NUMERICAL METHODS FOR ENGINEERING APPLICATIONS

MECHANICAL ENGINEERING

A Series of Textbooks and Reference Books

Editor

L. L. Faulkner

Columbus Division, Battelle Memorial Institute
and Department of Mechanical Engineering
The Ohio State University
Columbus, Ohio

1. *Spring Designer's Handbook,* Harold Carlson
2. *Computer-Aided Graphics and Design,* Daniel L. Ryan
3. *Lubrication Fundamentals,* J. George Wills
4. *Solar Engineering for Domestic Buildings,* William A. Himmelman
5. *Applied Engineering Mechanics: Statics and Dynamics,* G. Boothroyd and C. Poli
6. *Centrifugal Pump Clinic,* Igor J. Karassik
7. *Computer-Aided Kinetics for Machine Design,* Daniel L. Ryan
8. *Plastics Products Design Handbook, Part A: Materials and Components; Part B: Processes and Design for Processes,* edited by Edward Miller
9. *Turbomachinery: Basic Theory and Applications,* Earl Logan, Jr.
10. *Vibrations of Shells and Plates,* Werner Soedel
11. *Flat and Corrugated Diaphragm Design Handbook,* Mario Di Giovanni
12. *Practical Stress Analysis in Engineering Design,* Alexander Blake
13. *An Introduction to the Design and Behavior of Bolted Joints,* John H. Bickford
14. *Optimal Engineering Design: Principles and Applications,* James N. Siddall
15. *Spring Manufacturing Handbook,* Harold Carlson
16. *Industrial Noise Control: Fundamentals and Applications,* edited by Lewis H. Bell
17. *Gears and Their Vibration: A Basic Approach to Understanding Gear Noise,* J. Derek Smith
18. *Chains for Power Transmission and Material Handling: Design and Applications Handbook,* American Chain Association
19. *Corrosion and Corrosion Protection Handbook,* edited by Philip A. Schweitzer
20. *Gear Drive Systems: Design and Application,* Peter Lynwander

21. *Controlling In-Plant Airborne Contaminants: Systems Design and Calculations,* John D. Constance
22. *CAD/CAM Systems Planning and Implementation,* Charles S. Knox
23. *Probabilistic Engineering Design: Principles and Applications,* James N. Siddall
24. *Traction Drives: Selection and Application,* Frederick W. Heilich III and Eugene E. Shube
25. *Finite Element Methods: An Introduction,* Ronald L. Huston and Chris E. Passerello
26. *Mechanical Fastening of Plastics: An Engineering Handbook,* Brayton Lincoln, Kenneth J. Gomes, and James F. Braden
27. *Lubrication in Practice: Second Edition,* edited by W. S. Robertson
28. *Principles of Automated Drafting,* Daniel L. Ryan
29. *Practical Seal Design,* edited by Leonard J. Martini
30. *Engineering Documentation for CAD/CAM Applications,* Charles S. Knox
31. *Design Dimensioning with Computer Graphics Applications,* Jerome C. Lange
32. *Mechanism Analysis: Simplified Graphical and Analytical Techniques,* Lyndon O. Barton
33. *CAD/CAM Systems: Justification, Implementation, Productivity Measurement,* Edward J. Preston, George W. Crawford, and Mark E. Coticchia
34. *Steam Plant Calculations Manual,* V. Ganapathy
35. *Design Assurance for Engineers and Managers,* John A. Burgess
36. *Heat Transfer Fluids and Systems for Process and Energy Applications,* Jasbir Singh
37. *Potential Flows: Computer Graphic Solutions,* Robert H. Kirchhoff
38. *Computer-Aided Graphics and Design: Second Edition,* Daniel L. Ryan
39. *Electronically Controlled Proportional Valves: Selection and Application,* Michael J. Tonyan, edited by Tobi Goldoftas
40. *Pressure Gauge Handbook,* AMETEK, U.S. Gauge Division, edited by Philip W. Harland
41. *Fabric Filtration for Combustion Sources: Fundamentals and Basic Technology,* R. P. Donovan
42. *Design of Mechanical Joints,* Alexander Blake
43. *CAD/CAM Dictionary,* Edward J. Preston, George W. Crawford, and Mark E. Coticchia
44. *Machinery Adhesives for Locking, Retaining, and Sealing,* Girard S. Haviland
45. *Couplings and Joints: Design, Selection, and Application,* Jon R. Mancuso
46. *Shaft Alignment Handbook,* John Piotrowski

47. *BASIC Programs for Steam Plant Engineers: Boilers, Combustion, Fluid Flow, and Heat Transfer,* V. Ganapathy
48. *Solving Mechanical Design Problems with Computer Graphics,* Jerome C. Lange
49. *Plastics Gearing: Selection and Application,* Clifford E. Adams
50. *Clutches and Brakes: Design and Selection,* William C. Orthwein
51. *Transducers in Mechanical and Electronic Design,* Harry L. Trietley
52. *Metallurgical Applications of Shock-Wave and High-Strain-Rate Phenomena,* edited by Lawrence E. Murr, Karl P. Staudhammer, and Marc A. Meyers
53. *Magnesium Products Design,* Robert S. Busk
54. *How to Integrate CAD/CAM Systems: Management and Technology,* William D. Engelke
55. *Cam Design and Manufacture: Second Edition; with cam design software for the IBM PC and compatibles, disk included,* Preben W. Jensen
56. *Solid-State AC Motor Controls: Selection and Application*, Sylvester Campbell
57. *Fundamentals of Robotics,* David D. Ardayfio
58. *Belt Selection and Application for Engineers,* edited by Wallace D. Erickson
59. *Developing Three-Dimensional CAD Software with the IBM PC,* C. Stan Wei
60. *Organizing Data for CIM Applications,* Charles S. Knox, with contributions by Thomas C. Boos, Ross S. Culverhouse, and Paul F. Muchnicki
61. *Computer-Aided Simulation in Railway Dynamics,* by Rao V. Dukkipati and Joseph R. Amyot
62. *Fiber-Reinforced Composites: Materials, Manufacturing, and Design,* P. K. Mallick
63. *Photoelectric Sensors and Controls: Selection and Application,* Scott M. Juds
64. *Finite Element Analysis with Personal Computers,* Edward R. Champion, Jr., and J. Michael Ensminger
65. *Ultrasonics: Fundamentals, Technology, Applications: Second Edition, Revised and Expanded,* Dale Ensminger
66. *Applied Finite Element Modeling: Practical Problem Solving for Engineers,* Jeffrey M. Steele
67. *Measurement and Instrumentation in Engineering: Principles and Basic Laboratory Experiments,* Francis S. Tse and Ivan E. Morse
68. *Centrifugal Pump Clinic: Second Edition, Revised and Expanded,* Igor J. Karassik
69. *Practical Stress Analysis in Engineering Design: Second Edition, Revised and Expanded,* Alexander Blake

70. *An Introduction to the Design and Behavior of Bolted Joints: Second Edition, Revised and Expanded,* John H. Bickford
71. *High Vacuum Technology: A Practical Guide,* Marsbed H. Hablanian
72. *Pressure Sensors: Selection and Application,* Duane Tandeske
73. *Zinc Handbook: Properties, Processing, and Use in Design,* Frank Porter
74. *Thermal Fatigue of Metals,* Andrzej Weronski and Tadeuz Hejwowski
75. *Classical and Modern Mechanisms for Engineers and Inventors,* Preben W. Jensen
76. *Handbook of Electronic Package Design,* edited by Michael Pecht
77. *Shock-Wave and High-Strain-Rate Phenomena in Materials*, edited by Marc A. Meyers, Lawrence E. Murr, and Karl P. Staudhammer
78. *Industrial Refrigeration: Principles, Design and Applications*, P. C. Koelet
79. *Applied Combustion*, Eugene L. Keating
80. *Engine Oils and Automotive Lubrication*, edited by Wilfried J. Bartz
81. *Mechanism Analysis: Simplified and Graphical Techniques, Second Edition, Revised and Expanded*, Lyndon O. Barton
82. *Fundamental Fluid Mechanics for the Practicing Engineer*, James W. Murdock
83. *Fiber-Reinforced Composites: Materials, Manufacturing, and Design, Second Edition, Revised and Expanded*, P. K. Mallick
84. *Numerical Methods for Engineering Applications*, Edward R. Champion, Jr.
85. *Turbomachinery: Basic Theory and Applications, Second Edition, Revised and Expanded*, Earl Logan, Jr.
86. *Vibrations of Shells and Plates: Second Edition, Revised and Expanded*, Werner Soedel

Additional Volumes in Preparation

Steam Plant Calculations Manual: Second Edition, Revised and Expanded, V. Ganapathy

Industrial Noise Control: Second Edition, Revised and Expanded, Lewis H. Bell and Douglas H. Bell

Mechanical Properties of Polymers and Composites: Second Edition, Revised and Expanded, Lawrence E. Nielsen and Robert F. Landel

Mechanical Engineering Software

Spring Design with an IBM PC, Al Dietrich

Mechanical Design Failure Analysis: With Failure Analysis System Software for the IBM PC, David G. Ullman

NUMERICAL METHODS FOR ENGINEERING APPLICATIONS

EDWARD R. CHAMPION, JR.

Engineering Resources & Concepts
Kennesaw, Georgia

Marcel Dekker, Inc. New York • Basel • Hong Kong

Library of Congress Cataloging-in-publication Data

Champion, Edward R.
Numerical methods for engineering applications / Edward R. Champion, Jr.
p. cm. -- (Mechanical engineering; 84)
Includes bibliographical references and index.
ISBN 0-8247-9135-5
1. Engineering mathematics--Data processing. 2. Numerical analysis--Data processing. I. Title. II. Series: Mechanical engineering (Marcel Dekker, Inc.); 84.
TA335.C48 1993
620'.001'51--dc20 93-1189
CIP

The publisher offers discounts on this book when ordered in bulk quantities. For more information, write to Special Sales/Professional Marketing at the address below.

This book is printed on acid-free paper.

MARCEL DEKKER, INC.
270 Madison Avenue, New York, New York 10016

Current printing (last digit):
10 9 8 7 6 5 4 3 2 1

PRINTED IN THE UNITED STATES OF AMERICA

To My Family

Preface

This book is directed at undergraduate and perhaps first-year graduate students and practicing engineers who need a reference on numerical techniques. The purpose of this text is to provide a sampling of programs that have proven to be efficient and effective in performing numerical analysis. The theory behind the algorithms is kept to a minimum and the direction has been to provide workable routines that the user can readily implement on a personal computer. Each chapter contains problems that can be solved using the techniques described previously. There has been an effort not to delve into esoteric algorithms that often leave the reader asking, "How do I apply this to my problem ?" These areas are left for another volume.

Included in this edition are worked examples, in most cases, of the algorithms presented. In some areas, full application of the algorithms in the form of real-world problems is given. These examples show how to tie seemingly unrelated pieces of code into a solution for a specific problem.

Not all readers will be at the same level of expertise. The intent is for the text to have a broad appeal, from the undergraduate student to the practicing professional. A mixture of languages for algorithm exist in this text. You will find implementations of the algorithms in BASIC, C, FORTRAN, and PASCAL. The Microsoft versions of these languages were used in this text preparation.

In selecting routines, I reviewed the notes from many numerical analysis courses and seminars that I and others have attended. The programmers did likewise. Practically all the ideas behind the algorithms presented are in the open literature. The lineage of many programs in common circulation is often not clear. We have never intentionally attempted to copy implemented algorithms verbatim since these are copyrighted. These algorithms are implemented in this text through the synthesis of many ideas compiled and developed by the programmers based on a wide variety of public domain materials.

The topics covered in this text include the following:

Chapter 1 is an introduction to the remainder of the text because it contains definitions, numerical constants, and concepts used in the remainder of the text.

Chapter 2 covers interpolation and extrapolation.

Chapter 3 covers numerical integration and differentiation.

Chapter 4 discusses the vector basics, matrices, and linear algebraic equations.

Chapter 5 looks at solutions for roots of equations and nonlinear equations.

Chapter 6 expands on the material in Chapter 5 and focuses on eigensystems and eigensolutions.

Chapter 7 is an introduction to ordinary differential equations and includes one-dimensional initial-value problems and boundary-value problems.

Chapter 8 is devoted to partial differential equations and includes solution techniques for elliptic, parabolic, and hyperbolic differential equations.

Chapter 9 explores the application of previous chapter algorithms to finite element analysis.

Chapter 10 takes a cursory look at Fourier Transforms relative to data handling.

Chapter 11 reviews four commercially available math programs and packages with worked examples of their use.

Chapters 1 through 4 contain demonstration routines at the end of each chapter. Routines in these chapters are basically procedures required in the later chapters. The later chapters' programs are examples of specific problem solving and hence do not contain demonstrations per se.

No warranties, express or implied, are made that the programs contained in this volume are free of error or that they will meet any requirements for any particular application. They should not be relied on for solving a problem whose incorrect solution could result in injury to a person or loss of property. If you use the programs in such a manner, you do so at your own risk. The author, programmers, and publisher disclaim all liability for direct or consequential damages resulting from your use of the program.

All programs are available on disk in each of the four languages, FORTRAN, C, PASCAL, and BASIC. See Appendix C for details.

Acknowledgements

Four contributors (programmers) have helped make this text come together given the time and development constraints. Mr. Ben Landry is responsible for all of the BASIC programming included in the first four chapters. His text review helped clarify the draft text as it was originally presented.

Mr. Marvin Ziegler developed many of the C, FORTRAN, and PASCAL applications and contributed the problem in Chapter 7 on how numerical analysis was used to solve the problem of blood flow in a collapsing artery.

Miss Mary Chamlee, a PhD candidate at the Georgia Institute of Technology, reviewed programs and contributed to the writing of many of the included PASCAL programs.

Mr. Chris Baldwin a graduate student at the Georgia Institute of Technology in Computer Science and Electrical Engineering provided many of the program translations from one language to another.

Mr. S.M. Marathe, President of Universal Technical Systems, Inc., (1220 Rock St., Rockford, IL 61101; (815) 963-2220) provided the TKSolver software for review. He also provided insight to the workings of the program and how to best apply the software to problems.

Ms. Jane Rush, Customer Relations Representative for Wolfram Research, Inc. (100 Trade Center Drive, Champaign, IL 61820; (217) 398-0747) provided assistance in obtaining and using the Mathematica software.

Ms. Karen Stoutemyer of Soft Warehouse, Inc. (3660 Waialae Ave. Suite 304 Honolulu, HI 96816-3236; (808) 734-5801) was most helpful in obtaining the Derive software and providing reference material.

MathSoft, Inc. (201 Broadway St. Cambridge, MA 02139; (617) 577-1017) provided a copy of MathCad 3.1 for Windows. Technical support was representative of the quality of the product.

I also wish to thank the technical support staff at WordPerfect Corporation (1555 N. Technology Way, Orem, UT, 84057) and Bitstream, Inc. (215 First Street, Cambridge, MA, 02142-1270) for their assistance. This text was written entirely in WordPerfect V5.1 and WordPerfect for Windows using *Facelift for WordPerfect* fontware from Bitstream. Gratitude is extended to Microsoft Corporation for use of the compilers for the programming effort.

I want to thank my family: Ginger, Caroline, and Emily for their patience and support during this effort and especially my mother, Mrs. Joyce Champion, for her support and partial funding of the project.

Edward R. Champion, Jr.

Contents

List of Programs

1

Introduction to Numerical Analysis

1.1 Introduction

Numerical Analysis is, on the whole, a fascinating subject and one that a person could devote their entire career. The ability to solve difficult and sometimes intractable problems by approximating functions to the equations that describe a process or an approximation to the process and to arrive at meaningful solutions is the goal in any numerical endeavor. With the availability of modern computers (including personal computers) basic concepts as well as cutting edge concepts can be pursued in detail to levels heretofore deemed impossible. The power to perform such tasks lie within the reach of most scientists, engineers, numerical analysts, and anyone interested in mathematics in general. The purpose of this text is to provide a post-introductory numerical methods text that is somewhat void of detailed derivations and abstract theorems. This is not to say that the understanding of the fundamental concepts is not relevant. We have attempted to provide sound, fundamental numerical techniques and to provide a look into the algorithms and how the algorithms are implemented. To master the concepts presented, problems at the end of each chapter are given. Code is given in one of the four languages listed below. Sometimes the code is given in more than one language.

This volume of numerical methods in engineering presents a summary of the more widely used methods of basic numerical analysis. The thrust of this text is directed at the implementation of standard numerical methods in four of the common programming languages found in use today on personal computers, minis, and mainframes and the implementation of a commercially available program for solving like problems. These languages are Basic, C, Fortran, and Pascal. The implementation of these languages by Microsoft as

Microsoft Professional Basic, Microsoft C, Microsoft Fortran, and Microsoft Pascal are considered one of the de facto standard in programming languages. In addition, commercially available programs are reviewed later in the text to supplement the current discussion. Some of the problems given in the text are worked in the chapter on the review of commercially available programs selected for this text. With these programs, the thrust is to provide a tool to solve the problem at hand with the need for little or no understanding of the calculational procedure involved. The programs, Mathcad V3.1, TK Solver 2.0, Mathematica V2.04, and Derive V2.5, are all capable of solving problems from the most simple addition to the most sophisticated. There is a spread of costs and capabilities. MathCAD runs under Windows 3.x. Mathematica runs on a variety of platforms including PC's under Windows 3.x and DOS. TK Solver and Derive both run in the DOS environment.

All of the code written in these languages was developed using the Microsoft programs and is easily adaptable for use with other similar programs. The code for the algorithms developed in the Microsoft languages is presented at the end of each chapter. The algorithms given in the text are not implemented in each language version of the text. All subroutines and procedures referenced in the text are designated as Program x.yzN. The x.yz refers to the chapter and program number and the N suffix refers to the appropriate language (ie B - BASIC, C, P - PASCAL, F - FORTRAN).

Numerical calculations are best demonstrated by examples. This text not only includes source code for each algorithm but also an examples section at the end of each chapter to illustrate the application of the algorithms. Each chapter contains a set of problems for the student to solve using the routines developed in the text or by using any of the commercially available programs reviewed in the text.

The next section of the text summarizes the topics covered within this text. Although the material is not exhaustive and does not concentrate on all of the theory behind each algorithm, the text does try to balance the practical and theoretical with algorithms that work.

Chapter 1 is an introduction to the remainder of the text and contains definitions, numerical constants, and concepts used in the remainder of the text.

Chapter 2 covers interpolation and extrapolation.

Chapter 3 covers numerical integration and differentiation.

Chapter 4 discusses with the vector basics, matrices, and linear algebraic equations.

Chapter 5 looks at solutions for roots of equations and nonlinear equations.

Chapter 6 expands on the material in Chapter 5 and focuses on eigensystems and eigensolutions.

Chapter 7 is an introduction to ordinary differential equations and includes one-dimensional initial-value problems and boundary-value problems.

Chapter 8 is devoted to partial differential equations and includes solution techniques for elliptic, parabolic, and hyperbolic differential equations.

Chapter 9 explores the application of previous chapter algorithms to finite element analysis.

Chapter 10 takes a cursory look at Fourier Transforms relative to data handling.

Chapter 11 reviews four currently available PC-based math programs.

1.2 Basic Knowledge Required for This Text

To establish all readers on an equal basis, a review of fundamental approximations of both physical, and numerical, concepts and constants is in order. The basic concepts and constants given below are without proof and are for reference when used with the programs given in the following chapters. The reader is encouraged to pursue other texts for more in depth discussions on the derivation and definitions of these concepts and constants.

It is assumed that the reader has had at least an introduction to numerical programming and a Calculus course and a differential equations math course. With these assumptions, the following basic mathematical results are given below and will be used in various places in the ensuing text.

1.3 Sum of a Series

A series is a sequence of numbers given in a prescribed order and formed to a given set of rules. To represent the sum of the series of numbers a_1 through a_n the following equation format is used:

$$\sum_{i=1}^{n} a_i = a_1 + a_2 + a_3 + \dots + a_n = S_n \tag{1.1}$$

Series can be classified as either Monotonic or Alternating. In a Monotonic series, each term is summed together. Equation 1.1 above represents a finite monotonic series. An infinite monotonic series is represented by the following equation:

$$\sum_{i=1}^{\infty} a_i = a_1 + a_2 + a_3 + \dots + a_\infty = S_\infty \tag{1.2}$$

In an alternating series every other term is subtracted rather than added. The first number in the series to be subtracted can be either the first or the second term. An alternating finite series with the second term being the first subtraction term is represented by the following equation:

$$\sum_{i=1}^{n} \beta a_i = a_1 - a_2 + a_3 - \dots + \alpha a_n \tag{1.3}$$

In order to define the even terms as the terms to be subtracted, and to show whether the last term is to be added or subtracted, the following definitions are provided:

$$\begin{aligned} \beta &= (-1)^{(i+1)} \\ \alpha &= (-1)^{(n+1)} \end{aligned} \tag{1.4}$$

An alternating finite series with the first term being the first subtraction term is represented by the following equation along with the β and α definitions:

$$\sum_{i=1}^{n} \beta a_i = -a_1 + a_2 - a_3 \ldots + \alpha a_n$$
$$\beta = (-1)^{(i)} \qquad (1.5)$$
$$\alpha = (-1)^{(n)}$$

In computer implementation, a finite series is most easily handled by a one dimensional array. A program populates the array using the prescribed order and following the given set of rules that define the series. The subroutine, Program 1.1B (Equation 1.3) demonstrates how the computer can be used to perform a monotonic sum on an array that has been populated by another routine. It is followed by a subroutine that demonstrates how the computer can be used to perform an alternating sum on a populated array, Program 1.2B (Equation 1.5). Program 1.3B is a short program for -1 raised to any power and is used later in the text. Program 1.8B at the end of this chapter illustrates how to declare, populate, and then call these series routines.

1.3.1 Convergent and Divergent

A series is convergent if the sum of the series approaches a finite limit as the number of terms approaches infinity. A series is divergent if the value of the series increases without bounds as the number of terms in the series increases. An example of a convergent series would be the Taylor's series to compute sine or cosine, while an example of a divergent series would be the sum of all counting numbers. A series is said to be an absolutely convergent series if the absolute values of the individual terms in the series approaches a finite value.

To test for convergence, three commonly used tests can be conducted on a series.

1.3.1.1 Comparison Test

If a term a_i is the ith term to be tested and b_i is the ith term of another series which is known to be absolutely convergent and c is a constant that is independent of i, then if

$$|a_i| < c|b_i| \tag{1.6}$$

the series of *a* terms is also absolutely convergent.

1.3.1.2 Cauchy's Test

If a_i is the *i*th term of a series and

$$\lim_{i \to \infty} \sqrt[i]{a_i} = L \tag{1.7}$$

then the series is convergent if L < 1 and divergent if L > 1. The test fails if L = 1.

1.3.1.3 Ratio Test

If a_i and a_{i+1} are two successive terms of a series and

$$\lim_{i\to\infty}\left|\frac{a_{i+1}}{a_i}\right| = L \tag{1.8}$$

then the series is absolutely convergent for L < 1 and divergent for L > 1. The test fails for L=1.

1.3.2 Operations With Series of Numbers

The following apply to operations with series of numbers.

A. Grouping of terms in an absolutely or conditionally convergent series can be changed without affecting the sum of the series.

B. The order of terms of an absolutely convergent series may be rearranged without affecting the sum.

C. The rearrangement of the order of a conditionally convergent series may change the sum of the series.

D. The sum, difference, and product of two or more absolutely convergent series is an absolutely convergent series.

1.4 Products of Numbers

A finite product of a set of n factors is defined as

$$\prod_{i=1}^{n} a_n = a_1 * a_2 * a_3 \dots * a_n = P_n \qquad \textbf{(1.9)}$$

where if n<∞ the product is a finite product and if n=∞, the product is an infinite product. The subroutine corresponding to Equation 1.9 is given in Program 1.2.

The terms of section 1.2 apply to products of numbers. Familiar products of numbers used in numerical analyses are factorials, binomial coefficients, and nested products.

As in sums of series, products of finite series in computer work, is most easily handled by a one dimensional array. Again a program populates the array using the prescribed order and following the given set of rules that define the series. The subroutine 1.4B demonstrates how the computer can be used to perform a product on an array that has been populated by another routine. An example program, 1.8B, at the end of this chapter illustrates how to declare, populate, and then call the product series routine.

1.4.1 Factorials

A factorial of a positive integer is defined as the product of a given series of consecutive whole numbers beginning with 1 and ending with the specified number. The factorial of the positive integer *n* is defined as

$$n! = (1) * (2) * (3) * \cdots * (n-2) * (n-1) * n \tag{1.10}$$

The factorial, n!, can be computed from the integer n, using the subroutine found on the page following the product subroutine.

The factorial of <u>any</u> number K can be expressed as

$$K! = K * (K-1) * (k-2) * \ldots * (K-n+1) * u! \tag{1.11}$$

where K-n=u and n is the positive integer nearest to K yet smaller than K. The value of u! is calculated by the Gamma Function. Recall that the Gamma Function is defined by the following integral:

$$u! = \Gamma(u+1) = \int_0^\infty e^{-t} t^u dt. \tag{1.12}$$

The Program 1.5B illustrates the Integer Factorial.

Factorials for K>4 are approximated to a high degree of accuracy by Stirling's Expansion.

$$K! = e^{\lambda}\left(\frac{K}{e}\right)^K \sqrt{2K\pi} + \epsilon \tag{1.13}$$

Where λ is computed using the equation

$$\lambda = \frac{K^{-1}}{12} - \frac{K^{-3}}{360} + \frac{K^{-5}}{1260} - \frac{K^{-7}}{1680} \tag{1.14}$$

and π is Archimedes Constant (pi).

$\pi = 3.14159265358979323846 2642.......$

The quantity e is Euler's Constant, derived from the equation

$$e = \lim_{n \to \infty}(1 + \frac{1}{n})^n \tag{1.15}$$

or in a more useful form, Euler's Constant is

e = 2.71828182845904523536 0287.......

The epsilon term, the variant from the correct answer, is less than;

$$|\epsilon| < 5x10^{-10} \tag{1.16}$$

Factorials for K>100 can be approximated by Stirling's Formula

$$K! = (\frac{K}{e})^K \sqrt{2K\pi} + \epsilon \tag{1.17}$$

where pi, e, and epsilon are defined as in Stirling's Expansion.

The Program 1.6B demonstrates Stirling's Expansion to compute the factorial of any number. When using floating point numbers, most of today's computer languages can only carry a finite number of significant digits. Thus this implementation, like most computer implementation of Stirling's Expansion, can safely ignore epsilon.

1.5 Binomial Coefficients

The following section gives formulas for the *binomial coefficients* that are used throughout this text. Unless otherwise noted, x is the signed number and i,j, and k are positive integers.

$$\begin{aligned} \binom{x}{i} &= \frac{x(x-1)(x-2)\cdots(x-i+1)}{i!} \quad for\ i>1 \\ \binom{x}{i} &= 1 \quad for\ i=0 \\ \binom{x}{i} &= x \quad for\ i=1 \end{aligned} \tag{1.18}$$

For i > 0

$$\begin{aligned} \binom{x+1}{i} &= \binom{x}{i} + \binom{x}{i-1} \\ \binom{x+1}{i+1} &= \binom{x}{i} + \binom{x}{i+1} \end{aligned} \tag{1.19}$$

The Program 1.7B illustrates the above concept. This program can compute the above mentioned factorials and binomial coefficients.

For the positive integers i and j the following general relations hold:

$$\begin{aligned}
\binom{j}{i} &= \frac{j!}{(j-i)!i!} \ for\ j>i \\
\binom{j}{i} &= j\ for\ i=1\ ,\ i=j-1 \\
\binom{j}{i} &= 1\ for\ i=j\quad i=j-1 \\
\binom{j}{i} &= 0\ for\ i<0
\end{aligned} \tag{1.20}$$

Program 1.9B demonstrates the factorial algorithms given above.

1.6 Binomial Series

A binomial series is developed from expanding Newton's formula in a power series for the general expression a±x as follows:

$$\begin{aligned}
(a \pm x)^n = a^n(1 \pm u)^n &= a^n[1 \pm \binom{n}{1}u + \binom{u}{2}u^2 \pm \cdots \alpha \binom{n}{n}u^n] \\
&= a^n \sum_{i=0}^{n} \beta \binom{n}{i} u^i
\end{aligned} \tag{1.21}$$

where

$$\begin{aligned}
n &= signed\ number \\
\binom{n}{i} &= Binominal\ Coefficient \\
\alpha &= (\pm 1)^n \\
\beta &= (\pm 1)^i \\
u &= \frac{x}{a}
\end{aligned} \tag{1.22}$$

The series is finite if n = 0,1,2,3,.....z. The series is convergent if n ≠ 0,1,2,3,.....z and $u^2 < 1$. The series is divergent if n ≠ 0,1,2,3,.....z and $u^2 > 1$.

For computer implementation, the following expression is given and it introduces the concept of a nested series.

$$(a \pm x)^n = a^n(1 \pm n(u)(1 \pm \frac{n-1}{2}u(1 \pm \frac{n-2}{3}u(1 \pm \cdots)))) \qquad \textbf{(1.23)}$$

The error associated with this approximation is less than the absolute value of the first deleted term.

1.7 Integrals and Other Small Steps

1.7.1 Mean Value Theorem

For a given x_0 and h (step size), there exists a value ζ such that

$$f(x_0 + h) = f(x_0) + hf'(\zeta) \ \ with \ \ x_0 < \zeta < (x_0 + h) \qquad \textbf{(1.24)}$$

1.7.2 Taylor's Theorem

For a given x_0, h, and n there exists a value ζ $(x_0 < \zeta < (x_0+h)$ such that

$$f(x_0 + h) = f(x_0) + hf'(x_0) + \frac{h^2 f''(x_0)}{2!} + \dots + \frac{h^n f^{(n)}(x_0)}{n!} + \dots \tag{1.25}$$

1.7.3 Mean Value Theorem for Integrals

If the function f(x) is continuous on the interval [a,b] and the function g(x) is integrable and does not change sign on the interval [a,b], then there exists a value ζ ($a<\zeta<b$) such that

$$\int_a^b f(x)g(x)dx = f(\zeta)\int_a^b g(x)dx. \tag{1.26}$$

1.7.4 Chain Rule

If y is a function of x then the derivative of the function f(x,y(x)) is

$$\frac{d\,f(x,y(x))}{dx} = \frac{\partial f(x,y(x))}{\partial x} + \frac{\partial f(x,y(x))}{\partial y}\frac{dy(x)}{dx}. \tag{1.27}$$

1.7.5 Descartes' Rule of Signs

Given the polynomial equation

$$p(x) = a_n x^n + a_{n-1}x^{n-1} + \dots + a_1 x + a_0 = 0 \tag{1.28}$$

the number of positive real roots is equal to the number of sign changes in the sequence of coefficients, $a_n,\ldots,a_0$, or is smaller by an even integer. The number of negative real roots is related to the coefficients of p(-x) in a similar manner.

1.7.6 Legendre Polynomials

The Legendre polynomials $P_n(x)$ with n=0,1,2,..., are given by

$$\begin{aligned} P_0(x) &= 1 \\ P_1(x) &= x \\ P_{n+1}(x) &= \frac{2n+1}{n+1} x P_n(x) - \frac{n}{n+1} P_{n-1}(x) \end{aligned} \tag{1.29}$$

These are orthogonal with respect to the weight function w(x)=1 on the interval [-1,1].

1.8 Demonstrations of Developed Subroutines

The remainder of this chapter is devoted to complete working examples of routines that use the subroutines developed in this chapter.

As a prelude to the problems utilizing the algorithms in this chapter an **example routine** that follows demonstrates the three series subroutines and includes a sample output.

The second **example routine**, populates the series array with a series of consecutive whole numbers beginning with 1 and ending with the size of the array. When the product of this series is computed starting from one, the result is the same as a factorial.

1.9 Next Chapter

The next chapter is concerned with interpolation and extrapolation of data and builds on much of the material presented in this chapter. Practical problems are given to strengthen the concepts presented in this chapter.

1.10 Problems

The exercises below are intended to provide you with practice in using your computer to solve numerical problems. Using your preferred language and the programming techniques introduced in this section, create programs to compute the following:

1-1) Compute the sum of all positive integers from 12 to 104 inclusive.

Hint: Create any integer array an populate it with integers values equal to there index value. Then use the SumOfSeries technique introduced in this chapter to compute the answer.

1-2) Compute the product of all positive integers from 3 to 17.

1-3) Compute the integer factorials of 10, 20, 40, 80, and 160.

1-4) Compute the difference in the time it takes to compute the factorial of all positive integers 1 through 170 using Integer factorial and Stirling's Expansion (Equation 1.13)

1-5) Compute the magnitude of the error between sin(45) computed using the routine supplied with your compiler and sin(45) computed with the first 3 element of the series:

$$\sin(x) = x/1! - x^3/3! + x^5/5!$$

Hint: First convert the angle from degrees to radians by dividing by 57.29577952 radians per degree.

1-6) Find the difference in the computation time taken to generate cos(30) through cos(60) in 1/10 degree steps using Integer vs Stirling's Expansion for the factorials in the following 4 element series:

$$\cos(x) = -x^2/2! + x^4/4! - x^6/6! + x^8/8!$$

1-7) Find the maximum error between the two methods used for computing cos(x) in problem 1-6 over the range 1 to 89 degrees.

1-8) Compute the binomial coefficient of

$$\binom{170}{120}$$

1-9) Graph the binomial coefficients of

$$\binom{1^j}{40}$$

for all j integers from 1 to 40.

1-10) Graph the binomial coefficients of (i) for all i integers from 1 to 40.

Chapter 1 Programs

Program 1.1B Sum of a Series

```
SUB SumOfSeries (X#(), First%, Count%, Result#)
'
'THIS PROGRAM COPYRIGHT: (C) 1991 by LCPsoftware
'AUTHORS:   Benjamin J. Landry and Edward Champion
'TEXT BOOK: NUMERICAL METHODS FOR ENGINEERING APPLICATIONS
'
'EQUATION:  (01.03) EQU_0103 Sum Of Series
'
'PURPOSE:   Sums together any number of consecutive elements
'           in a numeric array.  May start from any element
'           in the array and sum together all, or a subset,
'           of the array elements.
'
'ARGUMENTS: X#(min)...X#(max)  Is an array of type DOUBLE,
'          , an array floating point numbers.
'
'           First%  Is an integer used as an indexer into
'           the X# array.  First% does NOT point to the
'           first element in the X# array, but points to the
'           first element in the X# array to be included
'           in the sum.  Note that (min) <= First% <= (max)
'
'           Count%  Is an integer used as the number
'           of elements in the X# array to be summed
'           together and includes the element First%.
'
'FORMULA:
'
'                         Count
'    Result# = Sum =        Σ   X(i)
'                          i=1
'
'
'COMMENTS:  Microsoft Professional Basic clears all NON
'           STATIC arguments when it enters a function or
'           subroutine, therefore it is NOT necessary to
'           clear the local variable, SumOf#, at the
'           beginning of the loop.
'
'
'
'---------------------- end introduction ------------------

FOR i% = First% TO (First% + Count% - 1)

   SumOf# = SumOf# + X#(i%)

NEXT i%

Result# = SumOf#

END SUB
```

Program 1.2B Alternating Sum of Series

```
SUB AltSumOfSeries (X#(), first%, Count%, Beta%, Result#)
'
'THIS PROGRAM COPYRIGHT: (C) 1991 by LCPsoftware
'AUTHORS:    Benjamin J. Landry and Edward Champion
'TEXT BOOK: NUMERICAL METHODS FOR ENGINEERING APPLICATIONS
'
'EQUATION:   (01.04) EQU_0104 Alternating Sum Of Series
'
'PURPOSE:    Similar to SumOfSeries, but alternates between
'            adding and subtracting any number of consecutive
'            elements in a numeric array.  May start from any
'            element in the array & performs the alternating
'            sum on all, or a subset, of the array elements.
'            The first term may be added or subtracted.
'
'ARGUMENTS: X#(min)...X#(max)  Is an array of type DOUBLE,
'            that is, an array of floating point numbers.
'
'            First%  Is an integer used as an indexer into
'            the X# array.  First% does NOT point to the
'            first element in the X# array, but points to the
'            first element in the X# array to be included
'            in the alternating sum. (min) <= First% <= (max)
'
'            Count%  Is an integer used as the number of
'            elements in the X# array to be alternately summed
'            together and includes the element First%.
'
'            Beta%  Is the sign of the first term and should
'            be set to either +1 or -1.
'
'FORMULA:                       Count
'            Result = Sum =   Σ  Beta * X(i)
'                              i=1
'
'--------------------- end introduction ------------------

FOR i% = first% TO (first% + Count% - 1)

   IF Beta% >= 0 THEN
      SumOf# = SumOf# + X#(i%)
      Beta% = -1
   ELSE 'Beta% < 0 THEN
      SumOf# = SumOf# - X#(i%)
      Beta% = 1
   END IF

NEXT i%

Result# = SumOf#

END SUB
```

Program 1.3B -1 Raised to Any Power

```
SUB MinusOnePower (Power%, Result%)
'
'THIS PROGRAM COPYRIGHT: (C) 1991 by LCPsoftware
'AUTHORS:   Benjamin J. Landry and Edward Champion
'TEXT BOOK: NUMERICAL METHODS FOR ENGINEERING APPLICATIONS
'
'EQUATION:  (1.5)    EQU_0105  Minus One Power
'
'PURPOSE: n:    Takes -1 raised to any Power, and returns +1 or -1
'              One raised to any power is still one.  Minus one raised
'              to a power takes on the value -1 for odd powers, and +1
'                for even power.
'
'KNOWNS:        Power        The power to raise -1 to
'
'FORMULA:                    Power
'               Result = -1^         i.e.; +1 or -1
'
'---------------------------end introduction------------------------

IF INT(Power% / 2) * 2 = Power% THEN
   Result% = 1  'Power is even, result is positive
ELSE
   Result% = -1 'Power is odd,  result is negative
END IF

END SUB
```

Program 1.3P Minus 1 Raised To An Integer Power

```
function MinusOnePower( Power : integer ) : integer;

{  EQUATION: (1.5) EQU_0105 Minus One Power                               }
{  PURPOSE:  Takes -1 raised to any power and returns +1 or -1.   }

begin
   if odd( Power ) then MinusOnePower := -1
   else MinusOnePower := 1;
end;
```

Program 1.4B Product of a Series

```
SUB ProductOfSeries (X#(), First%, Count%, Result#)
'
'THIS PROGRAM COPYRIGHT: (C) 1991 by LCPsoftware
'AUTHORS:   Benjamin J. Landry and Edward Champion
'TEXT BOOK: NUMERICAL METHODS FOR ENGINEERING APPLICATIONS
'
'EQUATION:  (01.09) EQU_0109 Product Of Series
'
'PURPOSE:   Multiplies together any number of consecutive
'           elements in a numeric array.  May start from any
'           element in the array and sum together all, or a
'           subset, of the array elements.
'
'ARGUMENTS: X#(min)...X#(max)  Is an array of type DOUBLE,
'           that is, an array of floating point numbers.
'
'           First%  Is an integer used as an indexer into
'           the X# array.  First% does NOT point to the
'           first element in the X# array, but points to the
'           first element in the X# array to be included
'           in the product.   (min) <= First% <= (max)
'
'           Count%  Is an integer used as the number
'           of elements in the X# array to be multiplied
'           together and includes the element First%.
'
'FORMULA:
'
'                               Count
'           Result = Sum =       π  X(i)
'                               i=1
'
'
'COMMENTS:  Microsoft Professional Basic clears all NON
'           STATIC arguments when it enters a function or
'           subroutine, therefore it IS necessary to seed
'           the local variable, ProductOf#, to 1 or a zero
'           result will be obtained.
'
'
'---------------------- end introduction ------------------

ProductOf# = 1               'Must be seeded with 1 upon entry
FOR i% = First% TO (First% + Count% - 1)

   ProductOf# = ProductOf# * X#(i%)

NEXT i%

Result# = ProductOf#

END SUB
```

Program 1.5B Integer Factorial

```
SUB IntegerFactorial (Argument%, Result#)
'
'THIS PROGRAM COPYRIGHT: (C) 1991 by LCPsoftware
'AUTHORS:   Benjamin J. Landry and Edward Champion
'TEXT BOOK: NUMERICAL METHODS FOR ENGINEERING APPLICATIONS
'
'EQUATION:  (01.11) EQU_0111 Integer Factorial
'
'PURPOSE:   Finds the factorial of the current argument.
'           The factorial is the product of all integers
'           from 1 through the current argument.
'
'ARGUMENTS: Argument (n in formula below)  The number whose
'           factorial is to be computed.
'
'FORMULA:   Result = n! = 1 * 2 * ... * n-1 * n
'
'COMMENT:   Note the use of STATIC on LastArgument% and
'           LastResult#, this allows this subroutine to save
'           the results from last call to IntegerFactorial.
'           This routine will resume the product sequence
'           from this last computed factorial if the new
'           Argument% is larger then the LastArgument%.
'
'           NOTE: Do NOT change the test from < to <=
'           If you do it will cause an error when
'           Argument% = LastArgument%
'
'           If LastArgument% = 0 then this is the first time
'           this subroutine has been called and the sequence
'           will start from 1.
'
'---------------------- end introduction ------------------

STATIC LastArgu%, LastResult#

IF (Argument% < LastArgument%) OR (LastArgument% = 0) THEN

   LastArgument% = 1
   LastResult# = 1

END IF

FOR LastArgument% = LastArgument% TO Argument%

   LastResult# = LastResult# * LastArgument%

NEXT LastArgument%

Result# = LastResult#

END SUB
```

Program 1.6B Stirling Factorial

```
SUB StirlingFactorial (K#, Result#)
'
'THIS PROGRAM COPYRIGHT: (C) 1991 by LCPsoftware
'AUTHORS:    Benjamin J. Landry and Edward Champion
'TEXT BOOK: NUMERICAL METHODS FOR ENGINEERING APPLICATIONS
'
'EQUATION:   (01.13) EQU_0113  Stirling Factorial
'
'PURPOSE:    Finds the factorial of the given argument K,
'            using the Stirling's expansion method.
'
'  Knowns:   K  The number whose factorial is to be computed
'
'FORMULAS:
'
'                       Lambda           K    __________
'    Result = K! = e^      * ( K / e)^   * √ 2 * K * π
'
'
'WHERE:
'
'                  -1         -3          -5           -7
'               K^         K^          K^           K^
'     Lambda =  -----  -  -----  +  ------   -   ------
'                12        360       1260         1680
'
'
'--------------------- end introduction ------------------

CONST Euler# = 2.71828182845905#
CONST Pi# = 3.14159265358979#


' ---------------- Computation of Lambda ------------------

LamdaArg1# = K# ^ -1 / 12
LamdaArg2# = K# ^ -3 / 360
LamdaArg3# = K# ^ -5 / 1260
LamdaArg4# = K# ^ -7 / 1680

Lambda# = LamdaArg1# - LamdaArg2# + LamdaArg3# - LamdaArg4#


' ----------- Computation of Sterling Factorial -----------

StirlingArg1# = Euler# ^ Lambda#
StirlingArg2# = (K# / Euler#) ^ K#
StirlingArg3# = SQR(2 * K# * Pi#)

Result# = StirlingArg1# * StirlingArg2# * StirlingArg3#

END SUB
```

Program 1.6F Stirling Factorial

```
procedure StirlingFactorial(K : real; var Result : real );

{  PURPOSE: Finds the factorial of the given argument K,     }
{           using Stirling's Expansion.                      }
{                                                            }
{  FORMULA:                                                  }
{              LAMBDA        K                               }
{      K! = e^        *(K/e)^  * sqrt(2*K*PI)                }
{                                                            }
{                         -1     -3     -5     -7            }
{      where LAMBDA = K^    - K^   + K^   - K^               }
{                     ----   ----   ----   ----              }
{                      12    360    1260   1680              }

const
     EULER = 2.71828182845905;
     PI    = 3.14159265358979;
var
     LambdaArg1 : real;
     LambdaArg2 : real;
     LambdaArg3 : real;
     LambdaArg4 : real;
     Lambda     : real;
     StirlingArg1 : real;
     StirlingArg2 : real;
     StirlingArg3 : real;

begin
   if K <> 0 then begin
     LambdaArg1 := 1 / ( K*12 );
     LambdaArg2 := 1 / ( K*K*K*360 );
     LambdaArg3 := 1 / ( K*K*K*K*K*1260 );
     LambdaArg4 := 1 / ( K*K*K*K*K*K*K*1680 );

     Lambda := LambdaArg1 - LambdaArg2 + LambdaArg3 - LambdaArg4;

     StirlingArg1 := Exp( Lambda );
     StirlingArg2 := Exp( Ln( K/EULER )*K );
     StirlingArg3 := Sqrt( 2*K*PI );
     Result := StirlingArg1 * StirlingArg2 * StirlingArg3;
   end else Result := 1;
end;
```

Program 1.7B Binomial Coefficient Determination

```
SUB BinomialCoefficient (TopTerm%, BottomTerm%, Result#)
'
'THIS PROGRAM COPYRIGHT: (C) 1991 by LCPsoftware
'AUTHORS:    Benjamin J. Landry and Edward Champion
'TEXT BOOK: NUMERICAL METHODS FOR ENGINEERING APPLICATIONS
'
'EQUATION:   (01.19) EQU_0119 Integer Binomial Coefficient
'
'PURPOSE:    Computes the binomial coefficient of the given
'            number pair.  Uses the IntegerFactorial routine
'            to the create the three factorial terms.
'
'
'KNOWNS:     ( j )  ( TopTerm% )
'            ( i )  ( BottomTerm% )
'
'
'FORMULA:    Result = ( j ) = j! / ((j-i)! * i!)
'                     ( i )
'
'
'COMMENTS:   TopTerm% variable represents j
'            BottomTerm% variable represents i.
'
'            Uses the IntegerFactorial routine to the create
'            the three factorial terms.
'
'
'            Order of Factorial CALLs takes advantage of the
'            Factorial routine that saves the last computed
'            value, as j > i.
'
'
'            No conditional tests are necessary as this code
'            generates the following conditional results:
'
'            When      i<0    i=j or i=0   i=1 or j-1
'
'            (j) =      0          1            j
'            (i)
'
'--------------------- end introduction ------------------

Difference% = TopTerm% - BottomTerm%
CALL IntegerFactorial(Difference%, differenceFactorial#)

CALL IntegerFactorial(BottomTerm%, iFactorial#)
CALL IntegerFactorial(TopTerm%, jFactorial#)

Result# = jFactorial# / (differenceFactorial# * iFactorial#)

END SUB
```

Program 1.8B Driver Program for Series Determination

```
SUB SeriesDemonstration
'
'THIS PROGRAM COPYRIGHT: (C) 1991 by LCPsoftware
'AUTHORS:   Benjamin J. Landry and Edward Champion
'TEXT BOOK: NUMERICAL METHODS FOR ENGINEERING APPLICATIONS
'
'EQUATION:  (01.24) EQU_0124 Series Demonstration
'
'PURPOSE:   This routine demonstrate the following:
'
'           (01.03) EQU_0103  Sum Of Series
'           (01.04) EQU_0104  Alternating Sum Of Series
'           (01.09) EQU_0109  Product Of Series
'
'--------------------- end introduction ------------------
DIM SeriesData#(50)
CLS                'Output sent to last listed device
OutputDevice$ = "DAT_0124.ASC"         'File name
OutputDevice$ = "LPT1:"                'Printer
OutputDevice$ = "SCRN:"                'Screen

SerFile% = FREEFILE
OPEN OutputDevice$ FOR OUTPUT AS #SerFile%

' Populate an array (series) with even numbers starting at 0
FOR Count% = 0 TO 50
   SeriesData#(Count%) = 2 * Count%
NEXT Count%

PRINT #SerFile%, " Array   Data    Sum Of    Alt   Product "
PRINT #SerFile%, " Index  Content          Sum Of    Of    "
PRINT #SerFile%, " -----  -------  ------  ------   ";
PRINT #SerFile%, "---------------------"

Start% = 1
FOR Count% = 1 TO 50 STEP 1

   PRINT #SerFile%, USING "  ##"; Count%;
   PRINT #SerFile%, USING "  ####"; SeriesData#(Count%);

   CALL SumOfSeries(SeriesData#(), Start%, Count%, S#)
   PRINT #SerFile%, USING "    #####"; S#;

   CALL AltSumOfSeries(SeriesData#(), Start%, Count%, 1, A#)
   PRINT #SerFile%, USING "   #####"; A#;

   CALL ProductOfSeries(SeriesData#(), Start%, Count%, P#)
   PRINT #SerFile%, "      "; STR$(P#)

NEXT Count%
CLOSE #SerFile%
END SUB
```

Program 1.9B Driver Program for Factorial Demonstration

```
SUB FactorialDemonstration
'
'THIS PROGRAM COPYRIGHT: (C) 1991 by LCPsoftware
'AUTHORS:   Benjamin J. Landry and Edward Champion
'TEXT BOOK: NUMERICAL METHODS FOR ENGINEERING APPLICATIONS
'
'EQUATION:  (01.25) EQU_0125 Factorial Demonstration
'
'PURPOSE:   This routine demonstrate the following:
'
'           (01.09) EQU_0109  Product Of Series
'           (01.11) EQU_0111  Integer Factorial
'           (01.12) EQU_0112  Stirling Factorial
'---------------------- end introduction ------------------
DIM SeriesData#(50)
CLS            'Output sent to last listed device
OutputDevice$ = "DAT_0125.ASC"          'File name
OutputDevice$ = "LPT1:"                 'Printer
OutputDevice$ = "SCRN:"                 'Screen

SerFile% = FREEFILE
OPEN OutputDevice$ FOR OUTPUT AS #SerFile%

' Populate an array (series) with even numbers starting at 0
FOR Count% = 0 TO 50 STEP 1
   SeriesData#(Count%) = Count%
NEXT Count%
PRINT #SerFile%, "Array   Data      Product        Integer";
PRINT #SerFile%, "        Stirling";
PRINT #SerFile%, "Index  Content                  Factorial";
PRINT #SerFile%, "       Factorial"
PRINT #SerFile%, "-----  -------  ------------  ";
PRINT #SerFile%, "------------  ------------"
Start% = 1
FOR Count% = 1 TO 50 STEP 1

   PRINT #SerFile%, USING "  ##"; Count%;
   PRINT #SerFile%, USING "   ####"; SeriesData#(Count%);

   CALL ProductOfSeries(SeriesData#(), Start%, Count%, P#)
   PRINT #SerFile%, USING "    ##########^^^^"; P#;

   CALL IntegerFactorial(Count%, Ifactorial#)
   PRINT #SerFile%, USING " ##########^^^^"; Ifactorial#;

   Count# = Count%
   CALL StirlingFactorial(Count#, Sfactorial#)
   PRINT #SerFile%, USING " ##########^^^^"; Sfactorial#

NEXT Count%
CLOSE #SerFile%
END SUB
```

2

Functional Interpolation and Extrapolation

2.1 Introduction

This chapter covers aspects of interpolation and extrapolation of data. In order to efficiently accomplish this task, a number of common routines were developed in Chapter 1 and the early part of this chapter that are called by the subroutines given at the end of the chapter. The combination of these routines form a complete method for finding the results sought.

If the value of a function is known at a given number of points, we do not *a priori* know the functional values at intermediary points. Such data may come from physical measurements or from a numerical calculation that cannot readily be put into a simple functional form. This chapter addresses the effort required to estimate the function f(x) at arbitrary x either between the maximum and minimum points (**interpolation**) or perhaps beyond the end points of the function definition (**extrapolation**).

An interpolating function is the function that passes through given data points that are typically sampled from a function table or directly from a known function. Several methods use polynomials to find the answers sought. Polynomial interpolation is one of the most fundamental subjects in numerical methods because many other numerical schemes are based on polynomial interpolation. The advantages of using polynomials would be of little consequence if no analytical justification existed for polynomials providing good approximations.

Any continuous function can be approximated to a given degree of accuracy on a closed interval by a polynomial. The *Weierstrass Approximation Theorem* states that

> If $f(x)$ is continuous in the closed interval $[a,b]$, then given any $\varepsilon > 0$, there is some polynomial $p_n(x)$ of degree $n \equiv n(\varepsilon)$ such that $|f(x) - p_n(x)| < \varepsilon, \quad a \le x \le b$.

Interpolation presumes a degree of smoothness for the function and hence an assumption of the existence of functional derivatives. Normally, you would think of this as a two-step process: (1) determine the function by fitting an interpolation polynomial to the specified data points and then (2) determine the value of the function at the required point. This process does work but it is computationally more intensive, less efficient, and may be subjected to intolerable roundoff errors than other methods. Interpolation methods presented in this chapter start at a point near the functional value required and add a sequence of decreasing corrections by incorporating the functional values from nearby data points.

If the continuity of derivatives is of concern, another interpolation function is provided by the *spline* function. A spline is a polynomial between each pair of adjacent data points. The most widely used type of spline is the **cubic spline** since it produces an interpolated function that is continuous through the second derivative. The splines are generally more stable than polynomials with less chance of eccentric behavior between the adjacent data points.

There is only one polynomial of degree n or less that assumes the exact values at n+1 distinct base points. Although many polynomial interpolation formulas may be found in the literature appear to be different, those that use the same base-point criterion for computing the polynomial coefficients are fundamentally the same. The basic interpolating polynomial has coefficients a_i that are the solutions of a set of n+1 simultaneous linear equations:

The determinant of the above matrix is known as the Vandermonde determinant and is non-zero if $x_i \neq x_j$, $i \neq j$. Therefore there is a unique solution

$$
\begin{aligned}
a_0 + a_1x_0 + a_2x_0^2 + \ldots + a_nx_0^n &= f(x_0) \\
a_0 + a_1x_1 + a_2x_1^2 + \ldots + a_nx_1^n &= f(x_1) \\
a_0 + a_1x_2 + a_2x_2^2 + \ldots + a_nx_2^n &= f(x_2) \\
\ldots\ldots\ldots\ldots\ldots\ldots &= \ldots \\
a_0 + a_1x_n + a_2x_n^2 + \ldots + a_nx_n^n &= f(x_n)
\end{aligned}
\tag{2.1}
$$

for the coefficients a_i. Vandermonde systems can be very ill-conditioned. If this is the case, no numerical technique will give an accurate answer.

The set of coefficients may be computed by solving the above set of simultaneous linear equations. This technique is a little more tedious for any system of size. The disadvantage of this method is not being able to estimate at least an upper bound on the error of approximation. The estimate on the error is important when evaluating a functional value at a given point. This section's purpose is to develop a substitute function ý(x) which closely approximates a more complicated function y(x) over a given interval and collocates with y(x) at certain points. The substitute function of a polynomial in terms of finite differences is called the *finite difference collocation polynomial*. Collocation polynomials of the finite difference type have general or equidistant spacing. If the function to be represented is a polynomial, the substitute formulas are exact. If the functions are not polynomials a remainder term is required to be added. Four types of formulas are given and discussed in the following subsections:

1. **Linear interpolation** followed by Finite Difference schemes

2. the **Lagrange** and **Newton Divided Difference** Interpolation formulas

3. the **Newton-Gregory** formulas for interpolation at or near the end of the interval and

4. the **Central Difference** formulas for interpolation over the intermediate range of the interval.

2.2 Linear Interpolation and Finite Differences

As an introduction let us look at the simplest type of interpolation, linear interpolation. The following formula for two points x_i and x_{i+1} with $y_i = f(x_i)$ and $y_{i+1} = f(x_{i+1})$ is basic linear interpolation. Given a value of x_j with $i \le j \le i+1$, the corresponding y_j is given below:

$$y_j = y_i + \frac{(y_{i+1} - y_i)}{(x_{i+1} - x_i)}(x_j - x_i) \tag{2.2}$$

This is illustrated in Program 2.3B.

Lets extend this concept to finite differences.

The finite difference of a function $y=f(x)$, where $f(x)$ is continuous and single valued in a given interval, can be represented by any of the following:

Divided Differences
Forward Differences
Backward Differences
Central Differences.

The following equations summarize to a primary level, finite differences:

A **divided difference of the first order** is given by

$$f(x_k,x_{k+1})=\frac{f(x_{k+1})-f(x_k)}{x_{k+1}-x_k}=\frac{y_{k+1}-y_k}{x_{k+1}-x_k} \quad (2.3)$$

Program 2.4B illustrates Equation 2.3.

The **forward difference** of first order is given in the above equation as

$$f(x_{k+1})-f(x_k)=y_{k+1}-y_k \quad (2.4)$$

$$\Delta y_k=y_{k+1}-y_k \quad (2.5)$$

and the Δ is the **forward difference operator.** Program 2.5B addresses the first order divided difference.

Higher order forward differences are given by the following formula

$$\Delta^n y_k=\sum_{r=0}^{n}(-1)^r\binom{n}{r}y_{(k+n-r)} \quad (2.6)$$

where the **binomial coefficient** is given by

$$\binom{n}{r} \quad (2.7)$$

Program 2.6B illustrates the higher order divided difference.

A **backward difference** of the first order is given as

$$f(x_k)-f(x_{k-1})=y_k-y_{k-1} \quad (2.8)$$

$$\nabla y_k=y_k-y_{k-1} \quad (2.9)$$

where ∇ is the **backward difference operator**. See Program 2.7B

The higher order backward difference expression is

$$\nabla^n y_k = \sum_{r=0}^{n} (-1)^r \binom{n}{r} y_{(k-r)} \tag{2.10}$$

A **central difference** of the first order is given by

$$f(x_{k+1/2}) - f(x_{k-1/2}) = y_{k+1/2} - y_{k-1/2} \tag{2.11}$$

where δ is the **central difference operator.**

$$\delta y_k = y_{k+1/2} - y_{k-1/2} \tag{2.12}$$

Expressions for the higher order central difference operator are given by

$$\delta^n y_k = \sum_{r=0}^{n} (-1)^r \binom{n}{r} y_{(k+1/2-r)} \tag{2.13}$$

Program 2.8B shows the central difference concept.

The operators given above are **LINEAR** operators and are governed by the following basic relations:

$$\begin{aligned} L[f_1(x)+f_2(x)] &= Lf_1(x)+Lf_2(x) \\ L^m[f(x)] &= L[L^{m-1}[f(x)]] \\ L^m L^n[f(x)] &= L^{m+n}[f(x)] \end{aligned} \tag{2.14}$$

From the basics given above, it is quite easy to show and develop the following difference tables:

Forward Difference Table

x_{-2}	y_{-2}				
		Δy_{-2}			
x_{-1}	y_{-1}		$\Delta^2 y_{-2}$		
		Δy_{-1}		$\Delta^3 y_{-2}$	
x_0	y_0		$\Delta^2 y_{-1}$		$\Delta^4 y_{-2}$
		Δy_0		$\Delta^3 y_{-1}$	
x_{+1}	y_{+1}		$\Delta^2 y_0$		
		Δy_{+1}			
x_{+2}	y_{+2}				

Backward Difference Table

x_{-2}	y_{-2}				
		∇y_{-1}			
x_{-1}	y_{-1}		$\nabla^2 y_0$		
		∇y_0		$\nabla^3 y_1$	
x_0	y_0		$\nabla^2 y_1$		$\nabla^4 y_2$
		∇y_1		$\nabla^3 y_2$	
x_{+1}	y_{+1}		$\nabla^2 y_2$		
		∇y_2			
x_{+2}	y_{+2}				

Central Difference Table

x_{-2}	y_{-2}				
		$\delta y_{-3/2}$			
x_{-1}	y_{-1}		$\delta^2 y_{-1}$		
		$\delta y_{-1/2}$		$\delta^3 y_{-1/2}$	
x_0	y_0		$\delta^2 y_0$		$\delta^4 y_0$
		$\delta y_{1/2}$		$\delta^3 y_{1/2}$	
x_{+1}	y_{+1}		$\delta^2 y_1$		
		$\delta y_{3/2}$			
x_{+2}	y_{+2}				

2.3 Lagrange and Newton Divided Difference Interpolation

Polynomial interpolations may be expressed in various alternative formats that can be transformed from one form to another. Among them are the power series, Lagrange interpolation and the Newton interpolations. An Nth order polynomial passing through N+1 data points is unique. Therefore, all polynomial interpolations that pass through these N+1 points are mathematically identical. The following algorithms are presented below for your consideration:

2.3.1 Lagrange Formula

$$\bar{y}(x) = y_0\lambda_0(x) + y_1\lambda_1(x) + \dots + y_n\lambda_n(x) + R_{n+1} \tag{2.15}$$

where

$$\lambda_i = \prod_{\substack{j=0 \\ j\neq i}}^{n} \frac{(x-x_i)}{(x_i-x_j)} = \dots\dots \\ \frac{(x-x_0)(x-x_1)(x-x_2)\dots(x-x_{i-1})(x-x_{i+1})\dots(x-x_n)}{(x_i-x_0)\dots\dots(x_i-x_{i-1})(x_i-x_{i+1})\dots(x_i-x_n)} \tag{2.16}$$

are the Lagrange multipliers and the remainder term is given by

$$R_{n+1} = \frac{(x-x_0)(x-x_1)\ldots(x-x_n)}{(n+1)!}[\max\ D^{n+1}f(\eta)] \tag{2.17}$$

This technique is illustrated in Program 2.9B and 2.9F.

2.3.2 Newton's Divided Difference Formula

$$\begin{aligned}\bar{y}(x) = y_0 + (x-x_0)\Delta_{11} + (x-x_0)(x-x_1)\Delta_{22}\ldots + \\ +[(x-x_0)(x-x_1)\ldots(x-x_{n-1})]\Delta_{nn}\end{aligned} \tag{2.18}$$

where

$$\begin{aligned}\Delta_{11} &= \frac{(y_1-y_0)}{(x_1-x_0)} \\ \Delta_{12} &= \frac{(y_2-y_0)}{(x_2-x_0)} \\ \Delta_{1n} &= \frac{(y_n-y_0)}{(x_n-x_0)} \\ \Delta_{mn} &= \frac{(\Delta_{m-1n}-\Delta_{m-1m-1})}{(x_n-x_{m-1})}\end{aligned} \tag{2.19}$$

Implementation of the above formula is given in Program 2.12B and

2.12F.

The advantages and disadvantages of using the Lagrange method are as follows: In using the Lagrange method, the difference table is not needed and is considered an advantage. However, the disadvantages are difficulty in subtraction of numbers that are almost equal for arguments that are closely spaced, and an excessive amount of computation is required when many interpolations are to be done using the same data set. In addition, the evaluation of error is not easy.

When a divided difference table for n+1 basepoints is prepared for all differences of order n or less, interpolating polynomials of any degree m where $m \leq n$ can be found. If the full difference table is not used, there is a choice of base points. The user should choose the base point nearest the argument to reduce the error associated with the approximation. When the argument is outside of the interval, extrapolation could result in large errors.

Inverse interpolation is the process of obtaining the argument of a function, x, if the value of the function, f(x), is known. For illustration, two types of inverse interpolation will be given. The first is the simple **inverse linear** interpolation and the second is the **inverse Lagrange interpolation** .

2.4 Inverse Interpolation

2.4.1 Inverse Linear Interpolation

For two points x_i and x_{i+1} with $y_i = f(x_i)$ and $y_{i+1} = f(x_{i+1})$, then given a functional value y_j with $i \leq j \leq i+1$, the corresponding x_j is determined by the following relation:

$$x_j = x_i + \frac{(y_j - y_i)}{(y_{i+1} - y_i)}(x_{i+1} - x_i) \tag{2.20}$$

2.4.2 Inverse Lagrange Interpolation

The Inverse Lagrange Interpolation formula is quite similar to the regular Lagrange formula given previously.

$$x_j = x_0 \lambda_0(y_j) + x_1 \lambda_1(y_j) + \dots + x_n \lambda_n(y_j) \tag{2.21}$$

where the λ's are given by the following formula:

$$\lambda_i(y_j) = \prod_{\substack{m=0 \\ m \neq i}}^{n} \frac{(y_j - y_m)}{(y_i - y_m)} \tag{2.22}$$

2.5 Newton-Gregory Formulas for Interpolation

A substitute function may be formulated in a polynomial form whose coefficients can be determined from forward, backward, or central differences of ascending order.

2.5.1 Forward Difference Interpolation Formula

$$y(x) = y_0 + \frac{u^{(1)}}{1!}\Delta y_0 + \frac{u^{(2)}}{2!}\Delta^2 y_0 + \dots + \frac{u^{(r)}}{r!}\Delta^r y_0 + R_{r+1} = \bar{y}(x) + R_{r+1} \tag{2.23}$$

where y(0)=f(a) with the remainder term given as

$$|R_{r+1}| \leq \left| \frac{u^{(r+1)}}{(r+1)!} h^{r+1} [max D^{r+1} f(\eta)] \simeq \frac{u^{(r+1)}}{(r+1)!} \Delta^{r+1} y_0 \right| \tag{2.24}$$

and η is the argument in [a,a+rh] that when inserted in the (r+1)st derivative of f(x) makes this derivative a maximum.

The computer implementation of the above formula is given in the following equation:

$$\bar{y}(x) = y_0 + u(\Delta y_0 - \frac{1-u}{2}(\Delta^2 y_0 - \frac{2-u}{3}(\Delta^3 y_0 - \dots - \frac{r-1-u}{r}\Delta^r y_0))) \tag{2.25}$$

Program 2.15B illustrates the above algorithm.

2.5.2 Backward Difference Interpolation Formula

The backward difference interpolation formula is given as

$$y(x) = y_0 + \frac{u^{(1)}}{1!}\nabla y_0 + \frac{(u+1)^{(2)}}{2!}\nabla^2 y_0 + \dots + \frac{(u+r+1)^{(r)}}{r!}\nabla^r y_0 + R_{(r+1)} \tag{2.26}$$

where y_0=f(a) and the remainder term is approximated by

$$|R_{(r+1)}| \le \left| \frac{(u+r)^{(r)}}{(r+1)!} \nabla^{r+1} y_0 \right| \tag{2.27}$$

and is as defined above. The $\nabla^n y_0$ are the backward differences of nth order.

For computer applications, this formula can be written as follows:

$$\bar{y}(x) = y_0 + u(\nabla y_0 + \frac{u+1}{2}(\nabla^2 y_0 + \frac{u+2}{3}(\nabla^3 y_0 + \dots + \frac{u+r+1}{r}\nabla^r y_0))) \tag{2.28}$$

See Program 2.17B for implementation.

2.6 Central Difference Formulas

To develop the central difference formulas, the following notation is used:

u=(x-a)/h v=1-u δ=central difference operator

$u^{(k)}$=u(u-1)(u-2) - - -(u-k+1) $v^{(k)}$=v(v-1)(v-2) - - -(v-k+1)

h = interval between points

2.6.1 Everett's Formula

For r=2,4,6,...

$$\begin{aligned} y(x) &= \frac{v}{1!}y_0 + \frac{u}{1!}y_1 + \frac{(v+1)^{(3)}}{3!}\delta^2 y_0 + \frac{(u+1)^{(3)}}{3!}\delta^2 y_1 + \\ &\frac{(v+2)^{(5)}}{5!}\delta^4 y_0 + \frac{(u+2)^{(5)}}{5!}\delta^4 y_1 + \dots + \\ &\frac{[v+1/2(r-2)]^{(r-1)}}{(r-1)!}\delta^{(r-2)} y_0 + \frac{[u+1/2(r-2)]^{(r-2)}}{(r-1)!}\delta^{(r-1)} y_1 + R_r \end{aligned} \tag{2.29}$$

The values of y_0, $\delta^2 y_0$,...., y_1, $\delta^2 y_1$,... are the difference values The expression for the remainder term is given by

$$|R_r| \leq \left| \frac{(u+1/2r)^r}{r} !h^r [\max D^r f(\eta)] \right| \tag{2.30}$$

where η has been defined earlier.

Implementation for computer analysis is given by the following formula with term definitions thereafter.

$$y(x) = vy_0 + uy_1 - C_0(\lambda_1 - C_1(\lambda_2 - C_2(\lambda_3 - C_3(\lambda_4 - \cdots)))) + R_r \tag{2.31}$$

$$\begin{aligned} C_k &= \frac{(u+k)(v+k)}{(2k+2)(2k+3)} \\ \lambda_k &= (v+k)\delta^{2k}y_0 + (u+k)\delta^{2k}y_1 \end{aligned} \tag{2.32}$$

See Program 2.18B for implementation.

2.6.2 Bessel's Formula

$$\begin{aligned} y(x) &= 1/2(y_0+y_1) + v\delta y_{1/2} + \frac{u^{(2)}}{2!}\lambda_2 + \cdots + \frac{(u+r-5)^{(r-2)}}{(r-2)!}\lambda_{r-2} \\ &+ \frac{(u+r-5)^{(r-2)}}{(r-1)!} v\delta^{r-1} y_{1/2} + R_r \end{aligned} \tag{2.33}$$

The remainder term is given as

$$|R_r| \leq \left| \frac{(u+1/2r)^{(r)} + [u+1/2(r-2)]^{(r)}}{r!} h^r [\max D^r f(\eta)] \right| \tag{2.34}$$

The above formula is rearranged and put into a form suitable for computer analysis as shown below:

$$\begin{aligned} y(x) &= 1/2(y_0+y_1) + v(\delta y_{1/2} + C_1(\lambda_2 + \\ &+ \frac{v}{3}(\delta^3 y_{1/2} + C_2(\lambda_4 + \frac{v}{5}(\delta^5 y_{1/2} + \cdots)))) + R_r \end{aligned} \tag{2.35}$$

where $C_k = -(k-1+u)(k-u)/2kv$ and $\lambda_{2k} = 1/2(\delta^{2k}y_0 + \delta^{2k}y_1)$ and $v = u - ½$.

2.6.3 Stirling's Formula

The formula for interpolation according to Stirling is given as follows for r=2,4,6,..:

$$\begin{aligned} y(x) &= y_0 + \frac{u^{(1)}}{1!}\omega_1 + \frac{u^{(1)}}{1!}u\delta^2 y_0 + \frac{(u+1)^{(3)}}{3!}\omega_3 + \frac{(u+1)^{(3)}}{3!}u\delta^4 y_0 \\ &+ \cdots + \frac{(u+r-5)^{(r-2)}}{(r-2)!}u\delta^{(r-1)}y_0 + \frac{(u+r-4)^{(r-1)}}{(r-1)!}\omega_{r-1} + R_r \end{aligned} \tag{2.36}$$

where y_0, $\delta^4 y_0$, ... , $\delta y_{1/2}$, $\delta^3 y_{1/2}$, ... , $\delta y_{-1/2}$, $\delta^3 y_{-1/2}$, are the difference values. The ω terms are given by the following:

$$\omega_{2k-1} = 1/2(\delta^{2k-1}y_{-1/2} + \delta^{2k-1}y_{1/2}) \tag{2.37}$$

The remainder term is given by

$$|R_r| \leq \left| \frac{(u+1/2r)^{(r)}}{r!} h^r [\max D^r f(\eta)] \right| \tag{2.38}$$

The above equation can be put in a form for computer aided implementation as given below:

$$y(x) = y_0 + u(\omega_1 + \frac{u}{2}(\delta^2 y_0 + C_2(\omega_3 + \frac{u}{4}(\delta^4 y_0 + C_4(\omega_5 + \frac{u}{6}(\delta^6 y_0 + \cdots)))))) + R_r \tag{2.39}$$

where $C_{2k} = (u^2 - k^2)/[(2k-1)u]$ and $\omega_{2k-1} = 1/2(\delta^{2k-1}y_{-1/2} + \delta^{2k-1}y_{1/2})$.

2.7 Rational Function Interpolation and Extrapolation

Rational functions are better than polynomial approximations in some cases. A rational function is comprised of quotients of polynomials as shown in the following equation:

$$f(x) \doteq g(x) = \frac{a_0 + a_1 x + a_2 x^2 + \cdots + a_n x^n}{b_0 + b_1 x + b_2 x^2 + \cdots + b_n x^n} = \frac{p_n(x)}{p_m(x)} \tag{2.40}$$

Rational functions have the ability to handle functions with poles. Poles are zeros of the denominator of the above equation. The function f(x) is finite for all finite real x, but has an analytic continuation with poles in the complex x-plane. These poles will ruin a polynomial approximation even if the function is restricted to real x values.

Bulirsch and Stoer [18] found an algorithm of the Neville type that does the rational function interpolation on a given set of tabulated data. The Neville type algorithm is a recursive way of filling in the numbers in the tables one column at a time from left to right. The algorithm is summarized by the following recurrence relation:

$$R_{(i+1)\ldots(i+m)}+\frac{R_{(i+1)\ldots(i+m)}-R_{i\ldots(i+m-1)}}{\left(\frac{x-x_i}{x-x_{i+m}}\right)\left(1-\frac{R_{(i+1)\ldots(i+m)}-R_{i\ldots(i+m)}}{R_{(i+1)\ldots(i+m)}-R_{(i+1)\ldots(i+m-1)}}\right)-1} \tag{2.41}$$

The above recurrence relation generates the rational functions through (m+1) points from the ones through m and (m-1) points.

Equation (2.41) can be put in a more tractable form ready for computer implementation using the following notation:

$$\begin{aligned} C_{m,i} &\equiv R_{i\ldots(i+m)}-R_{i\ldots(i+m-1)} \\ D_{m,i} &\equiv R_{i\ldots(i+m)}-R_{(i+1)\ldots(i+m)} \end{aligned} \tag{2.42}$$

The following recurrence relations are useful in developing a computer code for implementation.

$$D_{m+1,i}=\frac{C_{m,i+1}(C_{m,i+1}-D_{m,i})}{\left(\frac{x-x_i}{x-x_{i+m+1}}\right)D_{m,i}-C_{m,i+1}} \tag{2.43}$$

A simpler version of the case of **iterated interpolation** is given in Program 2.21F.

$$C_{m+1,i}=\frac{(\frac{x-x_i}{x-x_{m+i+1}})D_{m,i}(C_{m,i+1}-D_{m,i})}{(\frac{x-x_i}{x-x_{i+m+1}})D_{m,i}-C_{m,i+1}} \tag{2.44}$$

2.8 Cubic Spline Interpolation

Polynomial interpolation for a set of N+1 points is, as pointed out earlier, sometimes not satisfactory. A polynomial of degree N can have N-1 relative maxima and minima. Even though the polynomial may pass through the points, the curve between the points may not resemble the true curve. Another method is to piece together lower-degree polynomials and interpolate between the successive nodes. The simplest polynomial to use is the polynomial of degree 1. It is required to have continuity in the first and second derivatives. This produces a path of straight line segments that pass through the required points. A quadratic spline function can be constructed on each subinterval; however the curvature of the graph at the even nodes changes abruptly which can produce a distorted graph. If a piecewise cubic polynomial is used then both the first and second derivatives can be made continuous.

The fitting of a smooth curve to a set of data points is directly applicable to the area of Computer Aided Drafting (CAD). Mathematically, it is possible to construct cubic functions on each interval $[x_k, x_{k+1}]$ such that the resulting curve y=f(x) and the first and second derivatives, y' and y", are all continuous on the larger interval $[x_0, x_n]$. The continuity of the first derivative indicates that the graph of the points will not have sharp corners and the continuity of the second derivative indicates that the radius of curvature at each point is defined.

The proof of the existence of cubic splines and their ability to accurately model the data points is reserved for other texts. By definition, a function, say f(x), is called a cubic spline if there exists n cubic polynomials, $f_k(x)$ such that the following conditions are met:

A. $f(x) = f_k(x) = f_{k,0} + f_{k,1}(x-x_k) + f_{k,2}(x-x_k)^2 + f_{k,3}(x-x_k)^3$
for x in $[x_k, x_{k+1}]$ and $k = 0,1,..., n-1$.

B. $f(x_k) = y_k$ for $k = 0,1,..., n$.
The spline passes through each data point.

C. $f_k(x_{k+1}) = f_{k+1}(x_{k+1})$ for $k = 0,1,..., n-2$
The spline forms a continuous function.

D. $f'_k(x_{k+1}) = f'_{k+1}(x_{k+1})$ for $k = 0,1,..., n-2$.
The spline forms a smooth function.

E. $f''_k(x_{k+1}) = f''_{k+1}(x_{k+1})$ for $k = 0,1,..., n-2$.
The second derivative is continuous.

Implementing the above conditions results in a set of n-2 linear equations when put into matrix form result in a matrix of the tridiagonal type. This feature makes the system of equations easier to solve since each second order derivative is coupled only to the nearest point at k ± 1. For a unique solution the boundary conditions are specified on each end to be either zero or some value. To explore this further, we will define the following:

$$\sigma_n = f_{n-1}''(x_n) \tag{2.45}$$

and substitute this definition into the general expression for the cubic spline

$$f_i(x)=a_i(x-x_i)^3+b_i(x-x_i)^2+c_i(x-x_i)+d_i \quad \text{for } i=0,1,\ldots,n-1. \tag{2.46}$$

The mathematical algorithm to implement is given in two parts: (1) evaluate the expressions for σ_i and (2) using the σ_i values, evaluate the spline coefficients and the function value at $\bar{x}$. This example program assumes that the grid is evenly spaced. The FORTRAN version is given in Program 2.22F. The XBAR value is the point at which the functional value is to be interpolated.

2.9 Multidimensional Interpolation

In multidimensional interpolation, the goal is to estimate the function $y(x_1,x_2,\ldots,x_n)$ from an n-dimensional grid of tabulated values of each of the independent variables x_1, x_2,..., x_n. The fundamental process is to break the problem into a number of one-dimensional interpolations. For this bilinear process the gradient of the interpolated function is discontinuous at the boundaries of the surrounding grid. However, very reasonable estimates of the desired function can be obtained using the techniques outlined in the earlier sections of this chapter.

When a function is of two variables, say x and y, the most straightforward method of interpolation is to hold one variable constant, perhaps y, while interpolating on the other variable, x. From this a suitable of values for constant values of y is developed around the point of interest. The answer will then be obtained from application of another interpolation in the variable x.

It is not necessary to use the same form of interpolation for each variable. Depending on the point spacing, one may wish to use a Lagrange interpolating polynomial in one direction and a cubic spline interpolation in the other direction. The Programs 2.25B, 2.26B, and 2.27B illustrate the technique.

2.10 Next Chapter

The following chapters use some of the routines developed both in Chapter 1 and Chapter 2. Chapter 3 covers numerical differentiation and integration and is based on Finite Difference approximations. The covering of these areas leads to examples in the later chapters of how to use these routines in engineering problem solving.

2.11 Problems

The exercises below are intended to provide you with practice in using your computer to solve numerical problems. Using your preferred language and the programming techniques introduced in this section, create programs to compute the following:

2-1) The time card history of the new modification is as follows:

7 units	227 hours
14 units	354 hours
28 units	608 hours

Two new orders are being considered, one for 10 units and one for 20 units. Using linear interpolation compute the number of anticipated hours need for each order.

2-2) It has taken 28 days to test 37 samples of a special material. If there are 43 samples left, generate the anticipated completion schedule using linear interpolation.

2-3) The resistance of a material is found to be as follows:

10 degrees	2736 ohms
20 degrees	4420 ohms
30 degrees	6001 ohms
40 degrees	7428 ohms
50 degrees	8660 ohms

Using linear interpolation, compute the conductivity at 45 degrees.

2-4) A new resistance test of the material in problem 2-3 has found that at 45 degrees the resistance was 8071 ohms. This is over 20 ohms different from what was computed using linear interpolation. Find which of the following methods would have generated the smallest error:

Lagrange Interpolation.
Newton Divided
Newton Gregory Forward
Newton Gregory Backward
Everett

2-5) Using the method found to have the least error in problem 2-4, compute the resistance of the material in problem 2-3 for all temperatures from 5 to 55 degrees.

2-6) A sagging of a cable was measured in several places where access was not a problem. The distances from the pole, and the height above the ground was found to be as follows:

Distance from pole	Height above the ground
0	40
10	38
20	37
40	37
50	39
60	41

The lowest point is observed to be 30 feet from pole, but due to existing conditions it can not be measured directly.

Computed the height above the ground for the wire at this point.

2-7) Where should the second pole be relocated such that the wire can been attached 40 feet above the ground without changing the clearance between the lowest point of the wire and the ground?

2-8) As part of environmental testing, a new aircraft is being cooled in a special hanger. Thermal probes around the hanger monitor the temperature. Two critical failures occurred during the cooling process. The reading and locations of the probes are as follows:

X	Y	Temp	X	Y	Temp
25	25	4.0	75	25	9.3
25	50	5.0	75	50	12.7
25	75	7.1	75	75	14.9
25	100	10.2	75	100	21.2
50	25	6.1	100	25	13.5
50	50	8.5	100	50	17.9
50	75	11.2	100	75	22.4
50	100	15.6	100	100	28.8

One of the items that failed was located at (50, 70). Using linear interpolation, what was the ambient temperature in the vicinity of the first failed item?

2-9) Contrast the results computed using Linear and Everett's interpolation for the first item failure described in problem 2-8.

2-10) The other item that failed in problem 2-8 was located at (60,70). Using

Everett's interpolation, solve this multi-dimension interpolation to determine what the ambient temperature in the vicinity of the second failed item was.

Chapter 2 Programs

Program 2.1B Interval Variables

```
SUB Interval (X#(), xTerm#, Zero%, h#, u#, v#)
'
'THIS PROGRAM COPYRIGHT: (C) 1991 by LCPsoftware
'AUTHORS:    Benjamin J. Landry and Edward Champion
'TEXT BOOK:  NUMERICAL METHODS FOR ENGINEERING APPLICATIONS
'
'EQUATION:   (02.02) EQU_0202 Interval Variables
'
'PURPOSE:    Returns Index (Zero) to the nearest x argument
'            that is less than xTerm.  Returns the interval
'            (h), the percentage through the interval (u), and
'            the percentage remaining in the interval (v).
'
'ARGUMENTS:  X(min)...X(max) The x array of known arguments
'            !!!!!      MUST BE IN ASCENDING ORDER      !!!!!
'
'            xTerm    An x value between X(min) and X(max)
'FORMULAS:
'            h  = X(1) - X(0)              interval between x
'            u  = ( xTerm - X(0) ) / h     percentage through
'            v  = 1 - u                    percentage remaining
'
'            Zero   Search all x arguments for the x argument
'                   that is closest to but less than xTerm
'
'COMMENTS:   When dealing with arrays where the results f(x)
'            are known for equally spaced x arguments, it is
'            useful to know the interval between the x
'            arguments.
'
'            When an f(x) values is required for an x
'            argument that is not in the array, it is useful
'            to know the percentage through the interval from
'            the nearest but smaller x argument that is in
'            the array to the current x argument, and
'            to know the percentage of the interval from
'            the current x argument to the next larger x
'            argument that is in the array.
'
'---------------------- end introduction ------------------

Zero% = BOUND(X#)
'Search for the pointer to x array argument nearest xTerm
WHILE X#(Zero% + 1) <= xTerm#
   Zero% = Zero% + 1
WEND

h# = X#(Zero% + 1) - X#(Zero%)
u# = (xTerm# - X#(Zero%)) / h#
v# = 1 - u#

END SUB
```

Program 2.2B Partial Factorial

```
SUB PartialFactorial (Term#, Power%, Result#)
'
'THIS PROGRAM COPYRIGHT: (C) 1991 by LCPsoftware
'AUTHORS:    Benjamin J. Landry and Edward Champion
'TEXT BOOK: NUMERICAL METHODS FOR ENGINEERING APPLICATIONS
'
'EQUATION:   (02.03) EQU_0203 Partial Factorial
'
'PURPOSE:    Returns the Partial Factorial of Term#
'            (See definition below)
'
'ARGUMENTS: Term  Any real number
'
'            Power The number of terms to use in computing
'                  the partial factorial.
'
'FORMULAS:
'
'            (Power - 1)
'   Result  =     π      (Term - i)  = Term * (Term - 1) *
'              i = 0                          (Term - 2) *
'                                                 :
'                                             (Term - Power-1)
'
'COMMENTS:
'
'            Similar to factorial, Partial Factorial is a
'            term that is used in many interpolations.
'            In the formulas section, it is expressed as:
'
'                   (Power)
'            Term ^
'
'            This is not the same as a number Term raised to
'            the power of Power.  In the formulas section,
'            raised to a power is expressed as:
'
'                   Power
'            Term ^
'
'---------------------- end introduction ------------------

'For product loops, accumulator must be seed with 1
Accumulator# = 1

FOR i% = 0 TO Power% - 1
   Accumulator# = Accumulator# * (Term# - i%)
NEXT i%

Result# = Accumulator#

END SUB
```

Program 2.2P Partial Factorial Function

```
procedure PartialFactorial( Term : real;
                            Power : integer;
                            var Result : real );

{  PURPOSE :  Returns the partial factorial of Term
}

var Accumulator : real;
    i           : integer;

begin
   Accumulator := 1;
   for i := 0 to ( Power-1 ) do begin
      Accumulator := Accumulator*(Term-i);
   end;
   Result := Accumulator;
end;
```

Program 2.3B Linear Interpolation

```
SUB LinearInterpolation (x#(), Y#(), xTerm#, Zero%, Result#)
'
'THIS PROGRAM COPYRIGHT: (C) 1991 by LCPsoftware
'AUTHORS:    Benjamin J. Landry and Edward Champion
'TEXT BOOK:  NUMERICAL METHODS FOR ENGINEERING APPLICATIONS
'
'EQUATION:   (02.04) EQU_0204 Linear Interpolation
'
'PURPOSE:    Given two arrays, an x array populated with x
'            values and a y array populated with f(x) values,
'            this sub routine computes f(x) for x values that
'            are not in the x array via Linear Interpolation
'
'ARGUMENTS:  X(min)...X(max) The x array of known arguments
'            Y(min)...Y(max) The y array of results of f(x)
'
'            xTerm    An x value between X(min) and X(max)
'
'            Zero     The array entry in the X and Y arrays
'                     to be used as the 0th terms
'
'FORMULA:                        Y(1) - Y(0)
'            f(xTerm) = Y(0) + ------------ * (xTerm - X(0))
'                                X(1) - X(0)
'
'COMMENTS:   Although only two consecutive point in each
'            array are used, this subroutine can be used with
'            arrays of any length. The pointer Zero selects
'            the  X(0) and Y(0) terms from the two arrays.
'            This feature can be used to populate a large
'            array with known values once, and then use this
'            sub-routine to linearly interpolate as needed.
'
'            Best results are obtained when Zero is selected
'            such that the value of xTerm falls between two
'            values X(0) and X(1).
'
'            Because MPBasic pass all information by pointer,
'            passing two pointers to the array, take no more
'            memory the passing in the four required entries
'            in the array.
'
'---------------------- end introduction ------------------

xPointDelta# = x#(Zero% + 1) - x#(Zero%)
yPointDelta# = Y#(Zero% + 1) - Y#(Zero%)
xTermDelta# = xTerm# - x#(Zero%)

yTermDelta# = ((yPointDelta# / xPointDelta#) * xTermDelta#)
Result# = Y#(Zero%) + yTermDelta#

END SUB
```

Program 2.4B First Order Divided Difference

```
SUB DividedDifference (X#(), Y#(), Zero%, Result#)
'
'THIS PROGRAM COPYRIGHT: (C) 1991 by LCPsoftware
'AUTHORS:   Benjamin J. Landry and Edward Champion
'TEXT BOOK: NUMERICAL METHODS FOR ENGINEERING APPLICATIONS
'
'EQUATION:  (02.05) EQU_0205 First Order Divided Difference
'
'PURPOSE:   Finds first order divided difference operator
'           for any element in the x argument array.
'
'ARGUMENTS: X(min)...X(max) Array of equally spaced x points
'           Y(min)...Y(max) Array of known values for f(x)
'
'           Zero     The element pair in the X and Y arrays
'                    to be used as the 0th element.
'
'  Formula:   Result  =  Y(1) - Y(0) / X(1) - X(0)
'
'---------------------- end introduction -------------------

yPointDelta# = Y#(Zero% + 1) - Y#(Zero%)
xPointDelta# = X#(Zero% + 1) - X#(Zero%)

Result# = yPointDelta# / xPointDelta#

END SUB
```

Program 2.5B Forward Differences

```
SUB ForwardDifference (Y#(), Order%, Zero%, Result#)
'
'THIS PROGRAM COPYRIGHT: (C) 1991 by LCPsoftware
'AUTHORS:    Benjamin J. Landry and Edward Champion
'TEXT BOOK: NUMERICAL METHODS FOR ENGINEERING APPLICATIONS
'
'EQUATION:   (02.06) EQU_0206 Higher Order Forward Difference
'
'PURPOSE:    Finds a higher order forward difference operator
'            for any term in an array of known values.
'
'ARGUMENTS: Y(min)...Y(max) The y array of results of f(x)
'
'            Order    The higher order level selected.
'
'            Zero     The array entry in the X and Y arrays
'                     to be used as the 0th elements
'FORMULA:
'            Order     Count
' Result =     Σ  (-1)^ * ( Order ) * Y(Zero + Order - Count)
'            Count=0      ( Count )
'
'COMMENTS:   ( Order )       Represents a binomial coefficient
'            ( Count )
'
'            Note the computation of SignMinusOne below.
'            The parenthesis around the -1 is mandatory,
'            otherwise the order of computation would be
'            POWER then UNITARY MINUS.  Thus the difference
'            in the following forms of the equation hold:
'
'                     0                               0
'            -1 = -1 ^                     1 = (-1) ^
'                     1                               1
'            -1 = -1 ^                    -1 = (-1) ^
'                     2                               2
'            -1 = -1 ^                     1 = (-1) ^
'                     3                               3
'            -1 = -1 ^                    -1 = (-1) ^
'
'---------------------- end introduction ------------------

FOR Count% = 0 TO Order%

      SignMinusOne% = (-1) ^ Count%
      CALL BinomialCoefficient(Order%, Count%, Coefficient#)
      yTerm# = Y#(Zero% + Order% - Count%)
      SumOf# = SumOf# + (SignMinusOne% * Coefficient# * yTerm#)

NEXT Count%
Result# = SumOf#
END SUB
```

Program 2.6B Higher Order Forward Differences

```
SUB ForwardDifference (Y#(), Order%, Zero%, Result#)
'
'THIS PROGRAM COPYRIGHT: (C) 1991 by LCPsoftware
'AUTHORS:    Benjamin J. Landry and Edward Champion
'TEXT BOOK: NUMERICAL METHODS FOR ENGINEERING APPLICATIONS
'
'EQUATION:   (02.07) EQU_0207 Higher Order Forward Difference
'
'PURPOSE:    Finds a higher order forward difference operator
'            for any term in an array of known values.
'
'ARGUMENTS: Y(min)...Y(max) The y array of results of f(x)
'
'            Order    The higher order level selected.
'
'            Zero     The array entry in the X and Y arrays
'                     to be used as the 0th elements
'FORMULA:
'            Order      Count
' Result =     Σ   (-1)^  * ( Order ) * Y(Zero + Order - Count)
'            Count=0        ( Count )
'
'COMMENTS:   ( Order )        Represents a binomial coefficient
'            ( Count )
'
'            Note the computation of SignMinusOne below.
'            The parenthesis around the -1 is mandatory,
'            otherwise the order of computation would be
'            POWER then UNITARY MINUS.  Thus the difference
'            in the following forms of the equation hold:
'
'                       0                              0
'            -1 = -1 ^                      1 = (-1) ^
'                       1                              1
'            -1 = -1 ^                     -1 = (-1) ^
'                       2                              2
'            -1 = -1 ^                      1 = (-1) ^
'                       3                              3
'            -1 = -1 ^                     -1 = (-1) ^
'
'--------------------- end introduction -------------------

FOR Count% = 0 TO Order%

      SignMinusOne% = (-1) ^ Count%
      CALL BinomialCoefficient(Order%, Count%, Coefficient#)
      yTerm# = Y#(Zero% + Order% - Count%)
      SumOf# = SumOf# + (SignMinusOne% * Coefficient# * yTerm#)

NEXT Count%
Result# = SumOf#
END SUB
```

Program 2.7B Backward Differences

```
SUB BackwardDifference (Y#(), Order%, Zero%, Result#)
'
'THIS PROGRAM COPYRIGHT: (C) 1991 by LCPsoftware
'AUTHORS:    Benjamin J. Landry and Edward Champion
'TEXT BOOK: NUMERICAL METHODS FOR ENGINEERING APPLICATIONS
'
'EQUATION:   (02.10) EQU_0210 Higher Order Backward Difference
'
'PURPOSE:    Finds a higher order forward difference operator
'            for any term in an array of known values.
'
'ARGUMENTS: Y(min)...Y(max) The y array of results of f(x)
'
'            Order     The higher order level selected.
'
'            Zero      The array entry in the X and Y arrays
'                      to be used as the 0th elements
'FORMULA:
'            Order        Count
'  Result =    Σ     (-1)^       * ( Order ) * Y(Zero - Count)
'           Count=0                ( Count )
'
'---------------------- end introduction ------------------
FOR Count% = 0 TO Order%
      SignOne% = (-1) ^ Count%
      CALL BinomialCoefficient(Order%, Count%, Coefficient#)
      yTerm# = Y#(Zero% - Count%)
      SumOf# = SumOf# + (SignOne% * Coefficient# * yTerm#)
NEXT Count%
Result# = SumOf#
END SUB
```

Program 2.8B Central Difference

```
SUB CentralDifference (Y#(), Order%, Zero%, Result#)
'
'COPYRIGHT: (C) 1992 by LCPsoftware                   Ver 03.00
'AUTHORS:   Benjamin J. Landry  and Edward Champion
'TOOL:      Microsoft Professional Basic
'
'EQUATION:  (02.13) EQU_0213 Higher Order Central Difference
'
'FUNCTION:  Finds a higher order forward difference operator
'           for any term in an array of known values.
'
'ARGUMENTS: Y(min)...Y(max) The y array of results of f(x)
'
'           Order    The higher order level selected.
'
'           Zero     The array entry in the X and Y arrays
'                    to be used as the 0th elements
'
'FORMULA:
'
'           Order     Count
' Result =    Σ  (-1)^  * ( Order ) * (Y(k-r) + Y(k-r+1))/2
'          Count=0        ( Count )
'
'                                      k=Zero%   r=Count
'
'
'COMMENT:   Y(1/2) does not exist in the table of known
'           f(x) values.  The central difference is the
'           difference between the average of the Y(0) +
'           Y(1) and the average of the Y(-1) + Y(0).
'
'---------------------- end introduction ------------------

FOR Count% = 0 TO Order%

     SignOne% = (-1) ^ Count%

     CALL BinomialCoefficient(Order%, Count%, Coefficient#)

     yIndex1% = Zero% - Count%
     yIndex2% = Zero% - Count% + 1

     yTerm# = (Y#(yIndex1%) + Y#(yIndex2%)) / 2

     SumOf# = SumOf# + (SignOne% * Coefficient# * yTerm#)

NEXT Count%

Result# = SumOf#

END SUB
```

Program 2.9B Lagrange Interpolation

```
SUB Lagrange (x#(), Y#(), xTerm#, Zero%, Count%, Result#)
'
'THIS PROGRAM COPYRIGHT: (C) 1991 by LCPsoftware
'AUTHORS:   Benjamin J. Landry and Edward Champion
'TEXT BOOK: NUMERICAL METHODS FOR ENGINEERING APPLICATIONS
'
'EQUATION:  (02.15) EQU_0215 Lagrange Interpolation
'
'PURPOSE:   Finds f(x) using Lagrange Interpolation for
'           any x value.  Requires an array of known values
'           for x and for f(x).
'
'ARGUMENTS: X(min)...X(max) Array of equally spaced points
'           Y(min)...Y(max) Array of f(x) function values
'
'           xTerm  An X value with X(min) < xTerm < X(max)
'
'           Zero   The element pair in the X and Y arrays to
'                  be used as the 0th element, the value of
'                  X(Zero) should be close to the value
'                  xTerm for which f(x) is sought.
'
'           Count  The number of elements to use computing.
'
'FORMULA:
'          Count
'  Result = Σ (Y(element) *Lambda(element)*xTerm)+R(Count +1)
'       element=0
'
'  Lambda()  See description in module Lambda()
'  R()      See description in module LagrangeRemainder()
'
'COMMENT:   The term Accumulator# not being STATIC,
'           clears it by default on entry when using
'           Microsoft Professional Basic.
'
'--------------------- end introduction ------------------

TermCount% = Zero% + Count% - 1

FOR i% = Zero% TO TermCount%

   CALL Lambda(x#(), i%, xTerm#, Zero%, Count%, LamdaTerm#)
   Accumulator# = Accumulator# + (Y#(i%) * LamdaTerm#)

NEXT i%

CALL LagrangeRemainder(x#(), xTerm#, Zero%, Count%, rTerm#)

Result# = Accumulator# + rTerm#

END SUB
```

Program 2.9F Lagrange Interpolation

```
C LAGRANGE INTERPOLATION
IMPLICIT REAL*8(A-H,O-Z)
CHARACTER*64 FNAME
DIMENSION X(100),Y(100)
WRITE (*,900)
900 FORMAT (' Enter File Name For Sample Data - ' \)
READ (*,'(A)') FNAME
OPEN (3,FILE=FNAME)
READ (3,100) N,(X(I),I=1,N)
READ (3,101) (Y(I),I=1,N)
WRITE (*, 200)
      DO 1 I=1,N
  1   WRITE (*, 201) I,X(I),Y(I)
C
C     **** READ INTERPOLATION ARGUMENTS, EVALUATE
      WRITE (*, 202)
  2   READ (3, 102) XARG,IDEG,MIN
      YINT=EVAL(X,Y,XARG,IDEG,MIN,N)
      WRITE(*, 203) XARG,IDEG,MIN,YINT
      GO TO 2
C     **** FORMATS FOR I/O ****


  100 FORMAT (4X,I3/(15X,5F10.4))
  101 FORMAT (15X,5F10.4)
  102 FORMAT (7X,F10.4,13X,I2,12X,I2)
  200 FORMAT (1X,'THE SAMPLE FUNCTIONAL VALUES ARE')
  201 FORMAT (1H,I4,2F13.4)
  202 FORMAT (1X,'THE DATA AND RESULTS ARE')
  203 FORMAT (1H,F9.4,I8,I8,F12.4)
C
      END
C     **** EVALUATION FUNCTION ****
      FUNCTION EVAL(X,Y,XARG,IDEG,MIN,N)
      IMPLICIT REAL*8(A-H,O-Z)
      REAL*8 X,Y,XARG,EVAL
      DIMENSION X(N),Y(N)
      FACT=1.0
      MAX=MIN+IDEG
      DO 2 J=MIN,MAX
      IF(XARG.NE.X(J)) GO TO 2
      EVAL=Y(J)
      RETURN
 2    FACT=FACT*(XARG-X(J))
C     **** EVALUATE INTERPOLATING POLYNOMIAL ****
      YEST=0.
      DO 5 I=MIN,MAX
      TERM=Y(I)*FACT/(XARG-X(I))
      DO 4 J=MIN,MAX
 4    IF(I.NE.J) TERM=TERM/(X(I)-X(J))
 5    YEST=YEST+TERM
      EVAL=YEST
      RETURN
      END
```

Program 2.10B Lambda Calculation for Lagrange Interpolation

```
SUB Lambda (x#(), i%, xTerm#, Zero%, Count%, Result#)
'
'THIS PROGRAM COPYRIGHT: (C) 1991 by LCPsoftware
'AUTHORS:    Benjamin J. Landry and Edward Champion
'TEXT BOOK: NUMERICAL METHODS FOR ENGINEERING APPLICATIONS
'
'EQUATION:   (02.16) EQU_0216 Lagrange Lambda Multiplier
'
'PURPOSE:    Finds the Lagrange Lambda Multiplier for any x
'            value using any element in an array of known
'            values as the base point and using any number of
'            known points for the computation.
'
'ARGUMENTS: X(min)...X(max) Array of equally spaced points
'
'            i%     The displacement element for which Lambda
'                   is to be computed.
'
'            xTerm An X value within X(min) < xTerm < X(max)
'
'            Zero  The element pair in the X and Y arrays
'                  used as the 0th element, the value of
'                  X(Zero) should be close to the value
'                  xTerm for which f(x) is sought.
'
'           Count  The number of elements used in computing.
'
'FORMULA:            Count
'           Result =  π    ( xTerm - X(i) ) / ( X(i) - X(j)
'                    j=0
'                    j<>i
'
'COMMENT:   As this is a product loop, the Accumulator must
'           be seed with 1.
'
'---------------------- end introduction ------------------
Accumulator# = 1
TermCount% = Zero% + Count% - 1

FOR j% = Zero% TO TermCount%

      IF j% <> i% THEN
            xjTerm# = xTerm# - x#(j%)
            ijTerm# = x#(i%) - x#(j%)
            Accumulator# = Accumulator# * (xjTerm# / ijTerm#)
      END IF

NEXT j%

Result# = Accumulator#

END SUB
```

Program 2.11B Lagrange Remainder Term

```
SUB LagrangeRemainder (x#(), xTerm#, Zero%, Count%, Result#)
'
'THIS PROGRAM COPYRIGHT: (C) 1991 by LCPsoftware
'AUTHORS:    Benjamin J. Landry and Edward Champion
'TEXT BOOK: NUMERICAL METHODS FOR ENGINEERING APPLICATIONS
'
'EQUATION:   (02.17) EQU_0217 Lagrange Remainder
'
'PURPOSE:    Finds the Lagrange Remainder for any x value
'            using any element in an array of known values
'            as the base point and using any number of the
'            known points for computation.
'
'ARGUMENTS: X(min)...X(max) Array of equally spaced points
'
'            xTerm  An X value with X(min) < xTerm < X(max)
'
'            Zero   The element pair in the X and Y arrays to
'                   be used as the 0th element, the value of
'                   X(Zero) should be close to the value
'                   xTerm for which f(x) is sought.
'
'            Count  The number of elements to use computing.
'
'
'FORMULA:
'
'            Count
'  Result =  π  ( xTerm - X(element) ) / (element+1)!
'           element=0
'
'
'COMMENT:   As this is a product loop, the Accumulator must
'           be seed with 1.
'
'---------------------- end introduction ------------------

Accumulator# = 1
TermCount% = Zero% + Count% - 1

FOR element% = Zero% TO TermCount%

   Accumulator# = Accumulator# * (xTerm# - x#(element%))

NEXT element%

CountPlusOne% = Count% + 1
CALL IntegerFactorial(CountPlusOne%, PlusOneFactorial#)

Result# = Accumulator# / PlusOneFactorial#

END SUB
```

Program 2.12B Newton Divided Difference

```
SUB NewtonDividedDif (X#(), Y#(), xTerm#, Zero%, Count%, R#)
'
'THIS PROGRAM COPYRIGHT: (C) 1991 by LCPsoftware
'AUTHORS:    Benjamin J. Landry and Edward Champion
'TEXT BOOK: NUMERICAL METHODS FOR ENGINEERING APPLICATIONS
'
'EQUATION:   (02.18) EQU_0218 Newton Divided Difference
'
'PURPOSE:    Finds f(x) using Newton Divided Difference for
'            any x value.  Requires an array of known values
'            for x and for f(x).
'
'ARGUMENTS: X(min)...X(max) Array of equally spaced points
'            Y(min)...Y(max) Array of f(x) function values
'
'            xTerm  An X value with X(min) < xTerm < X(max)
'
'            Zero   The element pair in the X and Y arrays to
'                   be used as the 0th element, the value of
'                   X(Zero) should be close to the value
'                   xTerm for which f(x) is sought.
'
'            Count  The number of elements to use computing.
'FORMULA:
'                  Count i-1
' f(xTerm) = Y(0) +Σ   ( π (xTerm-X(j))*NewtonDelta(i,i))
'                   i=1  j=0
'
'COMMENT:    Newton's Divided Difference is computed by using
'            two loops and the NewtonDelta term subroutine.
'            The inner loop is used to compute the multiply
'            terms for each element in the array.  The outer
'            loop is used to sum the results of the inner
'            loop multiplied by the NewtonDelta term. The
'            SumOf term is seeded with "Zero" Y element, and
'            MultiplyOf is set to a Newton Delta Term at the
'            top of each inner loop.
'
'---------------------- end introduction ------------------

SumOf# = Y#(Zero%)

FOR i% = 1 TO Count%
   CALL NewtonDelta(i%, i%, X#(), Y#(), Zero%, MultiplyOf#)
   FOR j% = 0 TO (i% - 1)
      MultiplyOf# = MultiplyOf# * (xTerm# - X#(Zero% + j%))
   NEXT j%
   SumOf# = SumOf# + MultiplyOf#
NEXT i%

R# = SumOf#
END SUB
```

Program 2.12F Newton Divided Difference

```
C NEWTON'S DIVIDED DIFFERENCE INTERPOLATION POLYNOMIAL
C
      DIMENSION X(20), Y(20), TAB(20,20), XARG(20), IDEG(20)
      CHARACTER*64 FNAME,ANS,ARGDATA
      REAL*8 ACTUAL, NEWTF
      NARG=4
      WRITE (*,900)
  900 FORMAT (/' Enter Filename For Sample Data - '/)
      READ (*, '(A)') FNAME
      OPEN (3,FILE=FNAME)
C ***** READ DATA *****
      WRITE (*,901)
  901 FORMAT (/' If Y Is Not A Given Functional Of X, ENTER N - ')
      READ (*,'(A)') ANS
      IF (ANS.EQ.'Y') GO TO 2
      READ (3,100) N,M,(X(I),I=1,N)
      READ (3,101) (Y(I),I=1,N)
      GO TO 3
    2 READ (3,100) N,M,(X(I),I=1,N)
      DO 4 I=1,N
C ***** Put In Functional Value Here Y=f(x). *****
    4 Y(I)=COS(X(I))
    3 WRITE (*,200)
      DO 1 I=1,N
    1 WRITE (*,201) I, X(I), Y(I)
C ****** EVALUATE and PRINT DIVIDED DIFFERENCES *****
      CALL DIFTAB(X,Y,TAB,N,M,CONSIST,20)
      IF (CONSIST.NE.0.0) GO TO 999
      WRITE (*,202) M
      NM1=N-1
      DO 6 I=1,NM1
      L=I
      IF (I.GT.M) L=M
    6 WRITE (*,203) (TAB(I,J),J=1,L)
C ***** READ XARG AND IDEG, CALL NEWTF TO INTERPOLATE *****
      WRITE (*,902)
  902 FORMAT (//' Enter Argument Data File Name - '/)
      READ (*,'(A)') ARGDATA
      OPEN (3, FILE=ARGDATA)
      DO 8 J=1,NARG
      READ (3,102) XARG (J), IDEG(J)
    8 CONTINUE
      WRITE (*,204)
      IF (ANS.EQ.'N') GO TO 7
      DO 9 K=1,NARG
      XA=XARG(K)
      ID=IDEG(K)
      YARG=NEWTF(X,Y,TAB,N,M,ID,XA,CONSIST,20)
      ACTUAL=COS(XARG(K))
      WRITE (*,205) XARG(K), IDEG(K), YARG, ACTUAL, CONSIST
    9 CONTINUE
      GO TO 999
    7 DO 10 K=1,NARG
      ACTUAL=0.0
      YARG=NEWTF(X,Y,TAB,N,M,ID,XA,CONSIST,20)
      WRITE (*,205) XARG(K), IDEG(K), YARG, ACTUAL, CONSIST
   10 CONTINUE
```

```
C ***** FORMAT FOR I/O *****
  100 FORMAT (4X,I3,4X,I3 / (15X,5F10.4))
  101 FORMAT (4X,F8.4,4X,I3)
  102 FORMAT (4X,F6.4,4X,I3)
  200 FORMAT (' THE SAMPLE FUNCTIONAL VALUES ARE ')
  201 FORMAT (1H,I4,2F13.6)
  202 FORMAT (///' FOR M = ',I2,'  THE DIVIDED DIFFERENCES ARE ')
  203 FORMAT (1H / (1H,8E16.7))
  204 FORMAT (' THE DATA AND RESULTS ARE ' //)
  205 FORMAT (1H,F9.4,I8,2F12.6,F7.1)
  999 END
      SUBROUTINE DIFTAB(X,Y,TAB,N,M,CONSIST,KM)
      DIMENSION X(N), Y(N), TAB(KM,KM)
C ***** Check For Argument Consistency *****
      IF (M.LT.N) GO TO 2
      CONSIST=1.0
      RETURN
C ***** Calculate First Order Differences *****
    2 NM1=N-1
      DO 3 I=1,NM1
    3 TAB(I,1)=(Y(I+1)-Y(I))/(X(I+1)-X(I))
      IF (M.LE.1) GO TO 6
C ***** Calculate Higher Order Differences *****
      DO 5 J=2,M
      DO 5 I=J,NM1
      ISUB=I+1-J
    5 TAB(I,J)=(TAB(I,J-1)-TAB(I-1,J-1))/(X(I+1)-X(ISUB))
    6 CONSIST=0.0
      RETURN
      END
      FUNCTION NEWTF(X,Y,TAB,N,M,ID,XA,CONSIST,KM)
      REAL*8 NEWTF
      DIMENSION X(N), Y(N), TAB(KM,KM)
C ***** Check For Argument Consistency *****
      IF (ID.LE.M) GO TO 2
      CONSIST=1.0
      NEWTF=0.0
      RETURN
C ***** Search X Vector For Element >= XARG *****
    2 DO 4 I=1,N
      IF (I.EQ.N .OR. XA.LE.X(I)) GO TO 5
    4 CONTINUE
    5 MAX=I+ID/2
C ***** Insure ALL Required Differences Are In Table *****
      IF (MAX.LE.ID) MAX=ID+1
      IF (MAX.GT.N) MAX=N
C ***** Compute Interpolated Values *****
      YEST=TAB(MAX-1,ID)
      IF (ID.LE.1) GO TO 13
      IDEGM1=ID-1
      DO 12 I=1,IDEGM1
      ISUB1=MAX-I
      ISUB2=ID-I
   12 YEST=YEST*(XA-X(ISUB1))+TAB(ISUB1-1,ISUB2)
   13 ISUB1=MAX-ID
      CONSIST=0.0
      NEWTF=YEST*(XA-X(ISUB1))+Y(ISUB1)
      RETURN
      END
```

Program 2.13B Newton Delta Term

```
SUB NewtonDelta (i%, j%, X#(), Y#(), Zero%, Result#)
'
'THIS PROGRAM COPYRIGHT: (C) 1991 by LCPsoftware
'AUTHORS:    Benjamin J. Landry and Edward Champion
'TEXT BOOK: NUMERICAL METHODS FOR ENGINEERING APPLICATIONS
'
'EQUATION:   (02.19) EQU_0219 Newton Delta term
'
'PURPOSE:    Finds the Newton Delta term for any i, j, value
'
'ARGUMENTS: X(min)...X(max) Array of equally spaced points
'            Y(min)...Y(max) Array of f(x) function values
'
'            Zero   The element pair in the X and Y arrays to
'                   be used as the 0th element, the value of
'                   X(Zero) should be close to the value
'                   xTerm for which f(x) is sought.
'FORMULA:
'  When i <> 1
'
'  Delta(i,j) = ( Delta(i-1,j) - Delta(i-1,i-1) )
'               ---------------------------------
'                      ( X(j) - X(i-1) )
'
'  When i = 1
'
'  Delta(1,j) = ( Y(j) - Y(0) ) / ( X(j) - X(0) )
'
'COMMENT:   For i>1 the Newton Delta Term includes a Newton
'           Delta Term of i minus one in its definition. This
'           is an example of recursion. None of the variable
'           in the routine are STATIC. None of the input
'           values are changed, and a recursion limit of
'           i=1 exists.
'
'---------------------- end introduction ------------------
XjZero# = X#(Zero% + j%)
YjZero# = Y#(Zero% + j%)

IF i% > 1 THEN

  CALL NewtonDelta(i% - 1, j%, X#(), Y#(), Zero%, ijD#)
  CALL NewtonDelta(i% - 1, i% - 1, X#(), Y#(), Zero%, iiD#)
  Result# = (ijD# - iiD#) / (XjZero# - X#(Zero% + i% - 1))

ELSE

  Result# = (YjZero# - Y#(Zero%)) / (XjZero# - X#(Zero%))

END IF

END SUB
```

Program 2.14B Newton-Gregory Forward Difference Remainder

```
SUB NewtonGregFwdRemain (Y#(), Zero%, Order%, u#, Result#)
'
'THIS PROGRAM COPYRIGHT: (C) 1991 by LCPsoftware
'AUTHORS:    Benjamin J. Landry and Edward Champion
'TEXT BOOK: NUMERICAL METHODS FOR ENGINEERING APPLICATIONS
'
'EQUATION:   (02.24) EQU_0224 Newton-Gregory Fwd Remainder
'
'PURPOSE:    Newton-Gregory Forward Difference Remainder
'
'ARGUMENTS: Y(min)...Y(max) Array of f(x) function values
'
'            Zero  The index into the Y array that points to
'                  the known f(x) value that is the result of
'                  the x argument that lies closest to the x
'                  value whose unknown f(x) is being sought.
'                  See sub-routine Interval
'
'            Order The degree (or count) of higher order
'                  forward difference operators to be used.
'
'            u     Percentage through the current equally spaced
'                  x argument interval. See routine Interval
'
'FORMULA:
'
'                          (Order + 1)
'                        u^
'  Remainder = (  ------------  ) * d(Order + 1)Y(Zero)
'                  (Order + 1)!
'
'
'COMMENTS:
'
'  d(.a.)Y(.b.) represents the .a. th order forward
'               difference operator for the .b. th Y element
'
'    (.c.)
'  u^           represent the .c. th Partial Factorial of u
'
'
'--------------------- end introduction ------------------

CALL PartialFactorial(u#, Order% + 1, rPartial#)

CALL IntegerFactorial(Order% + 1, rFactorial#)

CALL ForwardDifference(Y#(), Order% + 1, Zero%, rFwdDif#)

Result# = ABS((rPartial# / rFactorial#) * rFwdDif#)

END SUB
```

Program 2.15B Newton-Gregory Forward Difference

```
SUB NewtonGregFwdDif (X#(), Y#(), xTerm#, Order%, Result#)
'
'THIS PROGRAM COPYRIGHT: (C) 1991 by LCPsoftware
'AUTHORS:   Benjamin J. Landry and Edward Champion
'TEXT BOOK: NUMERICAL METHODS FOR ENGINEERING APPLICATIONS
'
'EQUATION:  (02.25) EQU_0225 Newton-Gregory Fwd Difference
'
'PURPOSE:   Newton-Gregory Forward Difference Interpolation
'
'ARGUMENTS: X(min)...X(max) Array equally spaced x arguments
'           Y(min)...Y(max) Array of values f(x) function
'
'           xTerm An x argument whose f(x) value is unknown
'
'           Order The degree (or count) of higher order
'                 forward difference operators to be used.
'FORMULA:
'
'   f(xTerm) =  Y(0) + ( ((  0)-u)/1 * ( d1Y(0) -
'                      ( ((  1)-u)/2 * ( d2Y(0) -
'                            :  :  :        :
'                      ( ((r-1)-u)/r * ( drY(0) )))...) + R
'
'COMMENTS:
'
'           Use Interval routine to compute the following:
'
'           Zero  The index into the Y array that points to
'                 the known f(x) value that is the result of
'                 the x argument that lies closest to the x
'                 value whose unknown f(x) is being sought.
'                 See sub-routine Interval
'
'           u     Percentage through the current equally spaced
'                 x argument interval. See routine Interval
'
'--------------------- end introduction ------------------
CALL Interval(X#(), xTerm#, Zero%, h#, u#, v#)

FOR r% = Order% TO 1 STEP -1

      CALL ForwardDifference(Y#(), r%, Zero%, drYzero#)
      rTerm# = (((r% - 1) - u#) / r%)
      Accumulator# = rTerm# * (drYzero# - Accumulator#)

NEXT r%

CALL NewtonGregFwdRemain(Y#(), Zero%, Order%, u#, Remain#)

Result# = Y#(Zero%) - Accumulator# + Remain#
END SUB
```

Program 2.16B Newton-Gregory Backward Difference Remainder

```
SUB NewtonGregBakRemain (Y#(), Zero%, Order%, u#, Result#)
'
'THIS PROGRAM COPYRIGHT: (C) 1991 by LCPsoftware
'AUTHORS:   Benjamin J. Landry and Edward Champion
'TEXT BOOK: NUMERICAL METHODS FOR ENGINEERING APPLICATIONS
'
'EQUATION:  (02.27) EQU_0227 Newton-Gregory Bak Remainder
'
'PURPOSE:   Newton-Gregory Backward Difference Remainder
'
'ARGUMENTS: Y(min)...Y(max) Array of f(x) function values
'
'           Zero  The index into the Y array that points to
'                 the known f(x) value that is the result of
'                 the x argument that lies closest to the x
'                 value whose unknown f(x) is being sought.
'                 See sub-routine Interval
'
'           Order The degree (or count) of higher order
'                 forward difference operators to be used.
'
'           u     Percentage through the current equally spaced
'                 x argument interval. See routine Interval
'
'FORMULA:
'
'                              (Order)
'                ( u + r) ^
'  Remainder = ( ------------ ) * bd(Order + 1)Y(Zero)
'                (Order + 1)!
'
'
'COMMENTS:
'
'  bd(.a.)Y(.b.) represents the .a. th order backward
'                difference operator for the .b. th Y element
'
'      (.d.)
'  .c.^        represent the .d. th Partial Factorial of .c.
'
'
'--------------------- end introduction ------------------

CALL PartialFactorial(u# + Order%, Order%, rPartial#)

CALL IntegerFactorial(Order% + 1, rFactorial#)

CALL BackwardDifference(Y#(), Order% + 1, Zero%, rBakDif#)

Result# = ABS((rPartial# / rFactorial#) * rBakDif#)

END SUB
```

Program 2.17B Newton-Gregory Backward Difference Algorithm

```
SUB NewtonGregBakDif (X#(), Y#(), xTerm#, Order%, Result#)
'
'THIS PROGRAM COPYRIGHT: (C) 1991 by LCPsoftware
'AUTHORS:    Benjamin J. Landry and Edward Champion
'TEXT BOOK: NUMERICAL METHODS FOR ENGINEERING APPLICATIONS
'
'EQUATION:   (02.28) EQU_0228 Newton-Gregory Bak Difference
'
'PURPOSE:    Newton-Gregory Backward Difference Interpolation
'
'ARGUMENTS: X(min)...X(max) Array equally spaced x arguments
'            Y(min)...Y(max) Array of values f(x) function
'
'            xTerm An x argument whose f(x) value is unknown
'
'            Order The degree (or count) of higher order
'                  forward difference operators to be used.
'FORMULA:
'
'   f(xTerm) =  Y(0) + ( ((  0)+u)/1 * ( bd1Y(0) -
'                       ( ((  1)+u)/2 * ( bd2Y(0) -
'                            :  :  :       :
'                       ( ((r-1)+u)/r * ( bdrY(0) )))...) + R
'
'COMMENTS:
'
'            Use Interval routine to compute the following:
'
'            Zero  The index into the Y array that points to
'                  the known f(x) value that is the result of
'                  the x argument that lies closest to the x
'                  value whose unknown f(x) is being sought.
'                  See sub-routine Interval
'
'            u     Percentage through the current equally spaced
'                  x argument interval. See routine Interval
'
'---------------------- end introduction -------------------
CALL Interval(X#(), xTerm#, Zero%, h#, u#, v#)

FOR r% = Order% TO 1 STEP -1

   CALL BackwardDifference(Y#(), r%, Zero%, bdrYzero#)
   rTerm# = (((r% - 1) + u#) / r%)
   Accumulator# = rTerm# * (bdrYzero# + Accumulator#)

NEXT r%

CALL NewtonGregBakRemain(Y#(), Zero%, Order%, u#, Remain#)

Result# = Y#(Zero%) + Accumulator# + Remain#
END SUB
```

Program 2.18B Everett's Interpolation

```
SUB Everett (X#(), Y#(), xTerm#, Count%, Result#)
'
'COPYRIGHT: (C) 1991 by LCPsoftware                    Ver 02.00
'AUTHORS:   Benjamin J. Landry  and Edward Champion
'TOOL:      Microsoft Professional Basic
'
'EQUATION:  (02.29) EQU_0229 Everett's Interpolation
'
'FUNCTION:  Interpolates using Everett's Formula
'
'ARGUMENTS: X(min)...X(max) Array of equally spaced points
'           Y(min)...Y(max) Array of f(x) function values
'
'           xTerm  An x argument whose f(x) value is unknow
'
'           Count  The number of elements to use computation
'FORMULA:
'             Count                  (r-1)
'  Result = R + Σ  (((v + (r-2)/2) ^ * d(r-2)Y(0) / (r-1)!)
'              r=0
'                                    (r-1)
'                 + ((u + (r-2)/2) ^ * d(r-2)Y(1) / (r-1)!))
'
'COMMENTS:         |               (r)              r  |
'           R <=   | ((u + (r)/2) ^      / r! ) * h ^  |
'
'           d(.a.)Y(.b.) is the .a. th order central
'                 difference operator of the .b. th Y term
'
'           See Interval for Zero, u and v discriptions
'
'--------------------- end introduction ------------------

CALL Interval(X#(), xTerm#, Zero%, h#, u#, v#)
FOR r% = 4 TO Count% STEP 2
   HalfrLess2% = (r% - 2) / 2
   CALL CentralDifference(Y#(), r% - 2, Zero%, dC2y0#)
   CALL CentralDifference(Y#(), r% - 2, Zero% + 1, dC2y1#)
   CALL IntegerFactorial(r% - 1, rLess1Fact#)
   CALL PartialFactorial(v# + HalfrLess2%, r% - 1, vTerm#)
   CALL PartialFactorial(u# + HalfrLess2%, r% - 1, uTerm#)
   vuTerm# = (vTerm# * dC2y0#) + (uTerm# * dC2y1#)
   Accum# = Accum# + (vuTerm# / rLess1Fact#)
NEXT r%
Accum# = Accum# + (Y#(Zero%) * v#) + (Y#(Zero% + 1) * u#)
'---------------- Compute R (RemainderTerm) ---------------

CALL IntegerFactorial(Count%, CountFact#)
CALL PartialFactorial(u# + (.5 * Count%), Count%, PartFact#)
RemainderTerm# = (PartFact# / CountFact#) * h# ^ Count%

Result# = Accum# + RemainderTerm#
END SUB
```

Program 2.19B Bessel's Formula

```
SUB Bessel (X#(), Y#(), xTerm#, Count%, Result#)
'
'COPYRIGHT: (C) 1992 by LCPsoftware                    Ver 02.00
'AUTHORS:   Benjamin J. Landry  and Edward Champion
'TOOL:      Microsoft Professional Basic
'
'EQUATION:  (02.35) EQU_0235 Bessel's Interpolation
'
'FUNCTION:  Interpolates using Bessel's Formula
'
'ARGUMENTS: X(min)...X(max) Array of equally spaced points
'           Y(min)...Y(max) Array of f(x) function values
'
'           xTerm  An x argument whose f(x) value is unknow
'
'           Count  The number of elements to use computation
'FORMULA:          (step -2)
'                    2                              (r-1)
'  Result = R + Y(0) + Σ C(k) * ( l(r)+(v/(r-1)) * d^ Y(1/2)
'                    r = Count
'
'  C(k) = ( k -1 + u ) * ( k - u) / (2 * k * v)
'
'                  (2k)       (2k)
'  l(2k) = 1/2 * (d^ Y(0) + d^ Y(1)
'
'COMMENTS:        |                (r)            r  |
'           R <=  | ((u + (r)/2)  ^     / r! ) * h ^  |
'             a
'           d^ Y(b) is a central difference operator
'           See Interval for Zero, u and v discriptions
'--------------------- end introduction ------------------
CALL Interval(X#(), xTerm#, Zero%, h#, u#, v#)
FOR r% = Count% TO 4 STEP -2
   k# = (r% - 2) / 2
   CALL CentralDifference(Y#(), r% - 1, Zero%, drY0#)
   CALL CentralDifference(Y#(), 2 * k#, Zero% + 0, DyM#)
   CALL CentralDifference(Y#(), 2 * k#, Zero% + 1, DyP#)
   l# = (DyM# + DyP#) / 2
   C# = -(((k# - 1 + u#) * (k# - u#)) / (2 * k# * v#))
   Accum# = (((Accum# + drY0#) * (v# / (r% - 1))) + l#) * C#
NEXT r%
CALL CentralDifference(Y#(), 0, Zero%, drY0#)
Accum# = (Accum# + drY0#) * v#
Accum# = Accum# + ((Y#(Zero%) + Y#(Zero% + 1)) / 4)
'--------------- Compute R (RemainderTerm) ---------------

CALL IntegerFactorial(Count%, CountFact#)
CALL PartialFactorial(u# + (.5 * Count%), Count%, PartFact#)
RemainderTerm# = (PartFact# / CountFact#) * h# ^ Count%
Result# = Accum# + RemainderTerm#
END SUB
```

Program 2.20B Stirling's Formula

```
SUB Stirling (X#(), Y#(), xTerm#, Count%, Result#)
'
'COPYRIGHT: (C) 1992 by LCPsoftware                       Ver 02.00
'AUTHORS:   Benjamin J. Landry  and Dr. Ed Champion
'TOOL:      Microsoft Professional Basic
'
'EQUATION:  (02.39) EQU_0239 Stirling's Interpolation
'
'FUNCTION:  Interpolates using Stirling's Formula
'
'ARGUMENTS: X(min)...X(max) Array of equally spaced points
'           Y(min)...Y(max) Array of f(x) function values
'           xTerm  An x argument whose f(x) value is unknow
'           Count  The number of elements to use computation
'FORMULA:          (step -2)
'                    2                                  r
'  Result = R + Y(0) + Σ C(r-2) * ( w(r-1) + (u/r) * d^ Y(0)
'                   r = Count
'
'
'             (2)     (2)
'  C(2k) = (u^    - k^    ) / ((2k-u) * u)
'
'
'                       (2k-1)       (2k-1)
'  w(2k-1) = 1/2 * (d^  Y(-1/2) + d^  Y(1/2)
'
'COMMENTS:         |             (r)              r  |
'           R <=   | ((u + (r)/2) ^    / r! ) * h ^  |
'             a
'           d^ Y(b) is a central difference operator
'           See Interval for Zero, u and v discriptions
'------------------------ end introduction ------------------
CALL Interval(X#(), xTerm#, Zero%, h#, u#, v#)
FOR r% = Count% TO 2 STEP -2
   k# = (r% - 2) / 2
   CALL CentralDifference(Y#(), r%, Zero%, drY0#)
   CALL CentralDifference(Y#(), r% - 1, Zero% + 0, DyM#)
   CALL CentralDifference(Y#(), r% - 1, Zero% - 1, DyP#)
   w# = (DyM# + DyP#) / 2
   C# = (u# ^ 2 - k# ^ 2) / (((2 * k#) - 1#) * u#)
   IF k# = 0 THEN C# = u#
   Accum# = (((Accum# + drY0#) * (u# / r%)) + w#) * C#
NEXT r%
Accum# = Accum# + Y#(Zero%)

'---------------- Compute R (RemainderTerm) ---------------

CALL IntegerFactorial(Count%, CountFact#)
CALL PartialFactorial(u# + (.5 * Count%), Count%, PartFact#)
RemainderTerm# = (PartFact# / CountFact#) * h# ^ Count%
Result# = Accum# + RemainderTerm#

END SUB
```

Program 2.21F Iterated Interpolation

```
C      Iterated Interpolation Program
       DIMENSION X(10), XM(10), P(10,10)
     1 READ (3,100) N,XX
        IF (N) 40,40,4
        WRITE (*,200) N,XX
        DO 10 I = 1,N
            READ (3,101) X(I), P(I,1)
             XM(I) = XX - X(I)
    10 WRITE (*,201) X(I), XM(I)
       DO 20 I = 2,N
        DO 20 J = 2,I
        T = XM(J-1)*P(I,J-1) - XM(I)*P(J-1,J-1)
    20 P(I,J)=T/(X(I)-X(I-J))
        DO 30 I = 1,N
    30 WRITE (*,201) (P(I,J),J=1,I)
        GO TO 1
C**** FORMAT STATEMENTS ***************

  100 FORMAT (I2,F10.4)
  101 FORMAT (2F10.4)
  200 FORMAT (I4,E20.10)
  201 FORMAT (6E19.10)
   40 STOP
      END
```

Program 2.22F Subroutine For Cubic Spline Evaluation

```
C ****** CUBIC SPLINE EVALUATION ******
      SUBROUTINE CSPLINE (N,X,F,SIGMA,AT,DT,CT,BT,VALUE)
      REAL H,HS
       INTEGER I,N,NM1,STATUS
       REAL AT(N-1),BT(N-1),CT(N-1),DT(N-1),F(0:N),SIGMA(0:N),
     *     X(0:N)
       EXTERNAL TRISOLVE2
       H=(X(N)-X(0))/N
       HS=(H^2)/6
       NM1=N-1
       DO 10 I=1,NM1
            DT(I)=4.0
            BT(I)=(F(I-1)-2*F(I)+F(I+1))/HS
   10 CONTINUE
       DO 20 I=1,N-2
            CT(I)=1
            AT(I+1)=1
   20 CONTINUE
       CALL TRISOLVE (1,NM1,0,N,AT,BT,CT,DT,STATUS,SIGMA)
       SIGMA(0)=0
       SIGMA(N)=0
       VALUE=SPLNEV(N,XBAR,X,F,SIGMA)
       END

C ****** SUBROUTINE FOR TRIDIAGIONAL EQUATION SOLVER ******
      SUBROUTINE TRISOLVE (LB,UB,LBX,UBX,A,B,C,D,STATUS,X)
       INTEGER I,LB,UB,LBX,PSMALL,SOLVE,STATUS,UB,UBX
       REAL MULT,PIVOT,SMALL
       PARAMETER (SOLVE=0,PSMALL=1)
       PARAMETER (SMALL=1E-10)
       REAL A(LB:UB),B(LB:UB),C(LB:UB),D(LB,UB),X(LBX:UBX)
       INTRINSIC ABS
       DO 10 I=LB,UB-1
            PIVOT=D(I)
            IF(ABS(PIVOT).GT.SMALL) THEN
                  MULT=A(I+1)/PIVOT
                  IF(ABS(MULT).GT.SMALL) THEN
                        A(I+1)=MULT
                        D(I+1)=D(I+1)-MULT*C(I)
                        B(I+1)=B(I+1)-MULT*B(I)
                  ELSE
                        A(I+1)=0
                  END IF
   10 CONTINUE
       IF(ABS(D(UB)).GT.SMALL) THEN
            STATUS=SOLVE
            X(UB)=B(UB)/D(UB)
            DO 20 I=UB-1,LB-1
                  X(I)=(B(I)-C(I)*X(I+1))/D(I)
    20    CONTINUE
       ELSE
            STATUS=PSMALL
            END IF
       END
       REAL FUNCTION SPLNEV(N,XBAR,X,F,SIGMA)
       INTEGER J,N
```

```
REAL AJ,BJ,CJ,DJ,DIFF,H,X0,XBAR
REAL F(0:N),SIGMA(0:N),X(0:N)
INTRENSIC INT
X0=X(0)
H=(X(N)-X(0))/N
J=INT((XBAR-X0)/H)
AJ=(SIGMA(J+1)-SIGMA(J))/(6*H)
BJ=SIGMA(J)/2
CJ=(F(J+1)-F(J))/H-(SIGMA(J+1)+2*SIGMA(J))*H/6
DJ=F(J)
DIFF=XBAR-X(J)
SPLNEV=((AJ*DIFF+BJ)*DIFF+CJ)*DIFF+DJ
END
```

Program 2.23B Multidimensional Demo Setup

```
SUB InterpolationSetup
'
'THIS PROGRAM COPYRIGHT: (C) 1991 by LCPsoftware
'AUTHORS:   Benjamin J. Landry and Edward Champion
'TEXT BOOK: NUMERICAL METHODS FOR ENGINEERING APPLICATIONS
'
'EQUATION:  (02.47) EQU_0247 Interpolation Demonstration
'
'PURPOSE:   This routine declares and populates two array
'           used for interpolation demonstration.  The input
'           argument is declared and TEST function defined.
'--------------------- end introduction -------------------
DIM X#(100)
DIM Y#(100)
CLS                'Output sent to last listed device
OutputDevice$ = "DAT_0250.ASC"          'File name
OutputDevice$ = "LPT1:"                 'Printer
OutputDevice$ = "SCRN:"                 'Screen
F = FREEFILE:  OPEN OutputDevice$ FOR OUTPUT AS #F

    'Populate X array with argument and Y array with results
FOR Count% = 0 TO 100
   X#(Count%) = Count% / 100:  Y#(Count%) = SIN(X#(Count%))
NEXT Count%
                                          'Print out table headers
PRINT #F, " Array", " Argument x", " Know f(x) values"
PRINT #F, " Index", "      "
PRINT #F, " -----", " ----------", " ---------------"

xTerm# = .515#             'Input argument,     unknown f(x)
dTerm# = SIN(xTerm#)       'Test function, correct test f(x)

CALL Interval(X#(), xTerm#, Zero%, h#, u#, v#)  'Get nearest
Range% = 10         'For print and file output purpose only
FOR Element% = Zero% - Range% TO Zero% + Range%
   PRINT #F, Element%, X#(Element%), Y#(Element%)
NEXT Element%
                         'Print out appropriate demonstration terms
PRINT #F, ""
PRINT #F, "Nearest lower  argument value   :"; X#(Zero%)
PRINT #F, "Argument whose f(x) is not know :"; xTerm#
PRINT #F, "Nearest higher argument value   :"; X#(Zero% + 1)
PRINT #F, ""
PRINT #F, "Nearest lower  argument's f(x) value:"; Y#(Zero%)
PRINT #F, "Required input argument's f(x) value:         ?"
PRINT #F, "Nearest higher argument's f(x) value:";
PRINT #F, Y#(Zero% + 1): PRINT #F, ""
PRINT #F, "For TEST, set known functions f(x)  :"; dTerm#
                                     'Run the demonstration program
CALL InterpolateDemonstration(X#(), Y#(), xTerm#, Zero%, F)
CLOSE #F
END SUB
```

Program 2.24B Multidimensional Demo

```
SUB InterpolateDemonstration (X#(), Y#(), xTerm#, Zero%, F)
'
'THIS PROGRAM COPYRIGHT: (C) 1991 by LCPsoftware
'AUTHORS:   Benjamin J. Landry and Edward Champion
'TEXT BOOK: NUMERICAL METHODS FOR ENGINEERING APPLICATIONS
'
'EQUATION:  (02.48) EQU_0248 Interpolation Demonstration
'
'PURPOSE:   This routine demonstrate the following:
'
'           (02.02) EQU_0202 Linear Interpolation
'           (02.15) EQU_0215 Lagrange Interpolation
'           (02.18) EQU_0218 Newton Divided Difference
'           (02.25) EQU_0225 Newton-Gregory Fwd Difference
'           (02.28) EQU_0228 Newton-Gregory Bak Difference
'           (02.29) EQU_0229 Everett's Interpolation
'
'--------------------- end introduction ------------------

PRINT #F, "": PRINT #F, STRING$(54, "-")

CALL LinearInterpolation(X#(), Y#(), xTerm#, Zero%, Result#)
PRINT #F, ""
PRINT #F, "Results of Linear Interpolation     :"; Result#

Count% = 6
CALL Lagrange(X#(), Y#(), xTerm#, Zero%, Count%, Result#)
PRINT #F, ""
PRINT #F, "Results of Lagrange Interpolation   :"; Result#

Count% = 6
CALL NewtonDividedDif(X#(), Y#(), xTerm#, Zero%, Count%, R#)
PRINT #F, ""
PRINT #F, "Results of Newton Divided Difference:"; R#

Order% = 6
CALL NewtonGregFwdDif(X#(), Y#(), xTerm#, Order%, Result#)
PRINT #F, ""
PRINT #F, "Results of Newton Gregory Fwd Dif   :"; Result#

Order% = 6
CALL NewtonGregBakDif(X#(), Y#(), xTerm#, Order%, Result#)
PRINT #F, ""
PRINT #F, "Results of Newton Gregory Back Dif  :"; Result#

Count% = 6
CALL Everett(X#(), Y#(), xTerm#, Count%, Result#)
PRINT #F, ""
PRINT #F, "Results of Evertt's Interpolation   :"; Result#

END SUB
```

Program 2.25B Multidimensional Demo 2

```
SUB MultidimensionalDemoSetup
'
'THIS PROGRAM COPYRIGHT: (C) 1991 by LCPsoftware
'AUTHORS:   Benjamin J. Landry and Edward Champion
'TEXT BOOK: NUMERICAL METHODS FOR ENGINEERING APPLICATIONS
'
'EQUATION:  (02.49) EQU_0249 Multidimension Interpolat Setup
'
'PURPOSE:   This routine declares and populates three arrays
'           used for interpolation demonstration.  The input
'           argument is declared and a TEST function defined
'---------------------- end introduction ------------------
DIM X#(1 TO 30): DIM Y#(1 TO 30): DIM Z#(1 TO 30, 1 TO 30)
CLS                'Output sent to last listed device
OutputDevice$ = "DAT_0249.ASC"          'File name
OutputDevice$ = "LPT1:"                 'Printer
OutputDevice$ = "SCRN:"                 'Screen
F = FREEFILE:  OPEN OutputDevice$ FOR OUTPUT AS #F
                                         'Populate:
FOR i% = 1 TO 30: X#(i%) = i% / 30: NEXT i%    ' x arguments
FOR j% = 1 TO 30: Y#(j%) = j% / 60: NEXT j%    ' y arguments
FOR i% = 1 TO 30
   FOR j% = 1 TO 30
      Z#(i%, j%) = SIN(X#(i%)) + COS(Y#(j%))   'f(xy) values
   NEXT j%
NEXT i%

xTerm# = .54#: yTerm# = .22# 'Input argument, unknown f(x,y)
zTerm# = SIN(xTerm#) + COS(yTerm#)     'Test function, f(x,y)

CALL Interval(X#(), xTerm#, xZero%, xh#, xu#, xv#)'Nearest x
CALL Interval(Y#(), yTerm#, yZero%, yh#, yu#, yv#)'Nearest y
                  'Print out nearest appropriate known terms
PRINT #F, "Nearest terms to the required f(x,y) value:"
PRINT #F, "X(0), Y(0), Z(0,0):": PRINT #F, X#(xZero%);
PRINT #F, Y#(yZero%); Z#(xZero%, yZero%)
PRINT #F, "X(0), Y(1), Z(0,1):": PRINT #F, X#(xZero%);
PRINT #F, Y#(yZero% + 1); Z#(xZero%, yZero% + 1)
PRINT #F, "X(1), Y(0), Z(1,0):": PRINT #F, X#(xZero% + 1);
PRINT #F, Y#(yZero%); Z#(xZero% + 1, yZero%)
PRINT #F, "X(1), Y(1), Z(1,1):": PRINT #F, X#(xZero% + 1);
PRINT #F, Y#(yZero% + 1); Z#(xZero% + 1, yZero% + 1)
PRINT #F, "": PRINT #F, ""
PRINT #F, "Required f(x,y) has x and y values of:"
PRINT #F, xTerm#; yTerm#: PRINT #F, ""
PRINT #F, "For TEST, unknown functions f(x,y) is:"
PRINT #F, zTerm#
                               'Run the demonstration programs
CALL MultidimensionDemo(X#(), Y#(), Z#(), xTerm#, yTerm#, F)
CALL MultidimensionDem2(X#(), Y#(), Z#(), xTerm#, yTerm#, F)
CLOSE #F
END SUB
```

```
Array          Argument x     Know f(x) values
 Index
 -----          ----------     ----------------
 41             .41            .398609327984423
 42             .42            .40776045305957
 43             .43            .416870802429211
 44             .44            .425939465066
 45             .45            .434965534111123
 46             .46            .443948106965552
 47             .47            .452886285379068
 48             .48            .461779175541483
 49             .49            .470625888171158
 50             .5             .479425538604203
 51             .51            .488177246882908
 52             .52            .496880137843737
 53             .53            .505533341204847
 54             .54            .514135991653113
 55             .55            .522687228930659
 56             .56            .531186197920883
 57             .57            .539632048733969
 58             .58            .548023936791874
 59             .59            .556361022912784
 60             .6             .564642473395035
 61             .61            .572867460100481

Nearest lower  argument value    : .51
Argument whose f(x)is not known : .515
Nearest higher argument value    : .52

Nearest lower  argument's f(x) value: .488177246882908
Required input argument's f(x) value:         ?
Nearest higher argument's f(x) value: .496880137843737

For TEST, set known functions f(x)  : .492534849036109

-----------------------------------------------------

Results of Linear Interpolation     : .492528692363322

Results of Evertt's Interpolation   : .492534794487871

Results of Stirling's Interpolation : .492547109041606

Results of Bessel's Interpolation   : .492566192363322

Results of Lagrange Interpolation   : .492534849036095

Results of Newton Divided Difference: .492534849036109

Results of Newton Gregory Fwd Dif   : .492534849036109

Results of Newton Gregory Back Dif  : .492534849036114
```

3

Numerical Differentiation and Integration

3.1 Numerical Differentiation

Earlier in this book we addressed finite differences and methods of approximating functions are represented by sets of given data. Numerical differentiation is used to evaluate derivatives of a function using the functional values given at grid points. Difference approximations are important in the numerical solution of both ordinary and partial differential equations. Approximations to the derivatives of the functions are obtained by differentiating one of the polynomial approximations. The procedure may or may not lead to an accurate answer since the gradients of the approximating function may not be well behaved.

The first derivative of y=f(x) with constant intervals on h is

$$\frac{dy}{dx} = Dy \simeq \frac{d\bar{y}}{dx} = \frac{1}{h}\frac{d\bar{y}}{du} = D\bar{y}. \tag{3.1}$$

The expression for the general nth derivative is

$$\frac{d^n y}{dx^n} = D^n y \simeq \frac{d^n\bar{y}}{dx^n} = \frac{1}{h^n}\frac{d^n\bar{y}}{du^n} = D^n\bar{y} \tag{3.2}$$

where n=1,2,3,...., D=derivative operator, $u=(x-x_0)/h$, and

$$\bar{y} = \bar{f}(x) \tag{3.3}$$

is an interpolation polynomial. Interpolation polynomials were discussed in Chapter 2.

One must be careful to avoid large errors in determining derivative values because differentiation will highlight small errors in the approximating function. Higher order differentiation will magnify the approximating function discrepancies further. If derivative values have to be computed and the results will be used in subsequent calculations, it is often better to use a smoothing technique on the data, such as a least-squares approximation before computing the derivatives.

3.1.1 Differentiating Tabulated Functions

We can estimate derivatives of a tabulated function defined at discrete points. The formulas can be derived by differentiating any of the interpolation formulas used in Chapter 2. This section will cover the topic based on the Lagrange interpolation polynomial. The following text and program are based on a second degree interpolation polynomial. The expression for the second degree Lagrange polynomial is

$$\begin{aligned} P(x) &= f_0 L_0 + f_1 L_1 + f_2 L_2 = f_0 [\frac{(x-x_1)(x-x_2)}{(x_0-x_1)(x_0-x_2)}] \\ &+ f_1 [\frac{(x-x_0)(x-x_2)}{(x_1-x_0)(x_1-x_2)}] + f_2 [\frac{(x-x_0)(x-x_1)}{(x_2-x_0)(x_2-x_1)}] \end{aligned} \tag{3.4}$$

Since differentiation is a linear operation,

$$
\begin{aligned}
P'(x) &= f_0\lambda'_0(x) + f_1\lambda'_1(x) + f_2\lambda'_2(x) \text{ where} \\
\lambda'_0 &= \frac{(2x-x_1-x_2)}{(x_0-x_1)(x_0-x_2)} \\
\lambda'_1 &= \frac{(2x-x_0-x_2)}{(x_1-x_0)(x_1-x_2)} \\
\lambda'_2 &= \frac{(2x-x_0-x_1)}{(x_2-x_0)(x_2-x_1)}
\end{aligned}
\tag{3.5}
$$

Program 3.1B and 3.2B address Lagrange differentiation and the Lambda calculations.

3.1.2 Differentiating Analytic Functions

To differentiate an analytic function and evaluate that function at a point x, a formula can be derived by application of Taylor's series to the function evaluated at x+h and x-h, both points which lie inside the interval [a,b] in which the function is defined.

At x+h, the function expanded is

$$f(x+h) = f(x) + hf'(x) + \frac{h^2}{2!}f''(x) + \frac{h^3}{3!}f^{(3)}(x) + \cdots \tag{3.6}$$

and at x-h, the function expanded is

$$f(x-h) = f(x) - hf'(x) + \frac{h^2}{2!}f''(x) - \frac{h^3}{3!}f^{(3)}(x) + \cdots \tag{3.7}$$

Subtracting the above two equations and dividing by 2h results in the following expression:

$$f'(x) = \frac{f(x+h)-f(x-h)}{2h} - \sum_{i=1}^{\infty} \frac{f^{(2i+1)}(x)}{(2i+1)!} h^{2i} \tag{3.8}$$

If h is chosen to be sufficiently small then

$$f'(x) = \frac{f(x+h) - f(x-h)}{2h} \tag{3.9}$$

The subroutine that computes the derivative using this method with tabulated data is given in Program 3.3B. Higher derivatives follow as easily.

To obtain the accuracy necessary for a particular application, a decrease in the size of h may not be sufficient. A technique called *Richardson extrapolation* can be used. Expanding the function first derivative given above for terms involving h^4 and replacing h by rh, where r is any value other than ±1, results in an expression for the first derivative and is given by

$$f'(x) = \frac{f(x+rh) - f(x-rh)}{2rh} - \frac{r^2h^2}{6} f^{(3)}(x) - \frac{r^4h^4}{120} f^{(5)}(x) + \cdots \tag{3.10}$$

If h is chosen to be sufficiently small, then the third and higher derivative terms can be ignored. Program 3.4B illustrates Equation (3.10). A similar expression is derived for (x-rh).

Subtracting these equations results in the following algorithm for the Richardson extrapolation:

$$f'(x) = -\frac{1}{2hr(r^2-1)}[r^3 f(x+h) - f(x+rh) - r^3 f(x-h) + f(x-rh)] + \frac{r^2 h^4}{120} f^{(5)}(x) + \cdots \tag{3.11}$$

Again, if h is chosen to be sufficiently small, then the higher derivative term can be ignored. The Program 3.5B listed at the end of the chapter contains this simplified version of *Richardson extrapolation*.

3.2 Numerical Integration

If a closed form integration of a function is not possible or is greatly involved, the function may be integrated numerically by means of quadratures. A quadrature is a formula that evaluates the definite integral as a linear combination of a set of integrands in the range of integration. Numerical integration methods are derived by integrating interpolation polynomials. Different interpolation formulas will lead to different numerical integration methods. Numerical integration is an accurate process since, in general, errors that are incurred in each separate subinterval tend to cancel out. For this very reason, integration is often thought of as a smoothing process. These functions may be used for tabulated and analytical data forms.

This section covers several means for numerical integration including the infamous *Simpson's Rule* and the *Trapezoidal Rule*. These basic techniques are further extended for *Composite Integration Formula, Romberg Integration, Gaussian Quadrature, Improper Integrals*, and *Adaptive Quadrature*. In general, for equally spaced base points, the Newton-Coates formulas are used for unequally spaced points, Gaussian Quadrature is used. Newton-Coates formulas are further divided into the closed type and the open type.

The advantages and disadvantages for the different methods are summarized below:

The Trapezoidal rule is simple and optimal for improper integrals. However, you need a large number of subintervals for good accuracy.

Simpson's rule has higher accuracy than the Trapezoidal rule but can only be performed with an even number of intervals.

Newton-Coats formulas use equispaced points and the higher-order formulas are not necessarily more accurate.

Gaussian quadrature is more accurate than Newton-Coates formulas and the data points are not equispaced.

3.2.1 Newton-Coates Formulas (Polynomial Based)

The most commonly used classification for equal-interval integration formulas are called **open** and **closed**. Integration limits are either coincident with or displaced from base points by integral multiples of the base point spacing, that is, $x_i=x_0+ih$. Closed integration formulas use information concerning the functional values at the limits of integration and the open integration formulas require no such information at the integration limits.

The open and closed integration formula can be derived by integrating any of the interpolating polynomials given in Chapter 2 with appropriate base points and integration limits. We can assume that the function can either be computed or is known at its base points. If we use one of the finite difference formulas for the interpolating polynomial, then the development of the integration formulas will be straightforward.

From the earlier discussions of the Lagrange Interpolating Formula and defining $x_0 \equiv a$ and $x_1 \equiv b$ it is easy to show

$$\int_{x_0}^{x_1} f(x)dx = \frac{(x_1 - x_0)}{2}[f(x_0) + f(x_1)] \tag{3.12}$$

Letting h = b - a, the above formula can be written as

$$\int_{x_0}^{x_1} f(x)dx \approx \frac{h}{2}(f_0 + f_1) \tag{3.13}$$

The Program 3.6B contains a subroutine that approximates the integral using the **Trapezoidal Rule.**

To improve the accuracy of the above formula, the interval [a,b] can be split into two equal subintervals of size h/2. The polynomial is integrated over the interval [a,b] based on the points and functional values (x_i,f_i) with i = 0,1,2.

$$\int_{a}^{b} f(x)dx = \frac{h}{3}[f_0 + 4f_1 + f_2] \tag{3.14}$$

Program 3.7B contains a subroutine that approximates the integral using the above derived formula known as **Simpson's Rule.**

The errors incurred for both of the above formulas are

$$E_{Trap} = -\frac{h}{12}f''(\eta) \quad \text{where } x_0 < \eta < x_1 \tag{3.15}$$

for the Trapezoidal rule and

$$E_{Simp} = -\frac{h^5}{90}f^{(4)}(\eta) \quad \text{where } x_0 < \eta < x_1 \tag{3.16}$$

for Simpson's rule.

An examination of the error term for the trapezoidal rule indicates that the value returned by the equation is exact if the function is a linear polynomial. The second derivative of the function is zero. For Simpson's rule, the equation returns an exact value for all polynomials of degree three since the remainder term is zero for the fourth derivative.

The Newton-Coates family of integration formulas that use all (n+1) points are termed *closed* formula. In contrast an *open* formula is one that does not involve the end points of given interval. Closed formula are more accurate than the open formula. However the open formula are used in the solution of ordinary differential equations and their use gives more accurate results in this application than the closed formula representation.

3.2.2 Composite Integration Rules

In order to increase the accuracy of the Newton-Coates family of integration formula, it is necessary to split the integration interval into a number of subintervals and integrate the function over each subinterval separately. If we divided the range of integration into a number of subintervals and integrated the interpolation polynomial defined on the nodes of the subintervals, we would arrive at a value with considerably more effort and without a high level of confidence that the answer is accurate. *A high order interpolating polynomial does not constitute a high accuracy level.* The interpolating polynomial may not behave properly between the node points as the original function may have been intended.

Consider an interval [a,b] that is divided into n equivalent subintervals. The nodes are given as $x_i = x_0 + ih$ with i=0,1,2,...n. With h defined as (b-a)/n, the integral can be written as

$$\int_a^b f(x)dx = \int_{x_0}^{x_n} f(x)dx$$
$$= \int_{x_0}^{x_1} f(x)dx + \int_{x_1}^{x_2} f(x)dx + \cdots + \int_{x_{n-1}}^{x_n} f(x)dx \tag{3.17}$$

Each integral can be approximated by a numerical scheme such as the trapezoidal rule discussed earlier. Therefore the approximation for the integral in Equation (3.17) is

$$\int_a^b f(x)dx = \frac{h}{2}(f_0 + 2f_1 + 2f_2 + \cdots + 2f_{n-1} + f_n) \tag{3.18}$$

Equation (3.18) can also be written as

$$\int_a^b f(x)dx = \frac{h}{2}[f(a) + f(b) + 2\sum_{i=0}^{n} f(x_i)] - \frac{mh^3}{12}f''(\eta) \tag{3.19}$$

The Program 3.8B contains a subroutine that illustrates this concept. The derivation can be found in [58].

The error associated with this formulation is the sum of the errors of each of the individual approximations. The above formula is known as the **Composite** form of the trapezoidal rule. By dividing the integration regions in a different way, we are able to derive the **Composite Simpson's Rule.**

$$\int_a^b f(x)dx = \frac{h}{3}(f_0 + 4f_1 + 2f_2 + 4f_3 + \cdots + +2f_{n-2} + 4f_{n-1} + f_n) \tag{3.20}$$

The error is approximated by

$$|E_{Simp}| \leq \frac{h^4}{180}(b-a)F \tag{3.21}$$

where F is a bound on the fourth derivative of the function f.
An alternate, equivalent expression is given by

$$\int_a^b f(x)dx = \frac{h}{3}[f(a) + f(b) + 2\sum_{i=1}^{m-1} f(x_{2i}) + 4\sum_{i=1}^{m} f(x_{2i-1})] - \frac{mh^5}{90}f^{(4)}(\eta) \tag{3.22}$$

Program 3.9B contains the subroutine for the Composite Simpson's Rule.

3.2.3 Romberg Integration Formula

Richardson's extrapolation as discussed in the Numerical differentiation section when applied to integration theory (the trapezoidal rule) provides an accurate algorithm for numerical integration. The algorithm is called **Romberg** integration. The composite form of the trapezoidal rule to the general integral is

$$\int_a^b f(x)dx \approx \frac{h}{2}(f_0 + 2f_1 + 2f_2 + \cdots + 2f_{n-1} + 2f_n) \qquad \textbf{(3.23)}$$

where $h=(b-a)/n$, $x_0 \equiv a$ and $x_n \equiv b$. The error approximation is given by

$$E_{Trap_c} = -\frac{h^2}{12}(b-a)f''(\eta) \qquad \textbf{(3.24)}$$

The Romberg method evaluates the above equation for a sequence of decreasing values of h. Usually, the initial value of $h=b-a$ is used. The subintervals are successively halved so that $h_1=h_0/2$, $h_2=h_1/2$, and for the general expression, $h_k=h_{k-1}/2=(b-a)/2^k$.

Setting the right hand side of Equation (3.23) to $S_{k,o}$ and $h=h_k$, a sequence of terms can be developed such that the following forms are derived:

$$S_{0,0} = (h_0/2)[f(a) + f(b)]$$

$$S_{1,0} = (h_1/2)[f(a) + 2f(a+h_1) + f(b)] = (1/2)S_{0,0} + h_1\, f(a+h_1)$$

$$S_{2,0} = (h_2/2)[f(a) + 2f(a+h_2) + 2f(a+2h_2) + 2f(a+3h_2) + f(b)]$$

$$= (1/2)S_{1,0} + h_2[f(a+h_2) + f(a+3h_2)]$$

The general expression is given by

$$S_{k,0} = 1/2 S_{k-1,0} + h_k \sum_{i=1}^{2^k-1} f(a+(2i-1)h_k) \quad \text{for } k=1,2,... \tag{3.25}$$

Approximations using the above method converges to the answer in a rather slow manner. To accelerate the convergence to a solution, we can apply Richardson's extrapolation to the above equation. Combining the error approximation for the composite trapezoidal rule and the equation that introduced this section gives

$$\int_a^b f(x)dx = S_{n-1,0} - \frac{h^2_{n-1}}{12}[f'(b) - f'(a)] + + \frac{(b-a)}{720}h^4{}_n{-1}f^{(4)}(\zeta_{n-1}) \quad \text{where } a< \zeta_{n-1} <b. \tag{3.26}$$

Halving the step size results in

$$\int_a^b f(x)dx = S_{n,0} - \frac{h^2_n}{12}[f'(b) - f'(a)] + + \frac{(b-a)}{720}h^4_n f^{(4)}(\zeta_n) \quad \text{where } a< \zeta_n <b. \tag{3.27}$$

Combining the above two equations to eliminate the h^2_{n-1} terms results in the following equation:

$$S_{k,1} = \frac{4S_{k,0} - S_{k-1,0}}{3} \qquad \text{for each } k = 1,2,.... \tag{3.28}$$

The general term for the Richardson extrapolation applied is

$$S_{k,j} = \frac{4^j S_{k,j-1} - S_{k-1,0}}{4^j - 1} \qquad \text{for } k=1,2,..., \ j=1,2,...,k \tag{3.29}$$

The subroutine in Program 3.10C illustrates the Romberg integration scheme.

3.3 Gaussian Quadrature

The above integration formula are based on equally spaced points for the following general equation:

$$\int_a^b f(x)dx \doteq \sum_{i=0}^{n} w_i f(x_i) \tag{3.30}$$

where w_i are the weights for the n+1 functional values, $f(x_i)$. The Gaussian quadrature formulas are also based on the above equation. The x_i values are not equally spaced and chosen such that the sum of the weighted functional values yields the integral exactly when f(x) is a polynomial of degree 2n+1 or less.

The basis for using Gaussian Quadrature is to be able not only to choose the weighting coefficients but also the locations at which the function is to be evaluated. Therefore, we have twice the number of degrees of freedom in which to arrive at a high accuracy evaluation of the function. A high order function is only highly accurate when the integrand is very smooth. The weighting functions can be chosen to remove singularities from the integral.

We will consider four cases of Gaussian quadrature; (1) Legendre, (2) Laguerre, (3) Hermite, and (4) Chebyshev polynomials. These are well-known polynomial sets that are called orthogonal polynomials. A polynomial is orthogonal if, with respect to the weighting function w(x) on the interval [a,b], the following condition exists:

$$\begin{aligned} &\int_a^b w(x) g_n(x) g_m(x) dx = 0 \quad for \;\; n \neq m \\ &\int_a^b w(x) [g_n(x)]^2 dx = c(n) \neq 0 \quad for \;\; n = m \end{aligned} \tag{3.31}$$

General recursion relations for the four types of polynomials are given below. The associated weights for use in running the programs or subroutines/functions given in each subsection. The n+1 base points are the n+1 roots of the (n+1)th-degree polynomial.

Legendre Polynomials

Legendre Polynomials are orthogonal on the interval [-1,1] with respect to the weighting function

$$w(x) = 1 \tag{3.32}$$

$$P_n(x) = \frac{2n-1}{n} x P_{n-1}(x) - \frac{n-1}{n} P_{n-2}(x) \tag{3.33}$$

Laguerre Polynomials

Laguerre Polynomials are orthogonal on the interval $[0,\infty]$ with respect to the weighting function

$$w(x) = e^{-x} \tag{3.34}$$

$$L_n(x) = (2n-x-1) L_{n-1}(x) - (n-1)^2 L_{n-2}(x) \tag{3.35}$$

Chebyshev Polynomials

Chebyshev Polynomials are orthogonal on the interval $[-1,1]$ with respect to the weighting function

$$w(x) = \frac{1}{\sqrt{1-x^2}} \tag{3.36}$$

$$T_n(x) = 2xT_{n-1}(x) - T_{n-2}(x) \tag{3.37}$$

Hermite Polynomials

Hermite Polynomials are orthogonal on the interval [-∞,∞] with respect to the weighting function

$$w(x) = e^{-x^2} \tag{3.38}$$

$$H_n(x) = 2xH_{n-1}(x) - 2(n-1)H_{n-2}(x) \tag{3.39}$$

Each of the polynomials given above is an nth degree polynomial in x with real coefficients and n well-defined real roots within the integration interval. An arbitrary nth-degree polynomial

$$p_n(x) = \sum_{i=0}^{n} \beta_i Z_i \tag{3.40}$$

is a linear function of any of the above four orthogonal polynomial families where $Z_i(x)$ is an nth-degree polynomial of one of the families of orthogonal polynomials.

To illustrate the use of the Gaussian quadrature, an example is given in Program 3.11P. The routine is coded in PASCAL. Note that it is not necessary to calculate the roots or weighting functions since these are extensively tabulated elsewhere in the literature. In addition, there are subroutines that are written and can be written to tabulate the necessary parameters.

Consider the general integral given by

$$\int_a^b \omega(x) f(x) dx. \tag{3.41}$$

The variable $\omega(x)$ is a weighting function greater than zero.

The integral above can be replaced by an interpolation polynomial, $p_n(x)$. This leads to the following:

$$\int_a^b \omega(x) f(x) dx \simeq \int_a^b \omega(x) p_n(x) dx = \sum_{i=0}^{n} \alpha_i f_i. \tag{3.42}$$

The term α_i using Lagrangian interpolation is

$$\alpha_i = \int_a^b \omega(x) l_i(x) dx. \tag{3.43}$$

3.4 Problems

These exercises are intended to provide you with practice in using your computer to solve numerical problems. Using your preferred language and the programming techniques introduced in this section, create programs to compute the following:

3-1) A new device must be tested at a temperature of 0 degrees. During the cooling of the test chamber the temperature was monitored an shown to have the following values at the times indicated:

Time	Temperature
0 min	72 degrees
10 min	56 degrees
20 min	42 degrees
30 min	30 degrees
40 min	20 degrees
50 min	12 degrees
60 min	5 degrees

What is the rate of cooling 30 min into the cooling process?

3-2) What is the rate of cooling 70 min into the cooling process described in problem 3-1?

3-3) At what time will the temperature in the test chamber of problem 3-1 reach 0 degrees?

Hint: Compute the rate of cooling at several time intervals, then use an extrapolation techniques from chapter 2 to compute the time at which the temperature will reach 0 degrees.

3-4) During testing of a new rocket, the motor burned for 9 seconds and the speed was recorded at the following intervals:

Time	Speed
0 sec	0 kmph
1 sec	3 kmph
2 sec	8 kmph
3 sec	15 kmph
4 sec	24 kmph
5 sec	35 kmph
6 sec	48 kmph
7 sec	63 kmph
8 sec	80 kmph
9 sec	99 kmph

Approximately how far did the rocket travel during the acceleration phase of its flight?

Hint: Use the Sum of function from chapter 1 with an integration function from this chapter.

3-5) The approximation of the distance traveled in problem 3-4 has proven to be to inaccurate. Compute a more accurate distance traveled by using the interpolation techniques from chapter 2 to determine the speed of the rocket every tenth of a second. Then use this derived information to compute a more accurate distance traveled.

3-6) A similar rocket to the one in problem 3-4 is being build with a motor that will burn for 10 seconds. How far will the new rocket travel during the acceleration phase of its flight?

Hint: Use an extrapolation techniques from chapter 2 to compute the final rocket velocity at time 10 sec. You can increase the accuracy of the answer by using the same method as in problem 3-5.

3-7) What is the anticipated maximum acceleration that the test equipment to be used on this new rocket in problem 3-6 be required to withstand?

Hint: The derivative of the velocity is the acceleration.

3-8) A new land test vehicle is being accelerated on a timing track. The time that it took for the vehicle to reach each check point and the distance from the starting point of the check point is shown below:

Distance	Elapsed Time
0 feet	9.8 sec
10 feet	14.0 sec
25 feet	16.1 sec
50 feet	17.2 sec
85 feet	18.2 sec
130 feet	20.0 sec
185 feet	20.9 sec
250 feet	21.7 sec
325 feet	22.2 sec

What was the velocity of the vehicle as it passed the 100 foot check point?

3-9) What was the maximum velocity and the maximum acceleration of the test vehicle in problem 3-8?

3-10) How far does the test vehicle in problem 3-8 travel between the time it reaches its maximum acceleration and the time it reaches its maximum velocity?

Chapter 3 Programs

Program 3.1B Lagrange Differentiation

```
SUB LagrangeDif (X#(), Y#(), xTerm#, Zero%, Count%, Result#)
'
'THIS PROGRAM COPYRIGHT: (C) 1991 by LCPsoftware
'AUTHORS:    Benjamin J. Landry and Edward Champion
'TEXT BOOK: NUMERICAL METHODS FOR ENGINEERING APPLICATIONS
'
'EQUATION:   (03.05) EQU_0305 Lagrange Differentiation
'
'PURPOSE:    Finds f'(x) using Lagrange Differentiation for
'            any x value.  Requires an array of known values
'            for x and for f(x).
'
'ARGUMENTS: X(min)...X(max) Array of equally spaced points
'            Y(min)...Y(max) Array of f(x) function values
'
'            xTerm  An X value with X(min) < xTerm < X(max)
'
'            Zero   The element pair in the X and Y arrays to
'                   be used as the 0th element, the value of
'                   X(Zero) should be close to the value
'                   xTerm for which f'(x) is sought.
'
'            Count  The number of elements to use computing.
'
'FORMULA:
'
'            Count
'     Result = Σ (Y(element) * LamdaDif(element))
'            element=0
'
'            LamdaDif()  See description in module LamdaDif()
'
'COMMENT:
'            The term Accumulator# not being STATIC,
'            clears it by default on entry when using
'            Microsoft Professional Basic.
'
'---------------------- end introduction ------------------

TermCount% = Zero% + Count% - 1

FOR i% = Zero% TO TermCount%

   CALL LamdaDif(X#(), i%, xTerm#, Zero%, Count%, lResult#)

   Accumulator# = Accumulator# + (Y#(i%) * lResult#)

NEXT i%

Result# = Accumulator#

END SUB
```

Program 3.2B Lagrange Lambda Determination

```
SUB LamdaDif (X#(), i%, xTerm#, Zero%, Count%, Result#)
'
'THIS PROGRAM COPYRIGHT: (C) 1991 by LCPsoftware
'AUTHORS:   Benjamin J. Landry and Edward Champion
'TEXT BOOK: NUMERICAL METHODS FOR ENGINEERING APPLICATIONS
'
'EQUATION:  (03.06) EQU_0306 Lagrange Lambda Differentiation
'
'PURPOSE:   Finds the Lagrange Lambda Differentiator for any
'           x value using any element in an array of known
'           values as the base point and using any number of
'           known points for the computation.
'
'ARGUMENTS: X(min)...X(max) Array of equally spaced points
'
'           i%    The displacement element for which Lambda
'                 is to be computed.
'
'           xTerm An X value within X(min) < xTerm < X(max)
'
'           Zero  The element pair in the X and Y arrays
'                 used as the 0th element, the value of
'                 X(Zero) should be close to the value
'                 xTerm for which f(x) is sought.
'
'          Count  The number of elements used in computing.
'
'FORMULA:                        Count       Count
'          L'n = (2 * xTerm - Σ  X(j)) /  π ( X(i) - X(j))
'                             j =0         j =0
'                             j<>i         j<>i
'
'COMMENT:  As this has a product loop, the Denominator must
'          be seeded with 1.
'
'---------------------- end introduction ------------------
TermCount% = Zero% + Count% - 1
Numerator# = 2 * xTerm#
Denominator# = 1

FOR j% = Zero% TO TermCount%

      IF j% <> i% THEN
            Numerator# = Numerator# - X#(j%)
            Denominator# = Denominator# * (X#(i%) - X#(j%))
      END IF

NEXT j%

Result# = Numerator# / Denominator#

END SUB
```

Program 3.3B Small Derivative

```
SUB SmallDerivative (Y#(), X#(), Zero%, Result#)
'
'THIS PROGRAM COPYRIGHT: (C) 1991 by LCPsoftware
'AUTHORS:    Benjamin J. Landry and Edward Champion
'TEXT BOOK: NUMERICAL METHODS FOR ENGINEERING APPLICATIONS
'
'EQUATION:   (03.09) EQU_0309 Small Derivative
'
'PURPOSE:    Estimates the derivative of a function at any X
'            point based on the tabulated data in the two
'            arrays x AND f(x) by computing the linear slope
'            between the two know points either side of X(0)
'
'ARGUMENTS: X(min)...X(max) The x array of known arguments
'           Y(min)...Y(max) The y array of results of f(x)
'
'           Zero      The array entry in the X and Y arrays
'                     to be used as the 0th elements, i.e.;
'                     the element for which the derivative
'                     is being sought.
'
'FORMULA:   f'(x) = f(x+h) - f(x-h) / 2h
'
'COMMENTS:  As the points in the X array are equally spaced
'           at a fixed distance of h the following applies:
'
'            h  = X(1) - X( 0)
'      ^    2h  = X(1) - X(-1)
'      |
'      |    f(x+h) = Y( 1)
'      |    f(x)   = Y( 0)
'      |    f(x-h) = Y(-1)
'      |                                              *Y(n+5)
'      |                                                 :
'      |                                      *Y(n+4)   :
'      |    *Y(n)                                :       :
'      |    :       *Y(n+1)         *Y(n+3)      :       :
'      |    :       :       *Y(n+2) :            :       :
'      |    :       :       :       :            :       :
'      |    :       :       :       :            :       :
'      o ---:-------:-------:-------:-------:-------:------>
'           X(n)    X(n+1)  X(n+2)  X(n+3)  X(n+4)  X(n+5)
'
'           When Zero is passed in as n+3, then the slope
'           of the line from point n+2 to n+4 is returned.
'
'---------------------- end introduction ------------------

Twoh# = X#(Zero% + 1) - X#(Zero% - 1)
Result# = (Y#(Zero% + 1) - Y#(Zero% - 1)) / Twoh#

END SUB
```

Program 3.4B Richardson's Approximation

```
SUB RichardsonApproximation (Y#(), X#(), Zero%, r%, Result#)
'
'THIS PROGRAM COPYRIGHT: (C) 1991 by LCPsoftware
'AUTHORS:    Benjamin J. Landry and Edward Champion
'TEXT BOOK: NUMERICAL METHODS FOR ENGINEERING APPLICATIONS
'
'EQUATION:   (03.10) EQU_0310 Richardsons Approximation
'
'PURPOSE:    Estimates the derivative of a function at any X
'            point based on the tabulated data in the two
'            arrays x AND f(x) by computing the linear slope
'            between two know points that are r tabulated
'            entries from the selected X(0) point.
'
'ARGUMENTS: X(min)...X(max) The x array of known arguments
'           Y(min)...Y(max) The y array of results of f(x)
'
'           Zero      The array entry in the X and Y arrays
'                     to be used as the 0th elements, i.e.;
'                     the element for which the derivative
'                     is being sought.
'
'           r         Any integer value used to offset the
'                     selected extrapolation points either
'                     side of X(0)
'
'FORMULA:
'           f'(x) = f(x+rh) - f(x-rh) / 2rh
'
'COMMENTS:  As the points in the X array are equally spaced
'           at a fixed distance of h the following applies:
'
'            h  = X(1) - X( 0)
'           2h  = X(1) - X(-1)
'           2rh = X(r) - X(-r)
'
'           f(x+rh) = Y( r)
'           f(x)    = Y( 0)
'           f(x-rh) = Y(-r)
'
'           Referring to the graph in the comments section of
'           equation 03.09, if Zero is passed in as n+3 and
'           r is passed in as 2, then the slope of the line
'           from the point n+1 to n+5 is returned.
'
'--------------------- end introduction ------------------

h# = X#(Zero% + 1) - X#(Zero%)

Result# = (Y#(Zero% + r%) - Y#(Zero% - r%)) / (2 * r% * h#)

END SUB
```

Program 3.5B Richardson Extrapolation

```
SUB RichardsonExtrapolation (Y#(), X#(), Zero%, r%, Result#)
'
'THIS PROGRAM COPYRIGHT: (C) 1991 by LCPsoftware
'AUTHORS:   Benjamin J. Landry and Edward Champion
'TEXT BOOK: NUMERICAL METHODS FOR ENGINEERING APPLICATIONS
'
'EQUATION:  (03.11) EQU_0311 Richardsons Extrapolation
'
'PURPOSE:   Estimates the derivative of a function at any X
'           point based on the tabulated data in the two
'           arrays x AND f(x) by computing the weighted
'           average of a linear slope between 2 know points
'           that are r tabulated entries from selected X(0)
'           point and the tabulated entries adjacent to X(0)
'
'ARGUMENTS: X(min)...X(max) The x array of known arguments
'           Y(min)...Y(max) The y array of results of f(x)
'
'           Zero     The array entry in the X and Y arrays
'                    to be used as the 0th elements, i.e.;
'                    the element for which the derivative
'                    is being sought.
'
'           r        Any integer value used to offset the
'                    selected extrapolation points either
'                    side of X(0)
'
'FORMULA:
'                3
'          (r^*(f(x+h) - f(x-h)))  -  (f(x+rh) - f(x-rh))
'     f'(x) = ---------------------------------------------
'                                    2
'                    2 * h * r * (r^ -1)
'
'           Referring to the graph in the comments section of
'           equation 03.09, if Zero is passed in as n+3 and
'           r is passed in as 2, then a weighted average of
'           the slope between the points n+1, n+5 and of the
'           slope between the points n+2 and n+4 is returned
'
'--------------------- end introduction ------------------

h# = X#(Zero% + 1) - X#(Zero%)

Weight% = r% ^ 3
Slope1# = Y#(Zero% + 1) - Y#(Zero% - 1)
Slope2# = Y#(Zero% + r%) - Y#(Zero% - r%)
Averager# = 2 * h# * r% * (r% ^ 2 - 1)

Result# = ((Weight% * Slope1#) - Slope2#) / Averager#

END SUB
```

Program 3.6B Trapezoidal Integration

```
SUB Trapezoidal (Y#(), X#(), LowLimit%, HiLimit%, Result#)
'
'THIS PROGRAM COPYRIGHT: (C) 1991 by LCPsoftware
'AUTHORS:   Benjamin J. Landry and Edward Champion
'TEXT BOOK: NUMERICAL METHODS FOR ENGINEERING APPLICATIONS
'
'EQUATION:  (03.13) EQU_0313 Trapezoidal Integration
'
'PURPOSE:   Estimates the integral of a function between
'           the two points in array elements X0 and X1 by
'           computing the area of the trapezoid defined by
'           these two X points and their f(x) values.
'
'ARGUMENTS: X(min)...X(max) The x array of known arguments
'           Y(min)...Y(max) The y array of results of f(x)
'
'           nLimit% - The array entry in the X and Y arrays
'           to be used as the end point elements, i.e.; the
'           elements that bound the area being sought.
'
'FORMULA:   X(1)
'            |              x1 - x0
'               f(x)dx = --------- * [ f(x0) + f(x1) ]
'            J                 2
'           X(0)
'
'COMMENTS:  Many of the functions that have been described
'           in the previous section dealt with tabulated
'           results.  Although the Trapezoidal rule does
'           not depend on the tabulated entries, this method
'           provides a consistent method for passing many
'           argument required for this type of computation.
'           When NOT using tabulated data call this function
'           with the following code fragments:
'
'           Module level code:
'           DIM X(0 TO 1) AS DOUBLE, Y(0 to 1) AS DOUBLE
'
'           Sub-routine level code:
'           X(0) = (SelectedLowerLimit)
'           X(1) = (SelectedUpperLimit)
'           Y(0) = FunctionName(X(0))
'           Y(1) = FunctionName(X(1))
'           CALL Trapezoidal(Y#(), X#(), 0, 1, Result#)
'
'---------------------- end introduction ------------------

h# = X#(HiLimit%) - X#(LowLimit%)

Result# = (h# / 2) * (Y#(LowLimit%) + Y#(HiLimit%))

END SUB
```

Program 3.7B Simpson Integration

```
SUB Simpson (Y#(), X#(), LowLimit%, Delta%, Result#)
'
'THIS PROGRAM COPYRIGHT: (C) 1991 by LCPsoftware
'AUTHORS:    Benjamin J. Landry and Edward Champion
'TEXT BOOK: NUMERICAL METHODS FOR ENGINEERING APPLICATIONS
'
'EQUATION:   (03.14) EQU_0314 Simpson's Rule
'
'PURPOSE:    Estimates the integral of a function between
'            the two points in array elements X0 and X2 by
'            Simpson's Rule.  Requires three equally spaced
'            arguments in the X array and three f(x) values.
'
'ARGUMENTS: X(min)...X(max) The x array of known arguments
'           Y(min)...Y(max) The y array of results of f(x)
'
'           LowLimit% - The array entry in the X and Y
'           arrays where the low end point values are.
'
'           Delta% - The number of array entries above the
'           LowLimit where the mid point value is stored.
'
'FORMULA:   X(2)
'            |           h
'   Formula: |  f(x)dx = --- * [ f(x0) + 4f(x1) + f(x2) ]
'            |           3
'           X(0)
'
'COMMENTS:  When NOT using tabulated data call this function
'           with the following code fragments:
'
'           Module level code:
'           DIM X(0 TO 2) AS DOUBLE, Y(0 to 2) AS DOUBLE
'
'           Sub-routine level code:
'           X(0) = (SelectedLowerLimit)
'           X(1) = (SelectedLowerLimit + SelectedDelta)
'           X(2) = (SelectedLowerLimit + 2 * SelectedDelta)
'           Y(0) = FunctionName(X(0))
'           Y(1) = FunctionName(X(1))
'           Y(2) = FunctionName(X(1))
'           CALL Trapezoidal(Y#(), X#(), 0, 1, Result#)
'-------------------- end introduction -------------------

h# = X#(LowLimit% + Delta%) - X#(LowLimit%)

fZero# = Y#(LowLimit%)
fOne# = Y#(LowLimit% + Delta%)
fTwo# = Y#(LowLimit% + 2 * Delta%)

Result# = (h# / 3) * (fZero# + (4 * fOne#) + fTwo#)
END SUB
```

Program 3.8B Composite Trapezoidal Integration

```
SUB CompositeTrapezoidal (Y#(), x#(), a%, b%, Result#)
'
'THIS PROGRAM COPYRIGHT: (C) 1991 by LCPsoftware
'AUTHORS:    Benjamin J. Landry and Edward Champion
'TEXT BOOK:  NUMERICAL METHODS FOR ENGINEERING APPLICATIONS
'
'EQUATION:   (03.19) EQU_0319 Composite Trapezoidal Integrate
'
'PURPOSE:    Estimates the integral of a function between
'            the points contained in array element X(Alpha)
'            through X(Beta) using all tabulated values that
'            exist between these two limits via the Composite
'            Trapezoid Rule.
'
'ARGUMENTS:
'            X(min)...X(max) The x array of known arguments
'            Y(min)...Y(max) The y array of results of f(x)
'
'            a and b are the array entry in the X and Y
'                    arrays that contain the values of
'                    the end points to be used.  i.e.;
'                    the elements that bound the area
'                    being sought.
'
'FORMULA:
'            Xb
'            |          h                   b-1
'            |  f(x)dx = --- * [ f(Xa) + 2 * Σ f(Xi) + f(Xb)]
'            |          2                   i=a+1
'            Xa
'
'COMMENTS:
'            As the points in the X array are equally spaced
'            at a fixed distance of h, h can be computed by
'            subtracting the content on any two consecutive
'            values in the X array.  For simplicity, the
'            first two elements to be used in this equation
'            have been selected, thus the following applies:
'
'            h  = X(a+1) - X( a)
'
'--------------------- end introduction ------------------

h# = (x#(a% + 1) - x#(a%))

FOR i% = (a% + 1) TO (b% - 1)
   SumTerm# = SumTerm# + Y#(i%)
NEXT i%

Result# = (h# / 2) * (Y#(a%) + (2 * SumTerm#) + Y#(b%))

END SUB
```

Program 3.9B Composite Simpson Integration

```
SUB CompositeSimpson (Y#(), X#(), a%, b%, Result#)
'
'THIS PROGRAM COPYRIGHT: (C) 1991 by LCPsoftware
'AUTHORS:    Benjamin J. Landry and Edward Champion
'TEXT BOOK: NUMERICAL METHODS FOR ENGINEERING APPLICATIONS
'
'EQUATION:   (03.20) EQU_0320 Composite Simpson Integration
'
'PURPOSE:    Estimates the integral of a function between
'            the points contained in array element X(Alpha)
'            through X(Beta) using all tabulated values that
'            exist between these two limits via the Composite
'            Simpson Rule.
'
'ARGUMENTS:
'            X(min)...X(max) The x array of known arguments
'            Y(min)...Y(max) The y array of results of f(x)
'
'            a and b are the array entry in the X and Y
'                    arrays that contain the values of
'                    the end points to be used.  i.e.;
'                    the elements that bound the area
'                    being sought.
'
'FORMULA:
'            Xb
'             |
'             |         h             b-1
'                 f(x)dx = --- *[ f(Xa) +4 *Σ f(X(a+i) +
'             |                3             i=1,3,5...
'            Xa
'                                            b-1
'                                          2 *Σ f(X(a+j) +f(Xb)]
'                                            j=2,4,6...
'
'---------------------- end introduction ------------------

h# = (X#(a% + 1) - X#(a%))

FOR i% = (a% + 1) TO (b% - 1) STEP 2
   Sum4Times# = Sum4Times# + Y#(i%)
NEXT i%

FOR j% = (a% + 2) TO (b% - 1) STEP 2
   Sum2Times# = Sum2Times# + Y#(j%)
NEXT j%

SumOfSumTerm# = (4 * Sum4Times#) + (2 * Sum2Times#)
SumTerm# = Y#(a%) + SumOfSumTerm# + Y#(b%)

Result# = (h# / 3) * SumTerm#

END SUB
```

Program 3.10C Romberg Integration

```
/* Romberg integration.  The function fct is evaluated starting
at 0  until it reaches approximately zero. */

#include <stdio.h>
#include <stdlib.h>
#include <math.h>
#define tol 0.00001 /* Convergence criteria */
#define range_incr 0.2 /* Function division size */
#define maxit 10 /* Maximum iterations allowed for reaching
convergence */


float fct(float x)  /* Example function with an integral */
{                   /* value of 0.443112 */
     float f;
     f = x*x*exp(-x*x);
     return(f);
}


float romb_value(float a, float b)
/* This function carries out Romberg integration on the
function fct */
/* using limits a and b. */
{
     float h; /* Subinterval size */
     float sum,x;
     float extr[9][9]; /* Used to hold intermediate values and
the integral */
     /* increment */
     int i,j,k,l; /* Counters */

/* Trapezoidal rule is used first to compute the integral over
*/
/* 10 subintervals */
     h = (b-a)/10;
     sum = fct(a) + fct(b);
     x = a;
     for (i=2; i<=10; i++)
     {
          x = x+h;
          sum = sum + fct(x)*2;
     }
     extr[1][1] = h/2*sum;

/* The integral is now computed with half of h and then
extrapolated. */
/* This process is repeated until convergence or the maximum
number */
/* of iterations is reached. */
     for (i=1; i<=maxit; i++)
     {
          h = h/2;
          x = a+h;
          k = 10*pow(2,i);
          for (j=2; j<=k; j=j+2)
          {
```

```
                        sum = sum + fct(x)*2;
                        x = x + h + h;
                  }
                  extr[1][i+1] = h/2 * sum;
                  for (l=1; l<=i; l++)
                        extr[l+1][i+1] = extr[l][i+1]+1/(pow(4,l)-1)
                  *(extr[l][i+1]-extr[l][i]);

      /* Check for convergence */
                  if (fabs(extr[i+1][i+1]-extr[i][i+1]) - tol <= 0)
                        return(extr[i+1][i+1]);
            }
      /* If loop is complete then convergence was not met. */
            printf("\nTolerance not met after %d iterations.\n",i-1);
            printf("Convergence value =
%e\n",fabs(extr[i][i]-extr[i-1][i]));
            exit(0);
      }


      main()
      {
      float integral = 0; /* Value of integral */
      float a = 0; /* a is the lower range of integral, replaced by
      b */
      float b = range_incr; /* b is the upper range of integral,
                               /* incremented by range_incr */
      float increment; /* Increment to the integral value */
      int i;

      clrscr();
      printf("\t A \t B \t Integral \t Increment \n");
      for (i=1; i<=100; i++)
      {
            increment = romb_value(a,b);
            integral = integral + increment;

printf("\t%7.3f\t%7.3f\t%9.6f\t%9.8e\n",a,b,integral,increment);
            if (fabs(increment) < tol) break;
            a = b;
            b = b + range_incr;
      }
      }
```

Program 3.11P Gauss Legendre Quadrature

```
PROGRAM GaussInt;
{$N+}

USES
  crt;

CONST
  k = 6; { Order of polynomial used in finding solution + 1 }
         { eg. k=6 means a 5th order polynomial is used }
TYPE
  orderarray = ARRAY[1..6] OF DOUBLE;

VAR
  integral : REAL;
  epsilon,weight  : orderarray;
  i,midk : INTEGER;


FUNCTION F(T:Real):Real; { Example function }
  BEGIN
    F := EXP(-SQR(1+T)/4)/2
  END;


PROCEDURE Legendre (k : INTEGER;
                    VAR epsilon,weight : orderarray);
{ Returns the epsilon and weight values for Gauss-Legendre quadrature
}

VAR
  i : INTEGER; { counter for generating lower epsilons and weights }

BEGIN { procedure Legendre }
  IF k > 6 THEN
    BEGIN { error check }
      WRITELN('The given value of k is greater than 6');
      WRITELN('Execution stopped in procedure Legendre')
    END { error check }
  ELSE

    BEGIN { execution section }
      CASE k OF
        1 : BEGIN
              epsilon[1] := 0;
              weight[1]  := 2;
            END;
        2 : BEGIN
              epsilon[2] := 0.577350269189626;
              weight[2]  := 1;
            END;
        3 : BEGIN
              epsilon[2] := 0;
              epsilon[3] := 0.774596669241483;
              weight[2]  := 0.888888888888889;
              weight[3]  := 0.555555555555556;
            END;
```

```
        4 : BEGIN
              epsilon[3] := 0.339981043584856;
              epsilon[4] := 0.861136311594053;
              weight[3]  := 0.652145154862546;
              weight[4]  := 0.347854845137454;
            END;
        5 : BEGIN
              epsilon[3] := 0;
              epsilon[4] := 0.538469310105683;
              epsilon[5] := 0.906179845938664;
              weight[3]  := 0.568888888888889;
              weight[4]  := 0.478628670499366;
              weight[5]  := 0.236926885056189;
            END;
        6 : BEGIN
              epsilon[4] := 0.238619186083197;
              epsilon[5] := 0.661209386466265;
              epsilon[6] := 0.932469514203152;
              weight[4]  := 0.467913934572691;
              weight[5]  := 0.360761573048139;
              weight[6]  := 0.171324492379170;
            END
      END; { case of }
      FOR i:=1 TO k DIV 2  DO
        BEGIN
          epsilon[i] := -epsilon[k+1-i];
          weight[i]  := weight[k+1-i];
        END
    END { execution section }
END; { procedure Legendre }


BEGIN { main program }
      { The integral of F(x) over -1<x<1 is approximated by the }
      { sum F(epsilon[i])*weight[i], i=1 to k. }
  CLRSCR;
Legendre (k,epsilon,weight);
  midk := k DIV 2;
  integral := 0;
  FOR i:=1 TO midk DO
    integral:=integral+weight[i]*(F(epsilon[i])+F(epsilon[k+1-i]));
  IF ODD(k) THEN
    BEGIN
      midk := midk + 1;
      integral := integral + weight[midk]*F(epsilon[midk])
    END;
  WRITELN('Gauss quadrature integral = ',integral)
END. { main program }
```

Program 3.12B Differentiation Demonstration

```
SUB DifferentiatingDemonstration
'
'THIS PROGRAM COPYRIGHT: (C) 1991 by LCPsoftware
'AUTHORS:    Benjamin J. Landry and Edward Champion
'TEXT BOOK: NUMERICAL METHODS FOR ENGINEERING APPLICATIONS
'
'EQUATION:  Differentiation Demonstration
'
'---------------------- end introduction ------------------
DIM X#(0 TO 100), Y#(0 TO 100)
CLS               'Output sent to last listed device
OutputDevice$ = "DAT_0344.ASC"          'File name
OutputDevice$ = "LPT1:"                 'Printer
OutputDevice$ = "SCRN:"                 'Screen

SerFile% = FREEFILE
OPEN OutputDevice$ FOR OUTPUT AS #SerFile%
'                                                 3
FOR Count% = 0 TO 100          'Seed Arrays with f(x) = X^
   X#(Count%) = Count% / 10   'Use X from 0 to 10 step .1
   Y#(Count%) = X#(Count%) * X#(Count%) * X#(Count%)
NEXT Count%

PRINT #SerFile%, "Test data loaded per equation:"
PRINT #SerFile%, "": PRINT #SerFile%, "  3"
PRINT #SerFile%, "X^  with delta on X ="; X#(1) - X#(0)
PRINT #SerFile%, "": PRINT #SerFile%, ""
PRINT #SerFile%, "  X   f'(X)  LagrangeDif SmallDeriva ";
PRINT #SerFile%, "RichardAppr RichardExtr"
PRINT #SerFile%, "----- ------ ---------- ---------- ";
PRINT #SerFile%, "---------- ----------"

FOR Zero% = 30 TO 75 STEP 1
   PRINT #SerFile%, USING "##.#"; X#(Zero%);

   Result# = 3 * X#(Zero%) ^ 2
   PRINT #SerFile%, USING " ###.###"; Result#;

   CALL LagrangeDif(X#(), Y#(), X#(Zero%), Zero%, 3, r#)
   PRINT #SerFile%, USING " ###.#######"; r#;

   CALL SmallDerivative(Y#(), X#(), Zero%, Result#)
   PRINT #SerFile%, USING " ###.#######"; Result#;

   CALL RichardsonApproximation(Y#(), X#(), Zero%, 3, r#)
   PRINT #SerFile%, USING " ###.#######"; r#;

   CALL RichardsonExtrapolation(Y#(), X#(), Zero%, 3, r#)
   PRINT #SerFile%, USING " ###.#######"; r#
NEXT Zero%
CLOSE #SerFile%
END SUB
```

Program 3.13B Integration Demonstration

```
SUB IntegrationDemonstration
'
'THIS PROGRAM COPYRIGHT: (C) 1991 by LCPsoftware
'AUTHORS:   Benjamin J. Landry and Edward Champion
'TEXT BOOK: NUMERICAL METHODS FOR ENGINEERING APPLICATIONS
'
'EQUATION:  (03.45) EQU_0345 Integration Demonstration
'---------------------- end introduction ------------------
DIM X#(0 TO 100), Y#(0 TO 100)
CLS                'Output sent to last listed device
OutputDevice$ = "DAT_0345.ASC"          'File name
OutputDevice$ = "LPT1:"                 'Printer
OutputDevice$ = "SCRN:"                 'Screen
SerFile% = FREEFILE
OPEN OutputDevice$ FOR OUTPUT AS #SerFile%      '       3
FOR Count% = 0 TO 100          'Seed Arrays with f(x) = X^
   X#(Count%) = Count% / 10   'Use X from 0 to 10 step .1
   Y#(Count%) = X#(Count%) * X#(Count%) * X#(Count%)
NEXT Count%
PRINT #SerFile%, "Test data loaded per equation:"
PRINT #SerFile%, "": PRINT #SerFile%, "          3"
PRINT #SerFile%, "f(x) = X^  with tabulated delta of X =";
PRINT #SerFile%, X#(1) - X#(0):  PRINT #SerFile%, ""
PRINT #SerFile%, "               x2"
PRINT #SerFile%, "               |"
PRINT #SerFile%, " x1   x2      f(x)   Trapezod  Simpson";
PRINT #SerFile%, "  Composite Composite"
PRINT #SerFile%, "             x1|"; SPACE$(27);
PRINT #SerFile%, "Trapezod  Simpson "
PRINT #SerFile%, "---- ----  ---------  --------  ";
PRINT #SerFile%, "--------  --------  --------"

FOR Alpha% = 30 TO 73 STEP 1: Beta% = Alpha% + 10
   PRINT #SerFile%, Alpha%; " "; Beta%;
   Answer# = (X#(Beta%) ^ 4 / 4) - (X#(Alpha%) ^ 4 / 4)
   PRINT #SerFile%, USING "  ###.#### "; Answer#;

   CALL Trapezoidal(Y#(), X#(), Alpha%, Beta%, Result#)
   PRINT #SerFile%, USING "  ###.####"; Result#;

   Delta% = (Beta% - Alpha%) / 2
   CALL Simpson(Y#(), X#(), Alpha%, Delta%, Result#)
   PRINT #SerFile%, USING "  ###.####"; Result#;

   CALL CompositeTrapezoidal(Y#(), X#(), Alpha%, Beta%, R#)
   PRINT #SerFile%, USING "  ###.####"; R#;

   CALL CompositeSimpson(Y#(), X#(), Alpha%, Beta%, Result#)
   PRINT #SerFile%, USING "  ###.####"; Result#
NEXT Alpha%
CLOSE #SerFile%
END SUB
```

4

Vectors, Matrices, and Linear Algebraic Equations

4.1 Introduction

This chapter covers the basics required for manipulation of vectors (one column or row arrays), matrices (multi-column and row arrays), and the application of these techniques to the solving sets of linear algebraic equations. This material is fundamental to scientific and engineering numerical analysis. The concepts of this chapter using direct solving of the equation set are applicable to other areas of analysis that follow later in the book. For example, Chapter 9 on Finite Element Analysis relies on these techniques.

4.2 Vectors

A real N-dimensional vector **X** is an ordered set of N real numbers or complex numbers and can be written as

$$X = (x_1, x_2, \ldots, x_N) \tag{4.1}$$

The numbers x_N are called the components of **X**. The set that contains all N-dimensional vectors is called *N-dimensional space*. If the vector is used to specify a point or position in space it is referred to as a position vector. When the vector is used to specify movement between points it is called a displacement vector.

Define another vector **Y** by

$$Y = (y_1, y_2, \ldots, y_N) \quad \textbf{(4.2)}$$

The two vectors defined in Equations (**4.1**) and (**4.2**) are equivalent only if each corresponding coordinate is the same, i.e.

$$x_i = y_i \; for \; i=1,2,\ldots,N \quad \textbf{(4.3)}$$

The following definitions are given without proof and are illustrated through use of the BASIC algorithm given in Programs 4.1B, 4.2B, 4.3B, and 4.4B.

The sum of two vectors is

$$X + Y = (x_1 + y_1, x_2 + y_2, \ldots, x_N + y_N) \quad \textbf{(4.4)}$$

The difference of two vectors is known as a displacement vector and is given by

$$Y - X = (y_1 - x_1, y_2 - x_2, \ldots, y_N - x_N) \quad \textbf{(4.5)}$$

For vectors in N-dimensional space

$$Y - X = Y + (-X) \quad \textbf{(4.6)}$$

If a is a real number, then scalar multiplication is

$$aX = (ax_1, a_{x_2}, \ldots, ax_N) \quad \textbf{(4.7)}$$

If a and b are scalars, then the weighted sum of two vectors is given by

$$aX + bY = (ax_1 + by_1, ax_2 + by_2, \ldots, ax_N + by_N) \tag{4.8}$$

The *dot product* of two vectors is a **scalar** quantity and is given by

$$X \cdot Y = (x_1 y_1 + x_2 y_2 + \cdots + x_N y_N) \tag{4.9}$$

and is illustrated in Program 4.5B.

The length of a vector is defined by

$$|X| = (x_1^2 + x_2^2 + \cdots + x_N^2)^{1/2} \tag{4.10}$$

and is addressed in Program 4.6B.

From the above equations it is easy to show that the formula for distance between two points in N-dimensional space is

$$|Y - X| = [(y_1 - x_1)^2 + (y_2 - x_2)^2 + \cdots + (y_N - x_N)^2]^{1/2} \tag{4.11}$$

See Program 4.7B for implementation.

For vectors, the following properties hold. X, Y, and Z are N-dimensional vectors and a and b are scalars (real numbers).

$$\begin{gathered} Y + X = X + Y \\ \mathbf{0} + X = X + \mathbf{0} = X \\ X - X = X + (-X) = 0 \\ (X + Y) + Z = X + (Y + Z) \\ (a + b)X = aX + bY \\ a(X + Y) = aX + aY \\ a(bX) = (ab)X \end{gathered} \tag{4.12}$$

4.3 Matrices

A matrix is a rectangular array of numbers, either real or complex, arranged systematically in M rows and N columns type of format. We will assign the letter A to denote a matrix with the numbers forming the matrix elements and denoted by a_{ij} and similarly the letter B denotes a matrix with elements b_{ij}. The following statements and properties (similar to vectors) apply. The main diagonal of a square matrix **A** of order n consists of the elements a_{ii}. If all of the elements below the diagonal are zero, the matrix is *upper triangular* and if the elements above the diagonal are zero, the matrix is *lower triangular.*

Matrices that are associated with linear systems can be classified as dense or sparse. Dense matrices have very few, if any, zero elements. The solution of dense systems of equations are usually handled by direct methods. Sparse matrices have very few non-zero elements and matrices of this type arise from solving partial differential equations using finite difference approximations. These types of equation sets are best handled by iterative techniques. With iterative techniques the errors associated in this approximation are minimized since the previous solution is the initial value for the next solution step.

Two matrices are said to be equal only if

$$a_{ij} = b_{ij} \quad \textbf{(4.13)}$$

The sum of two MxN matrices A and B is computed element by element with the following definition:

$$\boldsymbol{A} + \boldsymbol{B} = (a_{ij} + b_{ij}) \; for \; 1 \le i \le M, \; 1 \le j \le N \quad \textbf{(4.14)}$$

Program 4.8B implements Equation (4.14).

The opposite of a matrix is obtained by replacing each element in the matrix with its opposite.

$$-\boldsymbol{A} = (-a_{ij}) \; for \;\; 1 \le i \le M, \;\; 1 \le j \le N \tag{4.15}$$

An expanded implementation of Equation (4.15) for any scalar multiplication is given in Program 4.9B.

The difference is computed element by element with the following definition:

$$\boldsymbol{A} - \boldsymbol{B} = (a_{ij} - b_{ij}) \; for \;\; 1 \le i \le M, \;\; 1 \le j \le N \tag{4.16}$$

See Program 4.10B for implementation.

The properties given in Equation (4.12) for vectors hold for matrices when addition, subtraction, and scalar multiplication are involved.

The product AB of two matrices A and B is defined only when the rows of A contain the same number of elements as the columns of B. The product is then a matrix C whose (i,j)th element is the inner product of the ith row of A and the jth column of B.

$$c_{ij} = \sum_{k=1}^{m} a_{ik} b_{jk} \tag{4.17}$$

Program 4.11B illustrates matrix multiplication.

It is imperative to note that the product AB is not necessarily equivalent to the product BA as it is for scalar quantities. Only the *Identity Matrix* commutes with all n x n matrices. The identity matrix is defined as follows:

$$\begin{aligned} I_n &= a_{ij} = 1 \textit{ for } i = j \\ I_n &= a_{ij} = 0 \textit{ for } i \neq j \end{aligned} \tag{4.18}$$

In shorthand notation for the Identity matrix, the *Kronecker* δ is used.

$$\begin{aligned} \delta_{ij} &= 1 \textit{ for } i = j \\ \delta_{ij} &= 0 \textit{ for } i \neq j \end{aligned} \tag{4.19}$$

The operation of matrix multiplication has several properties given without proof as follows:

$\mathbf{A} = (a_{ij})_{n \times m}$ $\mathbf{B} = (b_{ij})_{m \times k}$ $\mathbf{C} = (c_{ij})_{k \times r}$

$$\begin{aligned} (AB)C &= A(BC) \\ a(AB) &= (aA)B = A(aB) \end{aligned} \tag{4.20}$$

With both **A** and **B** m x k matrices then

$$(A + B)C = CA + CB \tag{4.21}$$

4.3.1 Determinants

A frequently quoted test for the invertibility of a matrix and hence the solution of sets of equations is based on the concept of a **determinant**. If det(**A**) $\neq$ 0, then it is possible to solve for the variables. Determinants are not of practical interest for the solution of linear systems since the calculation of the determinant value is of the same order of difficulty as the calculation of the solution of the system of equations.

Associated with every square matrix **A** of numbers is a number called the *determinant* of the matrix. If $\mathbf{A}=(a_{ij})$, then

$$\det(A) = \sum_p \sigma_p a_{1p_1} a_{2p_2} \cdots a_{np_n} \tag{4.22}$$

where the sum is taken over all n! permutations p of degree n, and σ_p is 1 or -1 depending on whether p is even or odd.

4.4 Linear Algebraic Equations - Direct Solve Methods

The material presented above is background for the main problem at hand encountered in everyday engineering analysis. That is the solving of linear sets of equations that describe some process. Nonlinear equation set solutions will be discussed in Chapter 5.

A set of linear algebraic equations has the following form:

$$\begin{aligned}
a_{11}x_1 + a_{12}x_2 + \cdots + a_1nx_n &= b_1 \\
a_{21}x_1 + a_{22}x_2 + \cdots + a_2nx_n &= b_2 \\
a_{31}x_1 + a_{32}x_2 + \cdots + a_3nx_n &= b3 \\
\cdots \quad \cdots \quad \cdots & \\
a_{m1}x_1 + a_{m2}x_2 + \cdots + a_{mn}x_n &= b_m
\end{aligned} \tag{4.23}$$

The n unknowns are x_j are related by m equations. The quantities a_{ij} and b_i are known values.

If n=m there are as many equations as there are unknowns and there is a high probability that there exists a unique solution to the set of equations.

A solution may fail to be unique if one or more of the equations in the set is a linear combination of the other equations or if all of the equations contain the same variables in exactly the same linear combination. If some of the equations are not exactly linear combinations but are close to being linear combinations of the other equations, the roundoff errors associated with the solution method used will preclude a numerical solution.

Equation (4.22) can be written in the following form:

$$[A]\cdot \boldsymbol{x} = \boldsymbol{b} \tag{4.24}$$

where [A] is matrix coefficients, **x** is the required solution vector, and **b** is the known value vector. Note that in order to solve for the **x** values it is necessary to invert the [A] matrix to arrive at a solution. Methods to accomplish this inversion and retain accuracy are required. The remainder of this chapter focuses on techniques and solution strategies to successfully and accurately solve the given set of linear equations.

4.4.1 Gaussian Elimination When Special Features Exist

A common method of directly solving a set of equations is known as *Gaussian Elimination*. To summarize the process, the matrix is reduced to an upper triangular matrix by a series adding equations (when multiplied by a ratio of the first coefficients of the two equations to be added) . The process requires the user to eliminate the first term in each equation from the second equation to the last until only one unknown is required to be found in the last equation. Once the value for this unknown is found, the remainder of the unknowns in the set may be found by **back substitution**.

Often it is possible to reduce the computational work that is necessary for solving the equation **Ax** = **B** by taking into account special features of the coefficient matrix **A** such as symmetry or sparseness. One example of this is a system of equations that are *Lower Triangular*. A matrix is lower triangular it all of the element above the diagonal are 0. Lower triangular matrices can be solved through the process of **forward** substitution using the following equations

$$x_n = \frac{b_n}{A_{nn}}$$
$$x_i = \frac{1}{A_{ii}} * [\, b_i - \sum_{j=i+1}^{n} A_{ij} * x_j \,] \quad i=n-1,\ n-2 \ldots 1 \tag{4.25}$$

A computer implementation in BASIC of this equation is contained in Program 4.12B.

Another example of this is a system of equations that are *Upper Triangular*. A matrix is upper triangular it all of the element below the diagonal are 0. Lower triangular matrices can be solved through the process of **backward** substitution using the following equations

$$x_1 = \frac{b_1}{A_{11}}$$
$$x_i = \frac{1}{A_{ii}} * [\, b_i - \sum_{j=1}^{i-1} A_{ij} * x_j \,] \quad i=2,3 \ldots n \tag{4.26}$$

A computer implementation is given in BASIC in Program 4.13B of Equation (4.26).

A third example of a matrix $\mathbf{A}=(a_{ij})$ of order n where it is possible to reduce the computational work that is necessary for solving the equation $\mathbf{Ax} = \mathbf{B}$ by taking into account special features of the coefficient matrix is the **tridiagonal** matrix. A tridiagonal matrix is defined as

A matrix $\mathbf{A}=(a_{ij})$ of order n is **tridiagonal** if

$$a_{ij} = 0 \; for \; |i - j|>1 \tag{4.27}$$

A is tridiagonal only if nonzero entries of **A** lie on the diagonal of **A**, a_{ii}, i=1,....,n, or on the subdiagonal of **A**, $a_{i,i-1}$, i=2,.....,n, or the superdiagonal of **A**, $a_{i,i+1}$, i=1,....,n-1. As an example, the following matrix is **tridiagonal**.

$$\begin{bmatrix} 3 & 1 & 0 & 0 \\ 1 & 3 & 1 & 0 \\ 0 & 2 & 2 & 1 \\ 0 & 0 & 1 & 6 \end{bmatrix} \tag{4.28}$$

For solution of tridiagonal systems the following subroutine is given. The diagonal, sub-diagonal, and super-diagonal arrays are stored in a, b, and c respectively. The computed solution vector is stored in the array v. Program 4.14B shows the solving of a matrix by the tridiagonal method.

The process may seem cumbersome but for implementation on a computer, the algorithm is easily implemented. The process works well for numbers of equations in the 40 to 50 range. For larger sets of equations, the user may wish to use iterative techniques outlined later in this chapter.

4.4.2 Gaussian Elimination With Pivoting

Gaussian elimination works quite well unless a diagonal element has a zero value. The problem of a diagonal zero value can be overcome by exchanging the particular row with another row. The first equation is called the **pivotal** equation and the first coefficient is called the pivot. Because problems arise if a pivot is zero, it is reasonable to assume that problems could arise if the pivotal element is very small compared to the other pivotal elements. The small pivot could lead to a large multiplier and the rounding errors in the coefficients of the pivotal equation would be magnified in the next elimination stage of the process. The process to use is to scan the rows below the zero pivot element to determine the element of maximum modules and if necessary interchange the rows so that this element becomes the pivot. The procedure is known as partial ROW pivoting. Other pivotal elements include the elements on the diagonal. Partial ROW pivoting of the complete **A** matrix is illustrated in the Program 4.15B and Program 4.16B for solution with Partial Column Pivoting.

$$\boldsymbol{A}\,x = \boldsymbol{b} \quad (\textit{Row pivots } \boldsymbol{A} \textit{ and } \boldsymbol{b} \textit{ elements}) \tag{4.29}$$

A similar procedure known as partial COLUMN pivoting scans the columns to the right of the pivotal diagonals to determine the element of maximum modules and if necessary interchanges the columns so that the dominant elements are on the diagonal.

$$\boldsymbol{A}\,x = \boldsymbol{b} \quad (\textit{Column pivots } \boldsymbol{A} \textit{ elements only}) \tag{4.30}$$

A FORTRAN program for performing Gaussian elimination with row pivoting is given in Program 4.15F.

4.4.3 Gauss-Jordan Elimination

A variation on the above theme that accomplishes the backsubstitution simultaneously as the subdiagonal elements are reduced is called the Gauss-Jordan Elimination method. Basically, the idea behind the method is to normalize the diagonal elements and proceed to reduce the equation set to a triangulated matrix. The coefficient matrix is augmented by the addition of the known values and an Identity matrix. Let **A** be the coefficient matrix (a_{ij}), **B** the known values (b_i), and **I** the identity matrix ( ).

The augmented matrix is **ABI** and can be written as

$$\begin{array}{l} a_{11}a_{12}\cdots a_{1}j\ b_1\ 1\cdots 0 \\ a_{21}a_{22}\cdots a_{2}j\ b_2\ 0\cdots 0 \\ \cdots\ \cdots\ \cdots \\ a_{nl}a_{n2}\cdots a_{nn}\ b_n\ 0\cdots 1 \end{array} \quad \mathbf{(4.31)}$$

The magnitudes of the coefficients of the main diagonal of the coefficient matrix should be as large as possible relative to the off-diagonal elements. If a main diagonal element is small there is a possibility that due to round-off errors, the matrix may appear to be singular and thus the inverse of the matrix would not exist. However the existence of a small diagonal element value does not necessarily mean that the coefficient matrix is singular. By reordering the rows (equations) or columns (variables), it is possible to avoid this problem in the algorithm by finding many possible large diagonal element terms. Therefore the coefficient matrix can be arranged to produce diagonal dominance by rearranging the rows and/or columns. The largest element magnitudes are then guaranteed to be on the main diagonal. This modification to the Gauss-Jordan algorithm is known as the **maximum pivot strategy**. This algorithm is shown in Programs 4.17B and 4.17F.

4.4.4 LU Decomposition

The procedure for LU Decomposition is to decompose the matrix **A** into the product of two matrices, **L** and **U**, where **L** is the lower triangular matrix and **U** is the upper triangular matrix. From our basic equation, **Ax=b**, then

$$LUx = b \tag{4.32}$$

A new column vector is introduced

$$y = Ux \tag{4.33}$$

so that Equation (4.29) is replaced by two equation systems that are easy to solve because of their triangular form. These equations are given below.

$$Ly = b \tag{4.34}$$

and

$$Ux = y \tag{4.35}$$

The solution of (4.33) gives the column vector **y** for the right hand side of (4.34). Equation (4.35) is then solved for x.

How do you factorize an **A** matrix for the LU Decomposition scheme? The most direct way is to multiply the general forms of **L** and **U** and compare the elements of the product matrix with those of **A**. A system of n^2 equations in the n^2+n unknowns l_{ij} and u_{ij} is produced so that factorization of **A** is not unique. In fact, arbitrary values can be assigned to n of the unknowns. In order to produce a unique set of factors, the diagonal elements of either **L** or **U** are set to the value of 1. If the diagonal elements of **L** are set to unity, the process is known as Doolittle's algorithm, and if the diagonal elements of the **U** matrix are set to 1, the process is known as Crout's algorithm.

The FORTRAN program in Program 4.18F illustrates the LU Decomposition method. Program 4.18B shows a BASIC form of the algorithm for LU decomposition.

4.4.5 Jacobi and Gauss-Seidel Iterative Methods

Iterative methods for solutions of systems of equations are those that start with an initial approximation and by applying a suitable algorithm, lead to successively better approximations. These methods work quite well when the system of equations is large and most of the coefficients are zero. This situation occurs most often when approximating partial differential equations using the finite difference technique. In addition, these techniques work for a matrix **A** that is *strictly diagonally dominant* as defined in the following equation.

$$|a_{ij}| \geq \sum_{\substack{i=1 \\ j\neq i}}^{n} |a_{ij}| \quad for\ i=1,2,\ldots,n \tag{4.36}$$

Being diagonally dominant insures convergence of the methods.

4.4.5.1 Jacobi's Method

The following algorithm is given for Jacobi's method followed by the programming of the algorithm.

> Given: n (order of the matrix A)
>
> $A = a_{ij}$ for i=1,...n and j=1,...n
>
> $b = b_j$ for j=1,...n
>
> Set initial values of solution vector $x_i = b^i/a_{ii}$ for i=1,...n.
>
> Repeat for k=0,1,2,.... until convergence for i=1,2,....n

The expression for the x_i value is given in Equation 4.35.

$$x_i^{k+1} = \frac{b_i - a_{i1}x_1^{\,k} - \dots - a_{ii-1}x_{i+1}^{\,k} - a_{ii-1}x_{i+1}^{\,k} - \dots - a_{in}x_n^{\,k}}{a_{ii}} \qquad \textbf{(4.37)}$$

Program 4.19F illustrates the implementation of the Jacobi Method.

4.4.5.2 Gauss-Seidel Method

The Gauss-Seidel method is a modification of the Jacobi method in which the most recent approximations to the solution are used immediately as they become available. In general, this method converges more rapidly to the solution than the Jacobi method. In order to achieve faster convergence a method of successive over-relaxation may be used. the method is based on the premise that instead of using the newly calculated value of x_i, a better value may result is a weighted average of the old and new values be used. In the following expression ω is an arbitrary parameter that is independent of i and k.

$$x_i^{k+1} = \omega \bar{x}_i^{k+1} + (1-\omega)x_i^k \tag{4.38}$$

When ω is greater than 1, then equation (4.33) defines the method of successive over-relaxation.

The Gauss-Seidel iteration is equivalent to that given for the Jacobi method and replacement of equation (4.32) with the following equation.

$$x_i^{k+1} = \frac{b_i - a_{i1}x_1^{k+1} - \dots - a_{ii-1}x_{i+1}^{k+1} - a_{ii+1}x_{i+1}^{k} - \dots - a_{in}x_n^{k}}{a_{ii}} \tag{4.39}$$

When i=1 the right hand side includes terms with superscript (k) only and when i=n, the terms with superscript (k+1) only. Program 4.20F illustrates the Gauss-Seidel Method.

Programs 4.19B through 4.27B are demonstration programs for vector and matrix topics covered in this chapter.

4.5 Problems

4-1) In a 3d computer game the velocity on an object is given in an arbitrary x,y,z co-ordinate system. If an object is moving with a velocity of 2,12,5 unit/sec and the another object is launched from the first with an additional velocity of 2,2,2, what is the total x,y,z velocity of the second object?

4-2) If the second object in problem 4-1 had been given a new velocity 3 times that of the first, what would the velocity of the second object be?

4-3) An aircraft is moving north at 300 mph. The jet stream is out of the west at 100 mph. What is the velocity of the aircraft over the ground?

4-4) Two mirrors being used for laser testing are mounted on an optical table. One mirror is 27 inches in, 14 over and 7 up.
The other is 72 inches in, 48 inches over and 16 up. Compute the straight line distance between the mirrors.

4-5) In a computer generated picture, the suns light ray is represented by an x,y,z vector as 1,1,1. A bridge support is represented by the vector 25,0,5. Use the dot product and compute the length of the shadow.

4-6) What is the length of the bridge support described in problem 4-5?

4-7) $[A]x=b \quad [A] = \begin{matrix} 5 & 0 & 0 & 0 \\ 6 & 3 & 0 & 0 \\ 2 & 1 & 3 & 0 \\ 4 & 1 & 7 & 9 \end{matrix} \quad b = \begin{matrix} 6 \\ 2 \\ 1 \\ 7 \end{matrix}$ Solve for x.

Hint: [A] is a lower triangular matrix

4-8) [A]x=b [A] = 5 6 2 4 b = 6 Solve for x

[A]	b
5 6 2 4	6
0 3 1 1	2
0 0 3 7	1
0 0 0 9	7

Hint: [A] is a upper triangular matrix

4-9) [A]x=b [A] = 5 6 0 0 b = 6 Solve for x

[A]	b
5 6 0 0	6
6 3 1 0	2
0 6 3 7	1
0 0 6 9	7

4-10) [A]x=b [A] = 5 6 2 4 b = 6 Solve for x

[A]	b
5 6 2 4	6
6 3 1 1	2
7 6 3 7	1
8 7 6 9	7

Chapter 4 Programs

Program 4.1B Vector Sums

```
SUB VectorSum (X#(), Y#(), Result#())
'
'THIS PROGRAM COPYRIGHT: (C) 1991 by LCPsoftware
'AUTHORS:    Benjamin J. Landry and Edward Champion
'TEXT BOOK: NUMERICAL METHODS FOR ENGINEERING APPLICATIONS
'
'EQUATION:   (04.04) EQU_0404 Sum of Two Vectors
'
'PURPOSE:    Finds the sum of the two vectors X and Y, where
'            the ordered set of numbers for each vector are
'            represented by the elements of the arrays
'            X() and Y().
'
'ARGUMENTS:
'            X(min)...X(max) The elements of the X array.
'            Y(min)...Y(max) The elements of the Y array.
'
'FORMULA:
'            In array terms:    R = X + Y
'
'
'            In element terms:  r(1) = x(1) + y(1)
'                               r(2) = x(2) + y(2)
'                                    :
'                               r(n) = x(n) + y(n)
'
'COMMENT:
'            This subroutine supports vectors of any length.
'            All three vectors must have been dimensioned
'            using the same index range, example:
'
'            DIM X#(5 TO 9), Y#(5 TO 9), Result#(5 TO 9)
'
'
'            This sub routine does not change the content of
'            the X or Y vector unless the vector is also
'            given as the result vector.  Thus if it is
'            desired to perform the function X = X + Y,
'            then this routine should be called using the
'            following format:
'
'            Call VectorSum (X#(), Y#(), X#())
'
'--------------------- end introduction ------------------

   FOR Element% = LBOUND(X#) TO UBOUND(X#)

      Result#(Element%) = X#(Element%) + Y#(Element%)

   NEXT Element%

END SUB
```

Program 4.1P Vector Sum Procedure

```
const MAXV = 50;
type
   Vector = array [1..MAXV] of real;
procedure VectorSum( var X : Vector;
                     var Y : Vector;
                     DIM   : integer;
                     var Result : Vector );

{  PURPOSE: Finds the sum of the two vectors X and Y,         }
{           where the ordered set of numbers for each vector   }
{           are represented by the arrays X and Y.             }
{                                                              }
{  ARGUMENTS:  X and Y are Vectors                             }
{              DIM is the dimension of vectors X and Y.        }
var
    Index : integer;
begin
   for Index := 1 to DIM do begin
      Result[Index] := X[Index] + Y[Index];
   end;
end;
```

Program 4.2B Vector Differences

```
SUB VectorDif (X#(), Y#(), Result#())
'
'THIS PROGRAM COPYRIGHT: (C) 1991 by LCPsoftware
'AUTHORS:    Benjamin J. Landry and Edward Champion
'TEXT BOOK: NUMERICAL METHODS FOR ENGINEERING APPLICATIONS
'
'EQUATION:  (04.05) EQU_0405 Difference of Two Vectors
'
'PURPOSE:   Finds the difference of the two vectors X and Y,
'           where the ordered set of numbers for each
'           vector are represented by the elements of
'           the arrays X() and Y().
'
'ARGUMENTS:
'           X(min)...X(max) The elements of the X array.
'           Y(min)...Y(max) The elements of the Y array.
'
'FORMULA:
'           In array terms:    R = X - Y
'
'
'           In element terms: r(1) = x(1) - y(1)
'                             r(2) = x(2) - y(2)
'                                  :
'                             r(n) = x(n) - y(n)
'
'COMMENT:
'           This subroutine supports vectors of any length.
'           All three vectors must have been dimensioned
'           using the same index range, example:
'
'           DIM X#(1 TO 3), Y#(1 TO 3), Result#(1 TO 3)
'
'
'           This sub routine does not change the content of
'           the X or Y vector unless the vector is also
'           given as the result vector.  Thus if it is
'           desired to perform the function X = X - Y,
'           then this routine should be called using the
'           following format:
'
'           Call VectorDif (X#(), Y#(), X#())
'
'--------------------- end introduction ------------------

   FOR Element% = LBOUND(X#) TO UBOUND(X#)

      Result#(Element%) = X#(Element%) - Y#(Element%)

   NEXT Element%

END SUB
```

Program 4.2P Vector Difference Procedure

```
const MAXV = 50;
type
   Vector = array [1..MAXV] of real;
procedure VectorDif( var X : Vector;
                     var Y : Vector;
                     DIM   : integer;
                     var Result : Vector );

{  PURPOSE: Finds the difference of the two vectors X and Y,       }
{           where the ordered set of numbers for each vector       }
{           are represented by the arrays X and Y.                 }
{                                                                  }
{  ARGUMENTS:  X and Y are Vectors                                 }
{              DIM is the dimension of vectors X and Y.            }
var
    Index : integer;
begin
   for Index := 1 to DIM do begin
      Result[Index] := X[Index] - Y[Index];
   end;
end;
```

Program 4.3B Vector Scalar Multiplication

```
SUB VectorScalarMul (Scalar#, X#(), Result#())
'
'THIS PROGRAM COPYRIGHT: (C) 1991 by LCPsoftware
'AUTHORS:    Benjamin J. Landry and Edward Champion
'TEXT BOOK: NUMERICAL METHODS FOR ENGINEERING APPLICATIONS
'
'EQUATION:  (04.07) EQU_0407 Difference of Two Vectors
'
'PURPOSE:   Finds the product of a vector X multiplied by
'           a Scalar where the ordered set of numbers for
'           the vector is represented by the elements of
'           the array X().
'
'ARGUMENTS: X(min)...X(max) The elements of the X array.
'
'           Scalar          The scalar "a" in the
'                           following equation:
'
'FORMULA:
'           In array terms:    R = aX
'
'
'           In element terms:  r(1) = a * x(1)
'                              r(2) = a * x(2)
'                                   :
'                              r(n) = a * x(n)
'
'COMMENT:
'           This subroutine supports vectors of any length.
'           Both X and Result must have been dimensioned
'           using the same index range, example:
'
'           DIM X#(3), Result#(3)
'
'
'           This sub routine does not change the content of
'           the X vector unless the vector is also given as
'           the result vector.  Thus if it is desired to
'           perform the function X = aX, then this routine
'           should be called using the following format:
'
'           Call VectorScalarMul(Scalar, X#(), X#())
'
'--------------------- end introduction ------------------

   FOR Element% = LBOUND(X#) TO UBOUND(X#)

      Result#(Element%) = Scalar# * X#(Element%)

   NEXT Element%

END SUB
```

Program 4.4B Vector Scalar Addition

```
SUB VectorScalarAddition (a#, X#(), b#, Y#(), Result#())
'
'THIS PROGRAM COPYRIGHT: (C) 1991 by LCPsoftware
'AUTHORS:   Benjamin J. Landry and Edward Champion
'TEXT BOOK: NUMERICAL METHODS FOR ENGINEERING APPLICATIONS
'
'EQUATION:  (04.08) EQU_0408 Vector Scalar Addition
'
'PURPOSE:   Finds the sum of the two vectors X and Y, after
'           each has been multiplied by a different scalar.
'           The ordered set of numbers for each vector are
'           represented by the elements of the arrays.
'
'ARGUMENTS: X(min)...X(max)  The elements of the array X.
'           Y(min)...Y(max)  The elements of the array X.
'
'           a and b          The scalars "a" and "b" in the
'                            following equation:
'
'FORMULA:   In array terms:    R = aX + bY
'
'           In element terms:  r(1) = a * x(1) + b * y(1)
'                              r(2) = a * x(2) + b * y(2)
'                                   :
'                              r(n) = a * x(n) + b * y(n)
'
'COMMENT:  This subroutine supports a vector of any length.
'  All 3 vectors must have been dimensioned using the same
'  index range.  If it is desired to perform the function
'  X = aX + bY, then this routine should be called using
'  the following format:
'
'  Call VectorSum (scalar#, X#(), scalar#, Y#(), X#())
'
'  From the code below, it should be obvious that this is
'  not one of the primitive functions for vector calculation
'  In Microsoft Professional Basic includes a Toolbox of
'  Matrix Functions.  The vector math primitives have been
'  coded here to give a clear picture of the functions
'  performed by the computer for each operation.
'
'--------------------- end introduction ------------------

DIM aX#(LBOUND(X#) TO UBOUND(X#))
DIM bY#(LBOUND(X#) TO UBOUND(X#))

CALL VectorScalarMul(a#, X#(), aX#())
CALL VectorScalarMul(b#, Y#(), bY#())
CALL VectorSum(aX#(), bY#(), Result#())

ERASE aX#: ERASE bY#
END SUB
```

Program 4.5B Vector Dot Product

```
SUB VectorDot (X#(), Y#(), ScalarResult#)
'
'THIS PROGRAM COPYRIGHT: (C) 1991 by LCPsoftware
'AUTHORS:    Benjamin J. Landry and Edward Champion
'TEXT BOOK: NUMERICAL METHODS FOR ENGINEERING APPLICATIONS
'
'EQUATION:   (04.09) EQU_0409 Dot Product of Two Vectors
'
'PURPOSE:
'            Finds the dot product of the two vectors X and Y
'            where the ordered set of numbers for each vector
'            are represented by the elements of X() and Y().
'
'ARGUMENTS:
'            X(min)...X(max) The elements of the array X.
'            Y(min)...Y(max) The elements of the array Y.
'
'FORMULA:
'            In array terms:    r = X dot Y
'
'            In element terms:  r = x(1) * y(1)
'                                 + x(2) * y(2)
'                                          :
'                                 + x(n) * y(n)
'
'COMMENT:
'            This subroutine supports vectors of any length.
'            Both vectors should have the same dimensions,
'            and use the same index range, example:
'
'            DIM X#(2), Result#(2)  'Defines 3 elements 0 - 2
'
'            This sub routine does not change the content of
'            the X or Y vector.  As the result of the dot
'            product is a  scalar, neither input term can
'            be used as the output.
'
'---------------------- end introduction ------------------

SumProduct# = 0                 '  Initialize the SumProduct
                                '  Done by MS Basic by default
one% = LBOUND(X#)
n% = UBOUND(X#)

FOR Element% = one% TO n%

   SumProduct# = SumProduct# + (X#(Element%) * Y#(Element%))

NEXT Element%

ScalarResult# = SumProduct#         '  Return the Dot product
END SUB
```

Program 4.6B Vector Length Calculation

```
SUB VectorLength (X#(), ScalarResult#)
'
'THIS PROGRAM COPYRIGHT: (C) 1991 by LCPsoftware
'AUTHORS:   Benjamin J. Landry and Edward Champion
'TEXT BOOK: NUMERICAL METHODS FOR ENGINEERING APPLICATIONS
'
'EQUATION:  (04.10) EQU_0410 Vector Length
'
'PURPOSE:   Finds the scalar length of the vector X where
'           the ordered set of numbers for the vector are
'           represented by the elements of the array X().
'           The scalar length of the vector X is returned
'           in the argument Result.
'
'ARGUMENTS: X(min)...X(max) The elements of the X array.
'
'FORMULA:   In array terms:     r = |X|
'                                       2              2  ½
'           In element terms:   r = ( x(1)^ + ... + x(n)^ )^
'
'COMMENT:  This subroutine supports vectors of any dimension
'
'  In Microsoft Professional Basic, unless a local variable
'  has been declared as a static, it's value is reset to 0
'  every  time the sub routine is called.  In this routine
'  the local variable SumSquares has NOT been declared as a
'  static, and is cleared automatically when this routine is
'  called.
'
' Reminder:  The # on the end of the matrix term causes this
'  routine to use double precision floating point math which
'  requires 8 bytes per value.  By changing all the #s to !s
'  this routine would use single precision floating point at
'  4 bytes per value.  By changing all the #'s to &'s, this
'  routine would use LONG integer at 4 bytes per value.
'  Finally, by changing all the #s to %s, this routine
'  would use INTEGER math at 2 bytes per value.
'
'--------------------- end introduction ------------------

one% = LBOUND(X#):
n% = UBOUND(X#)

FOR Element% = one% TO n%

   SumSquares# = SumSquares# + (X#(Element%) ^ 2)

NEXT Element%

ScalarResult# = SQR(SumSquares#)

END SUB
```

Program 4.6P Vector Length Procedure

```
const MAXV = 50;
type
   Vector = array [1..MAXV] of real;
procedure VectorLength( var X : Vector;
                        DIM   : integer;
                        var Result : real );

{  PURPOSE: Finds the length of the vector x.                       }
{                                                                   }
{  ARGUMENTS:  X and Y are Vectors                                  }
{              DIM is the dimension of vectors X and Y.             }
var
    Index : integer;
    SumSquares : real;
begin
   SumSquares := 0.0;
   for Index := 1 to DIM do begin
      SumSquares := SumSquares + ( X[Index] * X[Index] );
   end;
   Result := Sqrt( SumSquares );
end;
```

Program 4.7B Distance Between Points

```
SUB DistancePoints (X#(), Y#(), ScalarResult#)
'
'THIS PROGRAM COPYRIGHT: (C) 1991 by LCPsoftware
'AUTHORS:   Benjamin J. Landry and Edward Champion
'TEXT BOOK: NUMERICAL METHODS FOR ENGINEERING APPLICATIONS
'
'EQUATION:  (04.11) EQU_0411 Distance Between Two Points
'
'PURPOSE:   Finds the scalar straight line distance between
'           two points defined by the ordered set on numbers
'           in the X and Y arrays.
'
'ARGUMENTS: X(min)...X(max) The elements of the point X.
'           Y(min)...Y(max) The elements of the point Y.
'
'FORMULA:   In array terms:    r = |X - Y|
'
'                                                    2
'           In element terms:  r = ( ( x(0) - y(0) )^   +
'                                                    2
'                                    ( x(1) - y(1) )^   +
'                                          :         2     ½
'                                    ( x(n) - y(n) )^   ) ^
'
'COMMENTS:  This routine supports a vector of any dimension.
'
'  From the code below, it should be obvious that this is
'  NOT one of the primitive functions for vector calculation
'  It has been coded here to serve as an example of how the
'  primitive vector functions can be used.  In Microsoft
'  Professional Basic there is a Toolbox of Matrix Functions
'  that can be used to perform some of these operations on
'  any type of numeric data, or the #'s can be changed to
'  the required data type. The vector math primitives have
'  been coded here to give a clear picture of the functions
'  performed by the computer for each operation.
'
'  Note the use of the temporary vector YlessX, it
'  is created, used, and then eliminated from memory.
'
'--------------------- end introduction -------------------
one% = LBOUND(X#)
n% = UBOUND(X#)

DIM YlessX#(one% TO n%)

CALL VectorDif(X#(), Y#(), YlessX#())
CALL VectorLength(YlessX#(), ScalarResult#)

ERASE YlessX#

END SUB
```

Program 4.7P Distance Procedure

```
procedure DistancePoints( var X, Y : Vector;
                          DIM      : real;
                          Result   : real );

begin
   VectorDif( X,Y,DIM,YLessX );
   VectorLength( YLessX, Result );
end;
```

Program 4.8B Matrix Sums

```
SUB MatrixSum (A#(), B#(), Result#())
'
'THIS PROGRAM COPYRIGHT: (C) 1991 by LCPsoftware
'AUTHORS:    Benjamin J. Landry and Edward Champion
'TEXT BOOK: NUMERICAL METHODS FOR ENGINEERING APPLICATIONS
'
'EQUATION:   (04.14) EQU_0414 Sum of Two Matrices
'
'PURPOSE:    Finds the sum of the two matrices A and B where
'            the ordered set of numbers for each matrix are
'            represented by two dimensional arrays A() & B().
'
'ARGUMENTS: A(iMin,jMin)...A(iMax,jMax)
'           B(iMin,jMin)...B(iMax,jMax)
'                          The elements of these two arrays
'                          are the ordered set of numbers of
'                          the A and B matrix
'
'FORMULA     In matrix terms:     [R] = [A] + [B]
'
'            In element terms:  r(1,1) = a(1,1) + b(1,1)
'                               r(1,2) = a(1,2) + b(1,2)
'                                      :
'                               r(1,n) = a(1,n) + b(1,n)
'                               r(2,1) = a(2,1) + b(2,1)
'                                      :
'                               r(m,n) = a(m,n) + b(m,n)
'
'COMMENTS:
'
'  This subroutine supports two dimensional matrices with
'  any number of elements in each dimension.  All three
'  arrays must have the same number and range of elements
'  in each dimension.  Example:
'
'  DIM A#(1 TO 3, 0 TO 7), B#(1 TO 3, 7), Result#(1 TO 3, 7)
'
'  This sub routine does not change the content of the A or
'  B matrix unless the matrix is also given as the result
'  matrix.  In order to perform [A] = [A] + [B], call this
'  routine as follows:
'
'  Call MatrixSum (A#(), B#(), A#()
'---------------------- end introduction ------------------

   FOR j% = LBOUND(A#, 2) TO UBOUND(A#, 2)
      FOR i% = LBOUND(A#, 1) TO UBOUND(A#, 1)
         Result#(i%, j%) = A#(i%, j%) + B#(i%, j%)
      NEXT i%
   NEXT j%

END SUB
```

Program 4.9B Matrix Scalar Multiplication

```
SUB MatrixScalarMul (Scalar#, A#(), Result#())
'
'THIS PROGRAM COPYRIGHT: (C) 1991 by LCPsoftware
'AUTHORS:    Benjamin J. Landry and Edward Champion
'TEXT BOOK: NUMERICAL METHODS FOR ENGINEERING APPLICATIONS
'
'EQUATION:   (04.15) EQU_0415 Scalar Matrix Multiply
'
'PURPOSE:    Finds the product of a matrix A multiplied by
'            a Scalar where the ordered set of numbers for
'            the matrix is represented by the elements of
'            the array A().
'
'ARGUMENTS: A(iMin,jMin)...A(iMax,jMax)  The elements of
'                          these two arrays are the ordered
'                          set of numbers of the A matrix
'
'FORMULA
'           In matrix terms:     [R] = Scalar * [A]
'
'           In element terms:  r(1,1) = Scalar * a(1,1)
'                              r(1,2) = Scalar * a(1,2)
'                                     :
'                              r(1,n) = Scalar * a(1,n)
'                              r(2,1) = Scalar * a(2,1)
'                                     :
'                              r(m,n) = Scalar * a(m,n)
'
'COMMENTS:
'
'  This subroutine supports two dimensional matrices with
'  any number of elements in each dimension.  All three
'  arrays must have the same number and range of elements
'  in each dimension.  Example:
'
'  DIM A#(1 TO 3, 0 TO 7), Result#(1 TO 3, 7)
'
'  This sub routine does not change the content of the A
'  matrix unless the matrix is also given as the result
'  matrix.  In order to perform -[A] = call as follows:
'
'  Call MatrixSum (-1, A#(), A#()
'
'---------------------- end introduction ------------------

  FOR j% = LBOUND(A#, 2) TO UBOUND(A#, 2)
     FOR i% = LBOUND(A#, 1) TO UBOUND(A#, 1)
        Result#(i%, j%) = Scalar# * A#(i%, j%)
     NEXT i%
  NEXT j%

END SUB
```

Program 4.10B Difference of Two Matrices

```
SUB MatrixDif (A#(), B#(), Result#())
'
'THIS PROGRAM COPYRIGHT: (C) 1991 by LCPsoftware
'AUTHORS:   Benjamin J. Landry and Edward Champion
'TEXT BOOK: NUMERICAL METHODS FOR ENGINEERING APPLICATIONS
'
'EQUATION:  (04.16) EQU_0416 Difference of Two Matrices
'
'PURPOSE:   Finds the difference between the 2 matrices A &
'           B where the ordered set of numbers for each
'           matrix are represented by 2 dimensional arrays.
'
'ARGUMENTS: A(iMin,jMin)...A(iMax,jMax)
'           B(iMin,jMin)...B(iMax,jMax)
'                          The elements of these two arrays
'                          are the ordered set of numbers of
'                          the A and B matrix
'
'FORMULA    In matrix terms:    [R] = [A] - [B]
'
'           In element terms:  r(1,1) = a(1,1) - b(1,1)
'                              r(1,2) = a(1,2) - b(1,2)
'                                     :
'                              r(1,n) = a(1,n) - b(1,n)
'                              r(2,1) = a(2,1) - b(2,1)
'                                     :
'                              r(m,n) = a(m,n) - b(m,n)
'COMMENTS:
'
'  This subroutine supports two dimensional matrices with
'  any number of elements in each dimension.  All three
'  arrays must have the same number and range of elements
'  in each dimension.  Example:
'
'  DIM A#(1 TO 3, 0 TO 7), B#(1 TO 3, 7), Result#(1 TO 3, 7)
'
'  This sub routine does not change the content of the A or
'  B matrix unless the matrix is also given as the result
'  matrix.  In order to perform [A] = [A] - [B], call this
'  routine as follows:
'
'  Call MatrixDif (A#(), B#(), A#()
'
'--------------------- end introduction ------------------

   FOR j% = LBOUND(A#, 2) TO UBOUND(A#, 2)
      FOR i% = LBOUND(A#, 1) TO UBOUND(A#, 1)
         Result#(i%, j%) = A#(i%, j%) - B#(i%, j%)
      NEXT i%
   NEXT j%

END SUB
```

Program 4.11B Product of Two Matrices

```
SUB MatrixMul (A#(), B#(), Result#())
'
'THIS PROGRAM COPYRIGHT: (C) 1991 by LCPsoftware
'AUTHORS:    Benjamin J. Landry and Edward Champion
'TEXT BOOK: NUMERICAL METHODS FOR ENGINEERING APPLICATIONS
'
'EQUATION:   (04.17) EQU_0417 Product of two matrices
'
'PURPOSE:    Finds the product of the two matrices X and Y.
'
'ARGUMENTS: A(iMin,jMin)...A(iMax,jMax)   The elements of
'           B(iMin,jMin)...B(iMax,jMax)   these two arrays
'           are the ordered set of numbers of A & B matrix
'
'FORMULA:    In matrix terms:      [R] = [A] * [B]
'
'            In element terms:     r(1,1) = a(1,1) * b(1,1)
'                                         + a(1,2) * b(2,1)
'                                                  :
'                                         + a(1,k) * b(k,1)
'
'                                  r(1,2) = a(1,1) * b(1,2)
'                                         + a(1,2) * b(2,2)
'                                                  :
'                                         + a(1,k) * b(k,2)
'
'                                     :
'
'                                  r(2,1) = a(2,1) * b(1,1)
'                                         + a(2,2) * b(2,1)
'                                                  :
'                                         + a(2,k) * b(k,1)
'
'                                     :
'
'                                  r(i,j) = a(i,1) * b(1,j)
'                                         + a(i,2) * b(2,j)
'                                                  :
'                                         + a(i,k) * b(k,j)
'--------------------- end introduction ------------------

FOR i% = LBOUND(Result#, 1) TO UBOUND(Result#, 1)
   FOR j% = LBOUND(Result#, 2) TO UBOUND(Result#, 2)
      Sum# = 0
      FOR k% = LBOUND(A#, 1) TO UBOUND(A#, 1)
         Sum# = Sum# + (A#(i%, k%) * B#(k%, j%))
      NEXT k%
      Result#(i%, j%) = Sum#
   NEXT j%
NEXT i%

END SUB
```

Program 4.12B Forward Substitution

```
SUB ForwardSubstitution (A#(), b#(), Result#())
'
'THIS PROGRAM COPYRIGHT: (C) 1991 by LCPsoftware
'AUTHORS:    Benjamin J. Landry and Edward Champion
'TEXT BOOK: NUMERICAL METHODS FOR ENGINEERING APPLICATIONS
'
'EQUATION:   (04.25) EQU_0425 Forward Substitution
'
'PURPOSE:    Solves for the x matrix in the equation [A]x = b
'            when [A] has been found to be a lower triangular
'            matrix
'
'ARGUMENTS: [A]  is a two dimensional matrix that holds m
'                rows and n columns that are the terms of a
'                linear equation.
'
'            b  is a 1 dimensional matrix that hold n results
'
'            x  is a 1 dimensional matrix whose Results are
'               to be computed by this routine.
'
'FORMULAS:
'
'            x(1,1) = b(1,1) / A(1,1)
'
'                       1                i-1
'            x(i,1) = ------ * ( b(i,1) - Σ A(i,j) * x(j,1) )
'                     A(i,i)             j=1
'
'COMMENTS:
'            In keeping with the concept that a matrix is an
'            m by n array, both the b and x (Result) matrices
'            are treated here as m by 1 arrays.  They could
'            also have been coded as one dimensional arrays.
'
'---------------------- end introduction ------------------
CONST ByOne% = 1

Result#(1, ByOne%) = b#(1, ByOne%) / A#(1, 1)

FOR i% = 2 TO UBOUND(Result#)
   Sum# = 0
   FOR j% = 1 TO i% - 1
      Sum# = Sum# + (A#(i%, j%) * Result#(j%, ByOne%))
   NEXT j%
   BracketTerm# = b#(i%, ByOne%) - Sum#

   Result#(i%, ByOne%) = 1 / A#(i%, i%) * BracketTerm#

NEXT i%

END SUB
```

Program 4.13B Backward Substitution

```
SUB BackwardSubstitution (A#(), b#(), Result#())
'
'THIS PROGRAM COPYRIGHT: (C) 1991 by LCPsoftware
'AUTHORS:    Benjamin J. Landry and Edward Champion
'TEXT BOOK: NUMERICAL METHODS FOR ENGINEERING APPLICATIONS
'
'EQUATION:   (04.26) EQU_0426 Backward Substitution
'
'PURPOSE:    Solves for the x matrix in the equation [A]x = b
'            when [A] has been found to be a upper triangular
'            matrix
'
'ARGUMENTS: [A]  is a two dimensional matrix that holds m
'                rows and n columns that are the terms of a
'                linear equation.
'
'            b  is a 1 dimensional matrix that hold j results
'
'            x  is a 1 dimensional matrix whose Results are
'               to be computed by this routine.
'
'FORMULAS:
'
'            x(n,1) = b(n,1) / A(n,n)
'
'                       1                   n
'            x(i,1) = ------ * ( b(i,1) - Σ A(i,j) * x(j,1) )
'                     A(i,i)               j=i+1
'
'COMMENTS:
'            In keeping with the concept that a matrix is an
'            m by n array, both the b and x (Result) matrices
'            are treated here as m by 1 arrays.  They could
'            also have been coded as one dimensional arrays.
'
'---------------------- end introduction ------------------
CONST ByOne% = 1
N% = UBOUND(Result#)

Result#(N%, ByOne%) = b#(N%, ByOne%) / A#(N%, N%)

FOR i% = N% - 1 TO 1 STEP -1
   Sum# = 0
   FOR j% = i + 1 TO N%
      Sum# = Sum# + (A#(i%, j%) * Result#(j%, ByOne%))
   NEXT j%
   BracketTerm# = b#(i%, ByOne%) - Sum#

   Result#(i%, ByOne%) = 1 / A#(i%, i%) * BracketTerm#

NEXT i%
END SUB
```

Program 4.14B TriDiagonal Substitution

```
SUB TriDiagonal (A#(), b#(), Result#())
'
'THIS PROGRAM COPYRIGHT: (C) 1991 by LCPsoftware
'AUTHORS:    Benjamin J. Landry and Edward Champion
'TEXT BOOK: NUMERICAL METHODS FOR ENGINEERING APPLICATIONS
'
'EQUATION:   (04.28) EQU_0428 Tridiagonal Substitution
'
'PURPOSE:    Solves for the x matrix in the equation [A]x = b
'            when [A] has been found to be a tridiagonal
'            matrix
'
'ARGUMENTS: [A]  is a two dimensional matrix that holds m
'                rows and n columns that are the terms of a
'                linear equation.
'
'            b  is a 1 dimensional matrix that hold n results
'            x  is a 1 dimensional matrix whose Results are
'               to be computed by this routine.
'
'FORMULAS:   Decomposition and backsubstitution as coded
'
'COMMENTS:   In keeping with the concept that a matrix is an
'            m by n array, both the b and "Result" x matrices
'            are treated here as m by 1 arrays.  They could
'            also have been coded as one dimensional arrays.
'            Further, if memory is limited, the A#(m,n) matrix
'            can be replaced by 3 one dimensional arrays,
'            each holds one of the diagonals as follows:
'
'            a(j) = A(j,j-1); b(j) = A(j,j); c(j) = A(j,j+1)
'
'---------------------- end introduction ------------------
CONST ByOne% = 1
N% = UBOUND(Result#)
DIM Temp#(1 TO N%)

Alpha# = A#(1, 1)
Result#(1, ByOne%) = b#(1, ByOne%) / Alpha#

FOR j% = 2 TO N%   'Decomposition
   Temp#(j%) = A#(j% - 1, j%) / Alpha#
   Alpha# = A#(j%, j%) - A#(j%, j% - 1) * Temp#(j%)
   Beta# = A#(j%, j% - 1) * Result#(j% - 1, ByOne%)
   Result#(j%, ByOne%) = (b#(j%, ByOne%) - Beta#) / Alpha#
NEXT j%

FOR j% = N% - 1 TO 1 STEP -1     'Back substitution
   Offset# = (Temp#(j% + 1) * Result#(j% + 1, ByOne%))
   Result#(j%, ByOne%) = Result#(j%, ByOne%) - Offset#
NEXT j%
END SUB
```

Program 4.15B Partial Row Pivoting

```
SUB PartialPivotRow (A#(), b#())
'
'THIS PROGRAM COPYRIGHT: (C) 1991 by LCPsoftware
'AUTHORS:    Benjamin J. Landry and Edward Champion
'TEXT BOOK: NUMERICAL METHODS FOR ENGINEERING APPLICATIONS
'
'EQUATION:   (04.29) EQU_0429 Partial Pivot on Rows
'
'PURPOSE:    Performs a partial pivot of the matrix rows.
'
'ARGUMENTS: [A]  is a two dimensional matrix that holds n by
'                 m terms known terms of a linear equation.
'
'            b  is a 1 dimensional matrix that hold m results
'
'FORMULA:    [A]x = b  (Make this equation diagonal dominant)
'
'COMMENTS:   This subroutine supports matrix of any length
'            and the use of the Microsoft Basic array min TO
'            max syntax.  The 2nd dimension of the A array
'            should be the same as the b array.
'---------------------- end introduction ------------------
ElementX0% = LBOUND(A#, 1): ElementXn% = UBOUND(A#, 1)
ElementY0% = LBOUND(A#, 2): ElementYn% = UBOUND(A#, 2)
ElementZ0% = LBOUND(b#)
'-------- Move large value elements to the diagonal -------
FOR xIndex% = ElementX0% TO ElementXn%
    yIndexBase% = ElementY0% + (xIndex% - ElementX0%)
    zIndexBase% = ElementZ0% + (xIndex% - ElementX0%)
    yIndexPivot% = yIndexBase%
    yValue# = A#(xIndex%, yIndexPivot%)
'-------- Find term below diagonal > diagonal term ---------
    FOR yIndex% = yIndexPivot% + 1 TO ElementYn%
      IF A#(xIndex%, yIndex%) > yValue# THEN
         yValue# = A#(xIndex%, yIndex%)
         yIndexPivot% = yIndex%
      END IF
   NEXT yIndex%
'-------------------- b vector exchange -------------------
   zIndexPivot% = ElementZ0% + (yIndexPivot% - ElementY0%)
   Temp# = b#(zIndexBase%)
   b#(zIndexBase%) = b#(zIndexPivot%)
   b#(zIndexPivot%) = Temp#
'------------------- [A] matrix exchange ------------------
   FOR Element% = ElementX0% TO ElementXn%
      Temp# = A#(Element%, yIndexBase%)
      A#(Element%, yIndexBase%) = A#(Element%, yIndexPivot%)
      A#(Element%, yIndexPivot%) = Temp#
   NEXT Element%
'----------------------------------------------------------
NEXT xIndex%              'Do again from next diagonal element
END SUB
```

Program 4.15F Gaussian Elimination/Back Substitution/Row Pivoting

```
      PROGRAM GAUSS
C_______________________________________________________________________

C  THIS PROGRAM COPYRIGHT: (C) 1991 by Marvin N. Zeigler
C  AUTHORS:  Marvin N. Zeigler and Edward Champion
C  TEXT BOOK:  NUMERICAL METHODS FOR ENGINEERING APPLICATIONS
C  EQUATIONS:  4.25, 4.26, 4.29, 4.30

C  PURPOSE: The system of equations represented by matrix A and vector
C   B are solved simultaneously using Gauss elimination and backward
C   substitution.  Partial pivoting is implemented to reduce round-off
C    errors.

C  ARGUMENTS:
C   N       Number of equations.
C   A       Matrix of coefficients.
C   B       Right-hand-side vector.
C   X       Solution vector.
C_______________________________________________________________________

      PARAMETER (NN=10)
      REAL A(NN,NN),B(NN),X(NN)
      INTEGER N,NN,IPIV

C  Read coefficient matrix A and B vector from data file.
      OPEN (UNIT=1,FILE='DATA')
      READ(1,*) N
      READ(1,*) ((A(I,J),J=1,N),I=1,N)
      READ(1,*) (B(I),I=1,N)
      WRITE(*,*) ' '
      WRITE(*,*) '    INITIAL ARRAY A'
      DO 10 I=1,N
         WRITE(*,*)  (A(I,J),J=1,N)
 10   CONTINUE
      WRITE(*,*) ' '
      WRITE(*,*) '    INITIAL VECTOR B'
      WRITE(*,15) (B(I),I=1,N)


C  Begin main program.

      DO 80 I=1,N-1

C  Search for the largest element in the current column, I.
      IPIV = I
      DO 40 ITEMP=I+1,N
        IF ( ABS(A(ITEMP,I)) .GT. ABS(A(IPIV,I)) )  IPIV = ITEMP
 40   CONTINUE

C  If the pivot element is very near zero the matrix is singular and
a
C  solution is not feasible.  Program execution is stopped.
      IF (ABS(A(IPIV,I)) .LT. 1.0E-7) THEN
         WRITE(*,*)'! ERROR !  A(',IPIV,I,') =',A(IPIV,I)
         WRITE(*,*)'MATRIX A IS SINGULAR AND SOLUTION UNFEASIBLE.'
```

```
         STOP
      ENDIF

C  If pivot number is unchanged then the largest element is already
C  on the diagonal and no row switches are necessary.
      IF (IPIV .NE. I) THEN
      WRITE(*,*)' EXCHANGING ROW ',IPIV,' FOR ROW ',I

C  Exchange rows I and IPIV in matrix A and vector B.
      DO 50 JK=1,N
         HOLD = A(I,JK)
         A(I,JK) = A(IPIV,JK)
         A(IPIV,JK) = HOLD
 50   CONTINUE
      HOLD = B(I)
      B(I) = B(IPIV)
      B(IPIV) = HOLD

      ENDIF

C  Begin Gauss Elimination for current I.
         DO 70 J=I+1,N
         RATIO = A(J,I)/A(I,I)
           DO 60 K=I+1,N
              A(J,K) = A(J,K) - RATIO*A(I,K)
 60        CONTINUE
           B(J) = B(J) - RATIO*B(I)
 70      CONTINUE

 80   CONTINUE

C  Equations are solve now by backward substitution and the solution
C  is placed in vector X.

      X(N) = B(N)/A(N,N)

      DO 100 I=N-1,1,-1
      SUM = 0.0
       DO 90 J=I+1,N
        SUM = SUM + A(I,J)*X(J)
 90    CONTINUE
      X(I) = ( B(I) - SUM ) / A(I,I)
 100  CONTINUE

      WRITE(*,*) ' '
      WRITE(*,*) 'SOLUTION VECTOR'
      DO 110 I=1,N
        WRITE(*,25) I,X(I)
 110  CONTINUE
  15  FORMAT (2X,6F15.9)
  25  FORMAT (2X,'X(',I2,') = ',F10.4)
      END
```

Program 4.16B Partial Column Pivoting

```
SUB PartialPivotCol (A#())
'
'THIS PROGRAM COPYRIGHT: (C) 1991 by LCPsoftware
'AUTHORS:    Benjamin J. Landry and Edward Champion
'TEXT BOOK: NUMERICAL METHODS FOR ENGINEERING APPLICATIONS
'
'EQUATION:   (04.30) EQU_0430 Partial Pivot on Columns
'
'PURPOSE:    Performs a partial pivot of the matrix columns.
'
'ARGUMENTS: [A]  is a two dimensional matrix that holds n by
'                 m terms known terms of a linear equation.
'
'FORMULA:    [A]x = b  (Make this equation diagonal dominant)
'
'COMMENTS:   This subroutine supports matrix of any length
'            and the use of the Microsoft Basic array min TO
'            max syntax.  The 2nd dimension of the A array
'            should be the same as the b array.
'
'--------------------- end introduction ------------------

ElementX0% = LBOUND(A#, 1): ElementXn% = UBOUND(A#, 1)
ElementY0% = LBOUND(A#, 2): ElementYn% = UBOUND(A#, 2)

'-------- Move large value elements to the diagonal --------
FOR yIndex% = ElementY0% TO ElementYn%
   xIndexBase% = ElementX0% + (yIndex% - ElementY0%)
   zIndexBase% = ElementZ0% + (yIndex% - ElementY0%)

   xIndexPivot% = xIndexBase%
   xValue# = A#(xIndexPivot%, yIndex%)

'------- Find term right of diagonal > diagonal term -------
   FOR xIndex% = xIndexPivot% + 1 TO ElementXn%
      IF A#(xIndex%, yIndex%) > xValue# THEN
         xValue# = A#(xIndex%, yIndex%)
         xIndexPivot% = xIndex%
      END IF
   NEXT xIndex%

'------------------ [A] matrix exchange ------------------
   FOR Element% = ElementY0% TO ElementYn%
      Temp# = A#(xIndexBase%, Element%)
      A#(xIndexBase%, Element%) = A#(xIndexPivot%, Element%)
      A#(xIndexPivot%, Element%) = Temp#
   NEXT Element%

'----------------------------------------------------------
NEXT yIndex%           'Do again from next diagonal element

END SUB
```

Program 4.17B Maximum Pivot Strategy

```
SUB MaximumPivot (A#(), b#())
'
'THIS PROGRAM COPYRIGHT: (C) 1991 by LCPsoftware
'AUTHORS:    Benjamin J. Landry and Edward Champion
'TEXT BOOK: NUMERICAL METHODS FOR ENGINEERING APPLICATIONS
'
'EQUATION:   (04.31) EQU_0431 Maximum Pivotal (Almost)
'
'PURPOSE:    Performs a partial pivot of the matrix rows.
'
'ARGUMENTS: [A]  is a two dimensional matrix that holds n by
'                 m terms known terms of a linear equation.
'
'            b  is a 1 dimensional matrix that hold m results
'
'            x  would be the one dimensional matrix of m
'               unknown values that completes the linear
'               equation, it is not passed to this routine
'               as it is not effected by a partial pivot.
'
'FORMULA:    Performs a maximum pivotal of the columns and
'            rows in a linear equation set
'
'            In array terms:    [A]X = B
'
'       In element terms:
'       a(11) * x(1) + a(12) * x(2) ... + a(1n) * x(n) = b(1)
'       a(21) * x(1) + a(22) * x(2) ... + a(2n) * x(n) = b(2)
'         :        :       :        :          :       :     :
'       a(m1) * x(1) + a(m2) * x(2) ... + a(1n) * x(n) = b(m)
'
'COMMENTS:
'  This routine sorts the [A] matrix by using repetitive
'  calls to the partial row and column pivoting routines.
'  At the end of the sorting term a(11) is the largest term
'  in the array.  Term a(22) is the largest term not
'  contained in row or column 1.  This continues until
'  the diagonal is dominantly larger terms.
'
'---------------------- end introduction ------------------

Count0% = LBOUND(A#, 1) + LBOUND(A#, 2)
CountN% = UBOUND(A#, 1) + UBOUND(A#, 2) + 1

FOR Index% = Count0% TO CountN% STEP 2

   CALL PartialPivotRow(A#(), b#())
   CALL PartialPivotCol(A#())

NEXT Index%

END SUB
```

Program 4.17F Gauss-Jordan Elimination

```
            PROGRAM GAUSS_JORDAN
C_______________________________________________________________________
C  THIS PROGRAM COPYRIGHT: (C) 1991 by Marvin N. Zeigler
C  AUTHORS:  Marvin N. Zeigler and Edward Champion
C  TEXT BOOK:  NUMERICAL METHODS FOR ENGINEERING APPLICATIONS
C  EQUATIONS:  4.31

C  PURPOSE:  The system of equations represented by matrix A and
vector
C    B are solved simultaneously using Gauss Jordan elimination.
Partial
C   pivoting is implemented to reduce round-off errors.

C  ARGUMENTS:
C   N      Number of equations.
C   A      Matrix of coeFficients.
C   B      Right-hand-side vector.
C   X      Solution vector.
C   DIAG   Center diagonal elements.
C_______________________________________________________________________
      PARAMETER (NN=10)
      REAL A(NN,NN),B(NN),X(NN)
      INTEGER N,NN,IPIV

C  Read coefficient matrix A and B vector from data file.
      OPEN (UNIT=1,FILE='DATA')
      READ(1,*) N
      READ(1,*) ((A(I,J),J=1,N),I=1,N)
      READ(1,*) (B(I),I=1,N)
      WRITE(*,*) ' '
      WRITE(*,*) '    INITIAL ARRAY A'
      DO 10 I=1,N
         WRITE(*,*)  (A(I,J),J=1,N)
 10   CONTINUE
      WRITE(*,*) ' '
      WRITE(*,*) '    INITIAL VECTOR B'
      WRITE(*,15) (B(I),I=1,N)

C  Begin main program.

      DO 80 IDNUM=1,N

C  Search for the largest element in the current column, IDNUM.
      IPIV = IDNUM
      DO 40 ITEMP=IDNUM+1,N
        IF ( ABS(A(ITEMP,IDNUM)) .GT. ABS(A(IPIV,IDNUM))) IPIV = ITEMP
 40   CONTINUE

C If the pivot element is very near zero the matrix is singular and
C a solution is not feasible.  Program execution is stopped.
      IF (ABS(A(IPIV,IDNUM)) .LT. 1.0E-7) THEN
         WRITE(*,*)'! ERROR !  A(',IPIV,IDNUM,') =',A(IPIV,IDNUM)
         WRITE(*,*)'MATRIX A IS SINGULAR AND SOLUTION UNFEASIBLE.'
```

```
         STOP
      ENDIF

C  If pivot number is unchanged then the largest element is already
C  on the diagonal and no row switches are necessary.
      IF (IPIV .NE. IDNUM) THEN
      WRITE(*,*)' EXCHANGING ROW ',IPIV,' FOR ROW ',IDNUM

C  Exchange rows I and IPIV in matrix A and vector B.
      DO 50 JK=1,N
         HOLD = A(IDNUM,JK)
         A(IDNUM,JK) = A(IPIV,JK)
         A(IPIV,JK) = HOLD
 50   CONTINUE
      HOLD = B(IDNUM)
      B(IDNUM) = B(IPIV)
      B(IPIV) = HOLD

      ENDIF

C  Normalize by dividing by diagonal elements.
      DIAG = A(IDNUM,IDNUM)
      DO 55 IDXX=1,N
       A(IDNUM,IDXX) = A(IDNUM,IDXX)/DIAG
 55   CONTINUE
      B(IDNUM) = B(IDNUM)/DIAG

C  Perform row replacement and generate solution.
         DO 70 IRR=1,N
          IF (IRR .NE. IDNUM) THEN
            RATIO = A(IRR,IDNUM)/A(IDNUM,IDNUM)
             DO 60 ICC=1,N
                A(IRR,ICC) = A(IRR,ICC) - RATIO*A(IDNUM,ICC)
 60          CONTINUE
            B(IRR) = B(IRR) - RATIO*B(IDNUM)
          ENDIF
 70      CONTINUE

 80   CONTINUE

C  Solution is in the B vector.
      WRITE(*,*) ' '
      WRITE(*,*) 'SOLUTION VECTOR'
      DO 90 I=1,N
        WRITE(*,25) I,B(I)
 90   CONTINUE
 15   FORMAT (2X,6F15.9)
 25   FORMAT (2X,'X(',I2,') = ',F10.4)
      END
```

Program 4.18B LU Decomposition

```
SUB LuDecomposition (A#(), b#(), Result#())
'
'THIS PROGRAM COPYRIGHT: (C) 1991 by LCPsoftware
'AUTHORS:    Benjamin J. Landry and Edward Champion
'TEXT BOOK: NUMERICAL METHODS FOR ENGINEERING APPLICATIONS
'
'EQUATION:   (04.36) EQU_0436 LU Decomposition
'
'PURPOSE:    Solves for the x matrix in the equation [A]x = b
'
'ARGUMENTS: [A]x = b is a linear equation.
'
'FORMULAS:
'
'        for j = 1, 2 ... N  do ALL three of the following:
'           L(jj) = 1
'
'           for i = 1, 2 ... N do the following:
'                                i-1
'              U(ij) = A(ij) - Σ  L(ik) * U(kj)
'                                k=1
'
'           for i = j+1, j+2 ... N do the following:
'                                        j-1
'              L(ij) = 1/U(jj) * ( A(ij) - Σ  L(ik) * U(kj) )
'                                        k=1
'---------------------- end introduction ------------------
N% = UBOUND(Result#):  DIM y#(1 TO N%, 1 TO 1)
DIM L#(1 TO N%, 1 TO N%), U#(1 TO N%, 1 TO N%)

FOR j% = 1 TO N%          'Top of loop for all three --------
   L#(j%, j%) = 1         'L(jj) -----------------------
   FOR i% = 1 TO j%       'Top of U(ij) <---------------.
      Sum# = A#(i%, j%)                                 '
      FOR k% = 1 TO i% - 1                              '
         Sum# = Sum# - (L#(i%, k%) * U#(k%, j%))        '
      NEXT k%                                           '
      U#(i%, j%) = Sum#                                 '
   NEXT i%                'Bottom U(ij) ----------------'
   FOR i% = j% + 1 TO N% 'Top of L(ij) <---------------.
      Sum# = A#(i%, j%)                                 '
      FOR k% = 1 TO j% - 1                              '
         Sum# = Sum# - (L#(i%, k%) * U#(k%, j%))        '
      NEXT k%                                           '
      L#(i%, j%) = Sum# / U#(j%, j%)                    '
   NEXT i%                'Bottom L(ij) --------------- '
NEXT j%                   ' Bottom of loop for all three ----

CALL ForwardSubstitution(L#(), b#(), y#())
CALL BackwardSubstitution(U#(), y#(), Result#())

END SUB
```

Program 4.18F LU Decomposition - FORTRAN

```
      PROGRAM LUDEC
C_______________________________________________________________________

C  THIS PROGRAM COPYRIGHT: (C) 1991 by Marvin N. Zeigler
C  AUTHORS:  Marvin N. Zeigler and Edward Champion
C  TEXT BOOK:  NUMERICAL METHODS FOR ENGINEERING APPLICATIONS
C  EQUATIONS:  4.32, 4.33, 4.34, 4.35

C  PURPOSE:  A set of N linear equations is solved using LU
C   decomposition. The coefficient matrix A and the Right-Hand-Side
C   vector B are both scaled in subroutine PRECOND. Subroutine GELIM
C   uses Gaussian elimination and partial pivoting to decompose matrix
C   A into an LU compact form matrix. The solution vector X is
C   generated in subroutine FBSUB using backward substitution with the
C   ordered B vector. After the LU decomposition is complete, multiple
C   B vectors can be used by repeatedly calling FBSUB.

C  ARGUMENTS:
C   N       Number of equations.
C   A       Matrix of coefficients.
C   B       Right-hand-side vector.
C   X       Solution vector.
C   ORD     Vector that tracks row switches for use in FBSUB.
C_______________________________________________________________________

      PARAMETER (NN=4)
      REAL A(NN,NN),B(NN),X(NN)
      INTEGER N,ORD(NN)

C  Read coefficient matrix A and B vector from data file.
      OPEN (UNIT=1,FILE='DATA')
      READ(1,*) N
      READ(1,*) ((A(I,J),J=1,N),I=1,N)
      READ(1,*) (B(I),I=1,N)
      WRITE(*,*)
      WRITE(*,*) '    INITIAL ARRAY A'
      DO 10 I=1,N
         WRITE(*,*)  (A(I,J),J=1,N)
10    CONTINUE
      WRITE(*,*)
      WRITE(*,*) '    INITIAL VECTOR B'
      WRITE(*,15) (B(I),I=1,N)

C  Begin main program.
      CALL PRECOND (N,A,B)
      CALL GELIM (N,A,ORD)
      CALL FBSUB (N,A,B,ORD,X)

C  Output the LU matrix in compact form and the solution vector.
      WRITE(*,*)
      WRITE(*,*) '    COMPACT LU ARRAY'
      DO 20 I=1,N
         WRITE(*,*)  (A(I,J),J=1,N)
20    CONTINUE
      WRITE(*,*)
```

```
      WRITE(*,*) '    SOLUTION VECTOR '
      WRITE(*,15) (X(K),K=1,N)
 15   FORMAT (2X,6F15.9)
      END

C______________________________________________________________________
C______________________________________________________________________
      SUBROUTINE PRECOND (N,A,B)
C______________________________________________________________________
C Subroutine to pre-condition the A matrix and B vector.  The largest
C  element is found for each row in the A matrix, and all of the
C  elements of that row are divided by this element. This gives the
C  A matrix such that the largest value in any row is 1.0.

C  ARGUMENTS:
C   CHECK  Holds magnitude of largest element so far.
C   DMAX   Largest element of current row.
C______________________________________________________________________
      REAL A(N,N),B(N),CHECK,DMAX
      INTEGER N

      DMAX = 0.0
      DO 10 I=1,N
       CHECK = 0.0
       DO 20 J=1,N
        IF (CHECK .LT. ABS(A(I,J))) THEN
         CHECK = ABS(A(I,J))
         DMAX = A(I,J)
        ENDIF
 20    CONTINUE

C  If all of the values in a row are zero the routine stops.
       IF (DMAX .EQ. 0.0) THEN
        WRITE(*,*) '! ERROR !  ALL VALUES ARE ZERO FOR ROW ',I
        WRITE(*,*) 'PROGRAM STOPPED IN SUBROUTINE PRECOND'
        STOP
       ENDIF

C  Divide the A and B elements by the largest element in the Ith row
C  of A.
        DO 30 J=1,N
         A(I,J) = A(I,J)/DMAX
 30     CONTINUE
         B(I) = B(I)/DMAX
 10    CONTINUE
       RETURN
       END

C______________________________________________________________________
C______________________________________________________________________
       SUBROUTINE GELIM (N,A,ORD)
C______________________________________________________________________
C  Gaussian elimination is performed on the A matrix to give the LU
C  decomposition. The L and U matricies are stored in the A matrix in
C  compact form.
C______________________________________________________________________
       REAL A(N,N),SUM
       INTEGER N,ORD(N)
```

```
C  Initialize the vector ORD.
      DO 10 I=1,N
         ORD(I) = I
  10  CONTINUE

C  Perform pivoting for the first column.
      CALL PIVOT (A,ORD,N,1)

C Begin computing LU matrix. Pivoting is performed for each row after
C  a column of L's has been calculated.
      DO 20 J=2,N

C Calculate elements of the L matrix and store in A.  Recall that the
C  first column of L is the same as the first column of A.
         DO 30 I=J,N
         SUM = 0.0
           DO 40 K=1,J-1
              SUM = SUM + A(I,K)*A(K,J)
 40        CONTINUE
           A(I,J) = A(I,J)-SUM
 30      CONTINUE

C Calculate elements of the U matrix and store in A.  Computation of
C  the first row of the U matrix is not necessary because A(1,1)=1
C  after pre-conditioning pivoting.
      CALL PIVOT (A,ORD,N,J)
         DO 50 I=J+1,N
            SUM = 0.0
            DO 60 K=1,J-1
               SUM = SUM + A(J,K)*A(K,I)
 60         CONTINUE
            A(J,I) = (A(J,I)-SUM)/A(J,J)
 50      CONTINUE
 20   CONTINUE
      RETURN
      END

C_____________________________________________________________________
C_____________________________________________________________________
      SUBROUTINE PIVOT (A,ORD,N,J)
C_____________________________________________________________________
C  Elements of matrix A are pivoted about the largest element in
C  column J.  The order of changes is tracked through vector ORD.
C_____________________________________________________________________
      REAL A(N,N),HOLD
      INTEGER J,IR,ORD(N)

C  Search for largest element in column J.
      IR = J
      DO 33 I=J+1,N
        IF ( ABS(A(I,J)) .GT. ABS(A(IR,J)) )  IR = I
 33   CONTINUE

C  If the pivot element is near zero the matrix is singular and a
C  solution is not feasible.  Program execution is stopped.
      IF (ABS(A(IR,J)) .LT. 1.E-7) THEN
         WRITE(*,*)'! ERROR ! A(',IR,J,') =',A(IR,J),' MAKING MATRIX'
         WRITE(*,*)'A SINGULAR.'
         WRITE(*,*)'PROGRAM EXECUTION STOPPED IN SUBROUTINE PIVOT'
         STOP
```

```
      ENDIF
C  If now row switches are necessary return.
      IF (IR .EQ. J) RETURN
C  Switch J and IR rows in A.
      DO 44 JK=1,N
         HOLD = A(J,JK)
         A(J,JK) = A(IR,JK)
         A(IR,JK) = HOLD
 44   CONTINUE
C  Switch ORD(J) and ORD(IR).
      HOLD = ORD(J)
      ORD(J) = ORD(IR)
      ORD(IR) = HOLD
      RETURN
      END
C_______________________________________________________________
C_______________________________________________________________
      SUBROUTINE FBSUB (N,A,B,ORD,X)
C_______________________________________________________________
C  Equations are solved by backward substitution and the solution is
C  returned through X.
C_______________________________________________________________
      REAL A(N,N),B(N),X(N),SUM
      INTEGER N,ORD(N)
C  First the triangular system LY = B is solved,taking into account
C  the row switches that were made.  X(1)=B(1) because A(1,1)=1 from
C  pre-conditioning and pivoting.
      X(1) = B(ORD(1))
      DO 10 K=2,N
      SUM = 0.0
       DO 20 J=1,K-1
        SUM = SUM + A(K,J)*X(J)
  20   CONTINUE
      X(K) = (B(ORD(K)) - SUM)/A(K,K)
  10  CONTINUE
C  Now the triangular system UX = Y can be solved for X by back
C  substitution.  The solutions to the equations are stored in the
C  X vector.  X(N) is known because U(N,N) = 1.
      DO 30 K=N-1,1,-1
      SUM = 0.0
       DO 40 J=K+1,N
        SUM = SUM + A(K,J)*X(J)
  40   CONTINUE
      X(K) = X(K) - SUM
  30  CONTINUE
      RETURN
      END
```

Program 4.19F Jacobi's Method

```
      SUBROUTINE JACOBI(A,B,X,XWORK,TOL,N,NN,MAXIT)
C_______________________________________________________________________

C  THIS PROGRAM COPYRIGHT: (C) 1991 by Marvin N. Zeigler
C  AUTHORS:  Marvin N. Zeigler and Edward Champion
C  TEXT BOOK:  NUMERICAL METHODS FOR ENGINEERING APPLICATIONS
C  EQUATIONS:  4.37

C  PURPOSE:  Jacobi iteration is used to solve N linear equations.
C   The vector X initially contains guesses of solution values for all
C  X's. Rows are exchanged to put the largest element on the diagonal.
C  If the diagonal element is zero (or very near zero) then a solution

C   is not possible.  Approximations for each individual X value are
C  calculated before any of them are updated. Iterations are continued
C   until the largest change in any single X value is less than TOL.

C  ARGUMENTS:
C   N      Number of equations.
C   A      Coefficient matrix of size N X N.
C   B      Vector of right-hand-side values.
C   X     Vector holding initial guesses and returned solution vector.
C   MAXIT  Maximum number of iterations allowed.
C   TOL    Convergence criteria.
C_______________________________________________________________________

      REAL A(NN,NN),B(NN),X(NN),XWORK(NN),TOL
      INTEGER N,MAXIT

C Perform partial pivoting to put the largest element on the diagonal.

      DO 30 IDNUM=1,N
C  Search for the largest element in the current column, IDNUM.
      IPIV = IDNUM
      DO 10 ITEMP=IDNUM+1,N
        IF ( ABS(A(ITEMP,IDNUM)) .GT. ABS(A(IPIV,IDNUM)) )  IPIV =
ITEMP
 10   CONTINUE


C  If pivot number is unchanged then the largest element is already
C  on the diagonal and no row switches are necessary.
      IF (IPIV .NE. IDNUM) THEN
      WRITE(*,14) IPIV,IDNUM

C  Exchange rows I and IPIV in matrix A and vector X.
      DO 20 JC=1,N
         HOLD = A(IDNUM,JC)
         A(IDNUM,JC) = A(IPIV,JC)
         A(IPIV,JC) = HOLD
 20   CONTINUE
      HOLD = B(IDNUM)
      B(IDNUM) = B(IPIV)
      B(IPIV) = HOLD
```

```
      ENDIF

C  Check if diagonal element is too near zero.
      IF (ABS(A(IDNUM,IDNUM)) .LT. 1.0E-8) THEN
         WRITE(*,13) IDNUM,IDNUM
         STOP
      ENDIF

 30   CONTINUE

C  Normalize array A by the diagonal elements.
      DO 33 I=1,N
         DIAG = A(I,I)
         B(I) = B(I)/DIAG
         DO 36 J=1,N
            A(I,J) = A(I,J)/DIAG
36       CONTINUE
33    CONTINUE

C Approximations are generated using x values from the last iteration.
      DO 100 ITER=1,MAXIT
         CHECK = 0.0
         DO 80 I=1,N
            XWORK(I) = B(I)
            DO 70 J=1,N
               IF (J .NE. I) THEN
                  XWORK(I) = XWORK(I) - A(I,J)*X(J)
               ENDIF
70          CONTINUE
            IF (ABS(XWORK(I)-X(I)) .GT. CHECK) THEN
               CHECK = ABS ( XWORK(I)-X(I) )
            ENDIF
80       CONTINUE
         DO 90 I=1,N
            X(I) = XWORK(I)
90       CONTINUE
         IF (CHECK .LE. TOL) THEN
           WRITE(*,25)
           RETURN
         ENDIF
100   CONTINUE

C  If MAXIT has been reached then the tolerance value was not met.
      WRITE(*,35) TOL,ITER-1

 14   FORMAT(/' EXCHANGING ROW ',I3,' FOR ROW ',I3)
 13   FORMAT(/' DATA ERROR IN SUBROUTINE JACOBI.  DIAGONAL ELEMENT'
     + /' A(',I3,I3,') IS NEAR ZERO.'/)
 15   FORMAT(/' WARNING IN SUBROUTINE JACOBI. OW',I3,' OF MATRIX A IS'
     + /' NOT DIAGONALLY DOMINANT.  CONVERGENCE IS NOT GUARANTEED.')
 25   FORMAT(/' TOLERANCE MET IN JACOBI.  SOLUTION IN VECTOR X.')
 35   FORMAT(/' TOLERANCE VALUE ',E10.4,' WAS NOT MET IN ',I4,'
      ITERATIONS'
     + /' PROGRAM STOPPED IN SUBROUTINE JACOBI.')
      STOP
      END
```

Program 4.20F Gauss-Seidel Method

```
      SUBROUTINE GSEIDEL(NN,N,A,B,X,TOL,MAXIT)
C_______________________________________________________________________
C  THIS PROGRAM COPYRIGHT: (C) 1991 by Marvin N. Zeigler
C  AUTHORS:  Marvin N. Zeigler and Edward Champion
C  TEXT BOOK:  NUMERICAL METHODS FOR ENGINEERING APPLICATIONS
C  EQUATIONS:  4.38, 4.39

C  PURPOSE:Gauss-Seidel iteration is used to solve N linear equations.
C  The vector X initially contains guesses of solution values for all
C  X's. Rows are exchanged to put the largest element on the diagonal.
C  If the magnitude of the diagonal element is larger than the sum of
C   the magnitudes of the other elements in the row the Gauss-Seidel
C   method is guaranteed to converge.  Note that this is a sufficient
C   condition for convergence, but it is not a necessary condition.
C   Also, if the diagonal element is zero (or very near zero) then a
C   solution is not possible.  Approximations for each X value are
C   calculated using the most current calculated values.Iterations are
C   continued until the largest change in any single X value is less
C   than TOL.

C  ARGUMENTS:
C   N      Number of equations.
C   A      Coefficient matrix of size N X N.
C   B      Vector of right-hand-side values.
C   X    Vector holding initial guesses and returned solution vector.
C   MAXIT  Maximum number of iterations allowed.
C   TOL    Convergence criteria.
C_______________________________________________________________________

      REAL A(NN,NN),B(NN),X(NN),TOL
      INTEGER N,MAXIT


C Perform partial pivoting to put the largest element on the diagonal.

      DO 30 IDNUM=1,N
C  Search for the largest element in the current column, IDNUM.
      IPIV = IDNUM
      DO 10 ITEMP=IDNUM+1,N
        IF ( ABS(A(ITEMP,IDNUM)) .GT. ABS(A(IPIV,IDNUM)) )  IPIV =
ITEMP
 10   CONTINUE


C  If pivot number is unchanged then the largest element is already
C  on the diagonal and no row switches are necessary.
      IF (IPIV .NE. IDNUM) THEN
      WRITE(*,14) IPIV,IDNUM

C  Exchange rows IDNUM and IPIV in matrix A and vector X.
      DO 20 JC=1,N
         HOLD = A(IDNUM,JC)
         A(IDNUM,JC) = A(IPIV,JC)
         A(IPIV,JC) = HOLD
```

```
 20    CONTINUE

       ENDIF

       HOLD = B(IDNUM)
       B(IDNUM) = B(IPIV)
       B(IPIV) = HOLD

C  Check if diagonal element is too near zero.
       IF (ABS(A(IDNUM,IDNUM)) .LT. 1.0E-8) THEN
          WRITE(*,13) IDNUM,IDNUM
          STOP
       ENDIF

C  Now see if the magnitude of the diagonal element is larger than the
C  sum of the other elements on its row.
       SUM = 0.0
       DO 50 ICOL=1,N
          IF (ICOL .NE. IDNUM) THEN
             SUM = SUM + ABS(A(IDNUM,ICOL))
          ENDIF
50     CONTINUE
       IF (ABS(A(IDNUM,IDNUM)) .LT. SUM) THEN
          WRITE(*,15) IDNUM
       ENDIF

30     CONTINUE

C  Normalize array A by the diagonal elements.
       DO 33 I=1,N
          DIAG = A(I,I)
          B(I) = B(I)/DIAG
          DO 36 J=1,N
             A(I,J) = A(I,J)/DIAG
36        CONTINUE
33     CONTINUE

C  Generate approximations using current calculated X values.
       DO 90 ITER=1,MAXIT
          CHECK = 0.0
          DO 80 I=1,N
             DIAG = X(I)
             X(I) = B(I)
             DO 70 J=1,N
                IF (J .NE. I) THEN
                   X(I) = X(I) - A(I,J)*X(J)
                ENDIF
70           CONTINUE
             IF (ABS(X(I)-DIAG) .GT. CHECK) THEN
                CHECK = ABS ( X(I)-DIAG )
             ENDIF
80        CONTINUE
          IF (CHECK .LE. TOL) THEN
            WRITE(*,25)
            RETURN
          ENDIF
90     CONTINUE

C  If MAXIT has been reached then the tolerance value was not met.
       WRITE(*,35) TOL,ITER-1
```

```
15   FORMAT(/'WARNING IN SUBROUTINE GSEIDEL.ROW',I3,' OF MATRIX A IS'
    + /' NOT DIAGONALLY DOMINANT.  CONVERGENCE IS NOT GUARANTEED.')
14   FORMAT(/' EXCHANGING ROW ',I3,' FOR ROW ',I3)
13   FORMAT(/' DATA ERROR IN SUBROUTINE GSEIDEL.  DIAGONAL ELEMENT'
    + /' A(',I3,I3,') IS NEAR ZERO.'/)
25   FORMAT(/' TOLERANCE MET IN GSEIDEL.  SOLUTION IN VECTOR X.'/)
35   FORMAT(/'  TOLERANCE  VALUE  ',E10.4,'  WAS  NOT  MET  IN  ',I4,'
     +ITERATIONS'
    + /' PROGRAM STOPPED IN SUBROUTINE GSEIDEL.'/)
     STOP
     END
```

Program 4.21B Vector Demonstration Setup

```
SUB VectorDemoSetup
'
'THIS PROGRAM COPYRIGHT: (C) 1991 by LCPsoftware
'AUTHORS:   Benjamin J. Landry and Edward Champion
'TEXT BOOK: NUMERICAL METHODS FOR ENGINEERING APPLICATIONS
'
'EQUATION:  (04.41) EQU_0441 Vector Demonstration Setup
'
'PURPOSE:   Creates two random vectors [X] and [Y], and use
'           them to demonstrate the following routines:
'
'           VectorSum (X#(), Y#(), Result#())
'           VectorDif (X#(), Y#(), Result#())
'           VectorScalarMul (Scalar#, X#(), Result#())
'           VectorScalarAddition (a#, X#(), b#, Y#(), R#())
'
'           VectorDot (X#(), Y#(), Result#)
'           VectorLength (X#(), ScalarResult#)
'           DistancePoints (X#(), Y#(), ScalarResult#)
'
'---------------------- end introduction ------------------

DIM X#(1 TO 5), Y#(1 TO 5)

CLS             'Output sent to last listed device

OutputDevice$ = "DAT_0441.ASC"        'File name
OutputDevice$ = "LPT1:"               'Printer
OutputDevice$ = "SCRN:"               'Screen

SerFile% = FREEFILE
OPEN OutputDevice$ FOR OUTPUT AS #SerFile%

RANDOMIZE (TIMER)
RANDOMIZE (1) 'Make it repeatable

FOR xIndex% = LBOUND(X#) TO UBOUND(X#)

   X#(xIndex%) = INT(RND * 99)
   Y#(xIndex%) = INT(RND * 99)

NEXT xIndex%

PRINT #SerFile%, "                    Vector Test Results"
PRINT #SerFile%, "": PRINT #SerFile%, STRING$(60, "-")

CALL VectorDemo1(X#(), Y#(), SerFile%)
CALL VectorDemo2(X#(), Y#(), SerFile%)

CLOSE #SerFile%

END SUB
```

Program 4.22B Vector Demonstration 1

```
SUB VectorDemo1 (X#(), Y#(), SerFile%)
'
'THIS PROGRAM COPYRIGHT: (C) 1991 by LCPsoftware
'AUTHORS:    Benjamin J. Landry and Edward Champion
'TEXT BOOK: NUMERICAL METHODS FOR ENGINEERING APPLICATIONS
'
'EQUATION:   (04.42) EQU_0442 Vector Demonstration 1
'
'PURPOSE:    Demonstrates the vector functions that have
'            vector results.
'---------------------- end introduction ------------------

DIM R1#(1 TO 5), R2#(1 TO 5), R3#(1 TO 5), Text$(0 TO 5)
Text$(4) = "": Text$(5) = "-"          'Vector and Sum bar on

Text$(0) = "VectorSum"   '---------------------------------
Text$(1) = " X vector = | A random vector"
Text$(2) = " Y vector = | Another random vector"
Text$(3) = "    X + Y  = | Sum of X and Y vectors"

CALL VectorSum(X#(), Y#(), R1#())
CALL VectorOutput(Text$(), X#(), Y#(), R1#(), SerFile%)

Text$(0) = "VectorDif"   '---------------------------------
Text$(1) = "    X + Y  = | Start with computed X + Y"
Text$(2) = " Y vector = | Remove the Y vector"
Text$(3) = " X vector = | Returns X vector from above"

CALL VectorSum(X#(), Y#(), R1#())
CALL VectorDif(R1#(), Y#(), R2#())
CALL VectorOutput(Text$(), R1#(), Y#(), R2#(), SerFile%)

Text$(0) = "VectorScalarMul" '-----------------------------
Text$(1) = " X vector = | Starting with the X vector"
Text$(2) = "    2 * X  = | Scalar 2 times the X vector"
Text$(3) = "    X + X  = | Compare to adding to itself"

CALL VectorScalarMul(2, X#(), R1#())
CALL VectorSum(X#(), X#(), R2#())
CALL VectorOutput(Text$(), X#(), R1#(), R2#(), SerFile%)

Text$(0) = "VectorScalarAddition" '------------------------
Text$(1) = "    2 * X  = | Scalar 2 times the X vector"
Text$(2) = " + 3 * Y  = | Add to 3 times the Y vector"
Text$(3) = "  2X + 3Y = | Equals VectorScalarAddition"

CALL VectorScalarMul(2, X#(), R1#())
CALL VectorScalarMul(3, Y#(), R2#())
CALL VectorSum(R1#(), R2#(), R2#())
CALL VectorScalarAddition(2, X#(), 3#, Y#(), R3#())
CALL VectorOutput(Text$(), R1#(), R2#(), R3#(), SerFile%)
END SUB
```

Program 4.23B Vector Demonstration 2

```
SUB VectorDemo2 (X#(), Y#(), SerFile%)
'
'THIS PROGRAM COPYRIGHT: (C) 1991 by LCPsoftware
'AUTHORS:    Benjamin J. Landry and Edward Champion
'TEXT BOOK: NUMERICAL METHODS FOR ENGINEERING APPLICATIONS
'
'EQUATION:   (04.43) EQU_0443 Vector Demonstration 2
'
'PURPOSE:    Demonstrates the vector functions that have
'            scalar results.
'---------------------- end introduction ------------------

DIM R1#(1 TO 5), R2#(1 TO 5), R3#(1 TO 5), Text$(0 TO 5)
Text$(5) = ""          'Outputing Scalars, don't output Sum bar

Text$(0) = "VectorDot"  '---------------------------------
Text$(1) = " X vector = | A random vector"
Text$(2) = " Y vector = | Another random vector"
Text$(3) = " X dot Y  = | Dot of X and Y vectors"

CALL VectorDot(X#(), Y#(), Scalar#)
Text$(4) = STR$(Scalar#)     'Pass scalar for printing
CALL VectorOutput(Text$(), X#(), Y#(), Y#(), SerFile%)
Text$(0) = "VectorLength"  '------------------------------
Text$(1) = ""
Text$(2) = " X vector = | Our random vector"
Text$(3) = "            | Has this length "

CALL VectorLength(X#(), Scalar#)
Text$(4) = STR$(Scalar#)
CALL VectorOutput(Text$(), X#(), X#(), Y#(), SerFile%)

PRINT #SerFile%, " The above length squared ="; Scalar# ^ 2

Text$(0) = "": Text$(1) = "": Text$(2) = "": Text$(5) = "-"
Text$(3) = " Which is same as X dot X = | "

CALL VectorDot(X#(), X#(), Scalar#)
Text$(4) = STR$(Scalar#)     'Pass scalar for printing
CALL VectorOutput(Text$(), X#(), X#(), X#(), SerFile%)

Text$(0) = "VectorDistance" '----------------------------
Text$(1) = " X vector = | Our random vector": Text$(5) = ""
Text$(2) = " 0 vector = | X + (-1X)"
Text$(3) = " Distance = | from 0 to X, = length of X"

CALL VectorScalarMul(-1, X#(), R1#())
CALL VectorSum(X#(), R1#(), R2#())
CALL DistancePoints(R2#(), X#(), Scalar#)
Text$(4) = STR$(Scalar#)     'Pass scalar for printing
CALL VectorOutput(Text$(), X#(), R2#(), X#(), SerFile%)
END SUB
```

Program 4.24B Vector Output Program

```
SUB VectorOutput (Text$(), Vec1#(), Vec2#(), Vec3#(), Port%)
'
'THIS PROGRAM COPYRIGHT: (C) 1991 by LCPsoftware
'AUTHORS:    Benjamin J. Landry and Edward Champion
'TEXT BOOK: NUMERICAL METHODS FOR ENGINEERING APPLICATIONS
'
'EQUATION:   (04.44) EQU_0444 Vector Demo Output
'
'PURPOSE:    Common output support for Vector demonstrations
'
'--------------------- end introduction ------------------

IF Text$(0) <> "" THEN                  'Print Title if it exists
   PRINT #Port%, "Test of routine:  "; Text$(0)
   PRINT #Port%, ""
END IF


IF Text$(1) <> "" THEN       'Print Vector 1 if its commented
   PRINT #Port%, LEFT$(Text$(1), INSTR(Text$(1), "|") - 1);
   FOR Element% = LBOUND(Vec1#) TO UBOUND(Vec1#)
         PRINT #Port%, USING "### "; Vec1#(Element%);
   NEXT Element%
   PRINT #Port%, MID$(Text$(1), INSTR(Text$(1), "|") + 1)
END IF


IF Text$(2) <> "" THEN       'Print Vector 2 if its commented
   PRINT #Port%, LEFT$(Text$(2), INSTR(Text$(2), "|") - 1);
   FOR Element% = LBOUND(Vec2#) TO UBOUND(Vec2#)
         PRINT #Port%, USING "### "; Vec2#(Element%);
   NEXT Element%
   PRINT #Port%, MID$(Text$(2), INSTR(Text$(2), "|") + 1)
END IF

              'Text(5) is used to command print of a Sum Bar
IF Text$(5) <> "" THEN PRINT #Port%, " "; STRING$(30, "-")

PRINT #Port%, LEFT$(Text$(3), INSTR(Text$(3), "|") - 1);
IF Text$(4) = "" THEN                           'Vector Output
   FOR Element% = LBOUND(Vec3#) TO UBOUND(Vec3#)
      PRINT #Port%, USING "### "; Vec3#(Element%);
   NEXT Element%
ELSE                                            'Scalar Output
   PRINT #Port%, USING "##### "; VAL(Text$(4));
   PRINT #Port%, SPACE$(14);
END IF
PRINT #Port%, MID$(Text$(3), INSTR(Text$(3), "|") + 1)

PRINT #Port%, STRING$(60, "-")

END SUB
```

Program 4.25B Matrix Demonstration Setup

```
SUB MatrixDemoSetup
'
'THIS PROGRAM COPYRIGHT: (C) 1991 by LCPsoftware
'AUTHORS:   Benjamin J. Landry and Edward Champion
'TEXT BOOK: NUMERICAL METHODS FOR ENGINEERING APPLICATIONS
'
'EQUATION:  (04.45) EQU_0445 Matrix Demonstration Setup
'
'PURPOSE:   Set up two 3 x 3 matrices and one 3 x 1 matices.
'           These matrices will be used to demonstrate
'           the matrix routine described in this chapter
'
'---------------------- end introduction -------------------

DIM A#(1 TO 3, 1 TO 3)
DIM B#(1 TO 3, 1 TO 3)
DIM C#(1 TO 3, 1 TO 1)

CLS              'Output sent to last listed device

OutputDevice$ = "DAT_0445.ASC"          'File name
OutputDevice$ = "LPT1:"                 'Printer
OutputDevice$ = "SCRN:"                 'Screen

SerFile% = FREEFILE
OPEN OutputDevice$ FOR OUTPUT AS #SerFile%

RANDOMIZE (TIMER)
RANDOMIZE (1)        'Or we could make it repeatable

ElementX0% = LBOUND(A#, 1): ElementXn% = UBOUND(A#, 1)
ElementY0% = LBOUND(A#, 2): ElementYn% = UBOUND(A#, 2)

FOR yIndex% = ElementY0% TO ElementYn%
   FOR xIndex% = ElementX0% TO ElementXn%
      A#(xIndex%, yIndex%) = INT(RND * 99)
      B#(xIndex%, yIndex%) = INT(RND * 99)
   NEXT xIndex%
NEXT yIndex%

FOR yIndex% = LBOUND(C#, 2) TO UBOUND(C#, 2)
   FOR xIndex% = LBOUND(C#, 1) TO UBOUND(C#, 1)
      C#(xIndex%, yIndex%) = INT(RND * 99)
   NEXT xIndex%
NEXT yIndex%

PRINT #SerFile%, "                             Matrix Test Results"
PRINT #SerFile%, "": PRINT #SerFile%, STRING$(60, "-")

CALL MatrixDemo1(A#(), B#(), C#(), SerFile%)
CLOSE #SerFile%
END SUB
```

Program 4.26B Matrix Demonstration 1

```
SUB MatrixDemo1 (A#(), B#(), C#(), SerFile%)
'
'THIS PROGRAM COPYRIGHT: (C) 1991 by LCPsoftware
'AUTHORS:   Benjamin J. Landry and Edward Champion
'TEXT BOOK: NUMERICAL METHODS FOR ENGINEERING APPLICATIONS
'
'EQUATION:  (04.46) EQU_0446 Matrix Demonstration 1
'
'PURPOSE:   Demonstrates the matrix functions
'---------------------- end introduction -------------------

DIM R1#(1 TO 3, 1 TO 3), R2#(1 TO 3, 1 TO 3)
DIM R4#(1 TO 3, 1 TO 1), Text$(0 TO 3)

Text$(0) = "MatrixSum"
Text$(1) = "   + ": Text$(2) = "   = "
Text$(3) = "Add two random matrices [A] + [B] = [R]"
CALL MatrixSum(A#(), B#(), R1#())
CALL MatrixOutput(Text$(), A#(), B#(), R1#(), SerFile%)

Text$(0) = "MatrixDif"
Text$(1) = "   - ": Text$(2) = "   = "
Text$(3) = "Then  [R] - [B]  should get us  [A]  back"
CALL MatrixDif(R1#(), B#(), R2#())
CALL MatrixOutput(Text$(), R1#(), B#(), R2#(), SerFile%)

Text$(0) = "MatrixScalarMul"
Text$(1) = "     ": Text$(2) = "   = "
Text$(3) = "Start with [A], compute [A] + [A] should = 2[A]"
CALL MatrixSum(A#(), A#(), R1#())
CALL MatrixScalarMul(2, A#(), R2#())
CALL MatrixOutput(Text$(), A#(), R1#(), R2#(), SerFile%)

Text$(0) = "MatrixMul"
Text$(1) = "   * ": Text$(2) = "   = "
Text$(3) = "Of 2 n x n matrices [A] x [B] "
CALL MatrixMul(A#(), B#(), R1#())
CALL MatrixOutput(Text$(), A#(), B#(), R1#(), SerFile%)

Text$(0) = "MatrixMul"
Text$(1) = "   * ": Text$(2) = "   = "
Text$(3) = "Now compute [B] x [A] it should NOT = [A] x [B]"
CALL MatrixMul(B#(), A#(), R1#())
CALL MatrixOutput(Text$(), B#(), A#(), R1#(), SerFile%)

Text$(0) = "MatrixMul"
Text$(1) = "   * ": Text$(2) = "   = "
Text$(3) = "Of 1 n x n  times n * 1  (Linear Equation)"
CALL MatrixMul(A#(), C#(), R4#())
CALL MatrixOutput(Text$(), A#(), C#(), R4#(), SerFile%)

END SUB
```

Program 4.27B Linear Demonstration Sequencer

```
SUB LinearDemoSequencer
'
'THIS PROGRAM COPYRIGHT: (C) 1991 by LCPsoftware
'AUTHORS:   Benjamin J. Landry and Edward Champion
'TEXT BOOK: NUMERICAL METHODS FOR ENGINEERING APPLICATIONS
'
'EQUATION:  (04.47) EQU_0447 Linear Demonstration Setup
'
'PURPOSE:   Set up one 3 x 3 matrices and two 3 x 1 matrices.
'           These matrices will be used to demonstrate
'           the Linear routines described in this chapter
'--------------------- end introduction ------------------
DIM a#(1 TO 5, 1 TO 5), b#(1 TO 5, 1 TO 1)
DIM x#(1 TO 5, 1 TO 1): CLS
RANDOMIZE (TIMER):  RANDOMIZE (1) 'Make it repeatable
'For each of four tests, create a different [A] and x matrix
FOR Seed% = 1 TO 4   '1=Lower  2=Upper  3=Tridiagonal 4=Full
                        'Output sent to last listed device
OutputDevice$ = "DAT_044" + LTRIM$(STR$(Seed% + 6)) + ".ASC"
'OutputDevice$ = "LPT1:"                          'Printer
'OutputDevice$ = "SCRN:"                          'Screen

IF (Seed% = 1) OR (Seed% = 3) THEN
   SerFile% = FREEFILE 'When output to a file, put 2 in each
   OPEN OutputDevice$ FOR OUTPUT AS #SerFile%
   PRINT #SerFile%, "                   Linear Test Results"
   PRINT #SerFile%, "": PRINT #SerFile%, STRING$(60, "-")
END IF
  IF Seed% = 3 THEN
    Diag% = 1
  ELSE Diag% = 0
  END IF
  FOR iIndex% = LBOUND(a#, 1) TO UBOUND(a#, 1)
    FOR jIndex% = LBOUND(a#, 2) TO UBOUND(a#, 2)
      a#(iIndex%, jIndex%) = INT(RND * 98) + 1
      IF (iIndex% > jIndex% + Diag%) AND (Seed% AND 2) THEN
        a#(iIndex%, jIndex%) = 0
      END IF
      IF (jIndex% > iIndex% + Diag%) AND (Seed% AND 1) THEN
        a#(iIndex%, jIndex%) = 0
      END IF
    NEXT jIndex%
  NEXT iIndex%
  'Initialize x# array
  FOR iIndex% = LBOUND(x#, 1) TO UBOUND(x#, 1)
    x#(iIndex%, 1) = INT(RND * 98) + 1
  NEXT iIndex%

  CALL LinearDemo(Seed%, a#(), b#(), x#(), SerFile%)
IF (Seed% = 2) OR (Seed% = 4) THEN CLOSE SerFile%
NEXT Seed%
END SUB
```

Program 4.28B Linear Demonstration

```
SUB LinearDemo (Seed%, A#(), B#(), x#(), SerFile%)
'
'THIS PROGRAM COPYRIGHT: (C) 1991 by LCPsoftware
'AUTHORS:    Benjamin J. Landry and Edward Champion
'TEXT BOOK: NUMERICAL METHODS FOR ENGINEERING APPLICATIONS
'
'EQUATION:   (04.48) EQU_0448 Linear Demonstration Setup
'
'PURPOSE:    Demonstration of ForwardSubstitution
'                             BackwardSubstitution
'                             Tridiagonal Solution
'                             LU Decomposition
'
'---------------------- end introduction ------------------

DIM Text$(4)
Text$(1) = "    * ":  Text$(2) = "    = ": Text$(3) = ""

Text$(0) = ""
CALL MatrixMul(A#(), x#(), B#())
CALL MatrixOutput(Text$(), A#(), x#(), B#(), 1)

FOR iIndex% = LBOUND(x#, 1) TO UBOUND(x#, 1)
      x#(iIndex%, 1) = 0
NEXT iIndex%:

Text$(3) = "Confirm that the x matrix has be erased"
CALL MatrixOutput(Text$(), A#(), x#(), B#(), 1)
Text$(3) = ""

SELECT CASE Seed%

   CASE 1
      Text$(0) = "Forward Substitution"
      CALL ForwardSubstitution(A#(), B#(), x#())

   CASE 2
      Text$(0) = "Backward Substitution"
      CALL BackwardSubstitution(A#(), B#(), x#())

   CASE 3
      Text$(0) = "Tridiagonal Matrix"
      CALL Tridiagonal(A#(), B#(), x#())

   CASE 4
      Text$(0) = "LU Decomposition"
      CALL LuDecomposition(A#(), B#(), x#())
END SELECT:

CALL MatrixOutput(Text$(), A#(), x#(), B#(), 1)

END SUB
```

Program 4.29B Matrix Output

```
SUB MatrixOutput (Text$(), A#(), B#(), Result#(), Port%)
'
'THIS PROGRAM COPYRIGHT: (C) 1991 by LCPsoftware
'AUTHORS:   Benjamin J. Landry and Edward Champion
'TEXT BOOK: NUMERICAL METHODS FOR ENGINEERING APPLICATIONS
'
'EQUATION:  (04.49) EQU_0449 Vector Demo Output
'
'PURPOSE:   Common output support for Vector demonstrations
'
'---------------------- end introduction ------------------

IF Text$(0) <> "" THEN
   PRINT #Port%, "Test of routine:  "; Text$(0)
   PRINT #Port%, ""
END IF

RowMin% = LBOUND(B#, 1): RowMax% = UBOUND(B#, 1)
FOR Rows% = RowMin% TO RowMax%
   FOR Cols% = LBOUND(A#, 2) TO UBOUND(A#, 2)
      IF Rows% <= UBOUND(A#, 1) THEN
         PRINT #Port%, USING "####"; A#(Rows%, Cols%);
      ELSE PRINT #Port%, USING "&      "; "";
      END IF
   NEXT Cols%

   IF Rows% = INT((RowMax% - RowMin%) / 2) + RowMin% THEN
      PRINT #Port%, Text$(1);
   ELSE PRINT #Port%, SPACE$(LEN(Text$(1)));
   END IF

   FOR Cols% = LBOUND(B#, 2) TO UBOUND(B#, 2)
      PRINT #Port%, USING " ####"; B#(Rows%, Cols%);
   NEXT Cols%

   IF Rows% = INT((RowMax% - RowMin%) / 2) + RowMin% THEN
      PRINT #Port%, Text$(2);
   ELSE PRINT #Port%, SPACE$(LEN(Text$(2)));
   END IF

   FOR Cols% = LBOUND(Result#, 2) TO UBOUND(Result#, 2)
      IF Rows% <= UBOUND(B#, 1) THEN
        PRINT #Port%, USING " #####"; Result#(Rows%, Cols%);
      END IF
   NEXT Cols%
   PRINT #Port%, ""
NEXT Rows%

PRINT #Port%, ""
IF Text$(3) <> "" THEN PRINT #Port%, Text$(3)
PRINT #Port%, STRING$(60, "-")
END SUB
```

5

Roots and Non-Linear Equations

5.1 Introduction

This chapter highlights methods for finding roots of equations or functions. One of the most frequently occurring problems in analytical work is to find the roots of equations of the form

$$f(x) = 0 \tag{5.1}$$

The function may be given explicitly or it may be known implicitly. A basic premise is that over the interval of interest, the function is continuous and differentiable for the methods given in this chapter. In linear problems (functions of one variable) finding roots usually proceeds by iteration. For smoothly varying functions, good algorithms will converge given a good initial guess at the answer. The ability to provide a relatively good first guess is imperative for multi-dimensional systems.

The primary reason to solve non-linear equations by computer methods is that non-linear equations have no closed-form solutions except for a few exceptions. Analytical solutions exist for polynomial equations up to fourth order.

For one-dimensional root finding it is necessary to understand the function and know the interval where the root is required (ie bracket the interval such that you know where the function changes sign).

5.2 Methods Without Derivatives

5.2.1 Bracketing and Bisection

In order to find the root of a one-dimensional function it is necessary to determine where a root is bracketed in the interval (a,b) if f(a) and f(b) have different signs. By the *intermediate value theorem* of Calculus, at least one root must lie in the interval if the function is continuous. The following program examines a function over a given interval and returns the interval for the possible root.

The basic procedure is outlined below. The subroutine to accomplish this task is in Program 5.1F at the end of the chapter.

Guess initial range Subdivide over guessed interval Evaluate function at each interval Determine where function changes sign Use information on bracket for next programs.

The **bisection** method is a tried and true method for finding roots. A unique feature of this method is that it works for non-analytic as well as analytic functions. Given an interval where a root exists (eg from the above program), evaluate the function at the midpoint of the interval and examine the function's sign. Use the midpoint to replace whichever limit has the same sign. After each iteration decrease the bounds containing the root by a factor of 2. The algorithm for the bisection method can be listed as follows:

```
Given f(x), x0, x1
Repeat
        Set x2 = (x0 + x1)/2
        If f(x0) and f(x2) have opposite signs
                Set x1 = x2 otherwise
                Set x0 = x2.
Until Convergence to user specified epsilon.
```

The FORTRAN program is found in Program 5.2F.

5.2.2 Method Of False Position (Regula Falsi)

The bisection method will converge to an answer but it may do so at the expense of time. In addition while the bisection method notes the sign changes of the function, it does nothing with their values. Another method that converges faster than the bisection method is called the Regula Falsi position or the Method of False Position. In this method and in the Secant method discussed in Section 5.2.3, the function is assumed to be linear in the local region of interest and the next improvement in the root is at the point where the approximating line crosses the axis. After each iteration one of the previous boundary points is discarded in favor of the latest estimate of the root. The formula for the next position x_2 is given by

$$x_2 = x_1 - \frac{(x_1 - x_0)f_1}{f_1 - f_0} \tag{5.2}$$

The program for the Regula Falsi method is illustrated in Program 5.3F.

There are two disadvantages in using the Regula Falsi method. The first is that the estimate x_2 may not be a better estimate than the midpoint value.

The second disadvantage is that the size of the interval containing the root does not in all cases approach zero. The reason for this lies in the fact that most functions are strictly concave or convex which requires one of the end points to approach the root while the other end point remains fixed.

5.2.3 Secant Method

The Secant method is quite like the Regula Falsi method except that it removes the restriction that successive iterations must bracket the root. Successive function values can have the same sign.

An algorithm for the Secant method is given as follows.

```
Given f(x), x_0, x_1
    Repeat for n = 1,2,.....
        Set x_{n+1} = [(x_n - x_{n-1})/(f_n - f_{n-1})]f_n
    Until convergence.
```

The Secant method is given in Program 5.4F.

5.2.4 Brent's Method

Brent's method combines root bracketing, bisection, and inverse quadratic interpolation to converge to a zero crossing of the function. This method is based on a method developed by Van Wijngaarden, Dekker, and others at the Mathematical Center in Amsterdam in the 1960's. This version is outlined in Reference 67. The interpolation formula is similar to the Lagrange interpolation formula given in Chapter 2 with the three interpolated point pairs [a,f(a)], [b,f(b)], [c,f(c)].

$$\begin{aligned} x = & \frac{[y-f(a)][y-f(b)]c}{[f(c)-f(a)][f(c)-f(b)]} \\ & + \frac{[y-f(b)][y-f(c)]a}{[f(a)-f(b)][f(a)-f(c)]} \\ & + \frac{[y-f(c)][y-f(b)]b}{[f(b)-f(c)][f(b)-f(a)]} \end{aligned} \tag{5.3}$$

If we set y =0, we get the next root estimate.

$$x = b + \frac{P}{Q} \tag{5.4}$$

where

$$\begin{aligned} P &= S[T(R-T)(c-b) - (1-R)(b-a)] \\ Q &= (T-1)(R-1)(S-1) \\ R &\equiv \frac{f(b)}{f(c)} \\ S &\equiv \frac{f(b)}{f(a)} \\ T &\equiv \frac{f(a)}{f(c)} \end{aligned} \tag{5.5}$$

Brent's method is outlined in Program 5.5.

5.3 A Method Using Derivatives

If f(x), f'(x), and f"(x) are continuous near a root, then this extra information regarding the nature of f(x) can be used to develop algorithms that will produce sequences that converge faster than either the bisection or false position methods. The Newton-Raphson method is one of the more useful and better known algorithms that rely on the continuity of the function f(x).

Let x_n be an approximation to alpha, a root of f(x). Using Taylor's series expansion about the point x_n gives

$$f(\alpha) = f(x_n) + (\alpha - x_n)f'(x_n) + \frac{(\alpha - x_n)^2}{2!}f''(x_n) + \cdots \tag{5.6}$$

If x_n is close to α, then $(\alpha - x)$ is small and terms of order $(\alpha - x_n)^2$ and higher can be ignored. Therefore

$$f(\alpha) \simeq f(x_n) + (\alpha - x_n)f'(x_n) \tag{5.7}$$

from which

$$\alpha \simeq x_n - \frac{f(x_n)}{f'(x_n)} \tag{5.8}$$

$$x_{n+1} = x_n - \frac{f(x_n)}{f'(x_n)} \tag{5.9}$$

This is the basis for the iteration algorithm for the Newton-Raphson method given below:

Given: A function, initial guess X Compute function value Compute function derivative Compare new value of X with tolerance If within tolerance, end Else use computed x value for new function calculations Loop until converged

The program implementing the above equation is given in Program 5.6.

Near a root, the number of significant digits nearly doubles with each step. The Newton-Raphson method is the method of choice for any function whose derivative is evaluated efficiently and whose derivatives are continuous and non-zero in the near region of the root.

5.4 Polynomial Equations

5.4.1 Nested Multiplication

The ability to find the roots (or zeros) of a polynomial is required for a number of scientific and engineering disciplines. It is obvious that the methods discussed previously can be applied to polynomials; however the special structure of the function deserves to be examined in more detail. The reasons for this are: (1) it is known exactly how many zeros a polynomial has and also, given that the coefficients a_i in the polynomial are real, it is known that the zeros are real or occur in complex conjugate pairs; (2) polynomials can be evaluated very efficiently on a computer by using a technique called nested multiplication.

Nested multiplication requires the rewriting of a polynomial of the form

$$f(x) = a_0x^n + a_1x^{n-1} + \cdots + a_{n-1}x + a_n \tag{5.10}$$

to the following form:

$$f(x) = (\ldots((a_0x + a_1)x + \cdots + a_{n-1})x + a_0 \tag{5.11}$$

For a particular value of x = b, the polynomial expression is evaluated from the innermost bracket outward. Program 5.7 illustrates the above technique.

5.4.2 Newton's Method

The derivatives of polynomials are very straightforward to obtain and because of this the application of Newton's method is suited for this discussion. When seeking roots of a polynomial, the amount of effort required can be reduced by use of **deflation**. Deflation is, for all purposes, synthetic division.

Assume f(x) is a polynomial of degree n. If we divide f(x) by (x-k) where k is an arbitrary constant, we obtain

$$\frac{f(x)}{(x-k)} = g(x) + \frac{r}{(x-k)} \tag{5.12}$$

where g(x) is the quotient polynomial of degree one less than f(x) and r is the constant remainder. Therefore we can rewrite (5.12) as

$$f(x) = (x-k)g(x) + r \tag{5.13}$$

At the point b, f(k) = r. Therefore f(k) can be evaluated by finding the remainder when f(x) is divided by (x-k). Because g(x) is of degree (n-1), g(x) assumes the following form:

$$g(x) = b_0 x^{n-1} + b_1 x^{n-2} + \dots + b_{n-2} x + b_{n-1} \tag{5.14}$$

The b values are constants. If we substitute Equation (5.14) into Equation (5.13) and compare coefficients with those of Equation (5.10) gives

$$\begin{aligned} a_0 &= b_0 \\ a_1 &= b_1 - b_0 k \\ a_2 &= b_2 - b_1 k \\ \dots &= \dots \\ a_{n-1} &= b_{n-1} - b_{n-2} k \\ a_n &= r - b_{n-1} k \end{aligned} \tag{5.15}$$

Rearranging Equation (5.15) such that the unknown b appears on the left hand side of the equation gives

$$\begin{aligned} b_0 &= a_0 \\ b_1 &= b_0 k + a_1 \\ b_2 &= b_1 k + a_2 \\ \dots &= \dots \\ b_{n-1} &= b_{n-2} k + a_{n-1} \end{aligned} \tag{5.16}$$

and

$$r = b_{n-1}k + a_n \tag{5.17}$$

If r is replaced by b_n, then f(k) is b_n, the last value obtained from the recurrence relation

$$b_i = b_{i-1}k + a_i \tag{5.18}$$

where i=1,2,..., n and $b_0 = a_0$.

To obtain a similar formula for f'(k), Equation (5.13) is differentiated and this gives

$$f'(x) = (x-k)g'(x) + g(x) \tag{5.19}$$

Therefore

$$f'(k) = g(k) \tag{5.20}$$

To evaluate g(k), we apply the process of synthetic division as before to obtain

$$g(x) = (x-k)(c_0x^{n-2} + c_1x^{n-3} + \dots + c_{n-3}x + c_{n-2}) + s \tag{5.21}$$

Therefore g(k) is the remainder when g(x) is divided by (x-k). Comparing coefficients as before, we obtain

$$\begin{aligned} c_0 &= b_0 \\ c_1 &= c_0 k + b_1 \\ \ldots &= \ldots \\ c_{n-2} &= c_{n-3} k + b_{n-2} \end{aligned} \tag{5.22}$$

and

$$s = c_{n-2} k + b_{n-1} \tag{5.23}$$

If s is replaced by c_{n-1}, then f'(k) is c_{n-1}, the last value obtained from the recurrence relation

$$c_i = c_{i-1} k + b_i \tag{5.24}$$

where i=1,2,....n-1 and $c_0 = b_0$.

With these relations defined, the expression for Newton's method for finding a real zero of a polynomial is

$$x_{m+1} = x_m - \frac{b_n}{c_{n-1}} \tag{5.25}$$

In FORTRAN, the method is programmed as Program 5.8F and in PASCAL as Program 5.8P.

Deflation must be used with care. Since each new root is known with a finite accuracy, errors will build in the determination of the coefficients of the successively deflated polynomial. Therefore, the roots become more and more inaccurate. It does matter how the error builds, whether it is stable or unstable. In order to minimize the increasing error, it is better to treat the roots of a successively deflated polynomial as only tentative roots. These values can be used as initial guesses to solve by another method the undeflated polynomial.

5.4.3 Bairstow's Method

Newton's method may be used for complex roots by suitable, minor code changes. This approach has as a disadvantage on computers since complex arithmetic is usually slower than real number arithmetic. Complex roots of polynomials with real coefficients occurring in conjugate pairs. This leads to the use of quadratic factor rather than linear factor for the polynomial. Let (x^2 + ux + v) be the approximate quadratic factor of f(x). Following the logic in the previous section, we have

$$f(x) = (x^2 + ux + v)g(x) + rx + t \tag{5.26}$$

where (rx + t) is the remainder and g(x) is the quotient polynomial of degree (n-2). The above equation can be rewritten letting the value of s = t - ru.

$$f(x) = (x^2 + ux + v)g(x) + r(x+u) + s \tag{5.27}$$

In order to obtain an exact quadratic factor, increments of δu and δv have to be determined that when added to u and v respectively will make r and s equal to zero. The set of equations to solve for δu and δv are as follows.

$$\begin{aligned} r(u+\delta u, v+\delta v) &= 0 \\ s(u+\delta u, v+\delta v) &= 0 \end{aligned} \tag{5.28}$$

These equations are non-linear. Apply Taylor's expansion to each expression in Equation (5.28) to give a linear approximation.

$$\begin{aligned} r + \frac{\partial r}{\partial u}\delta u + \frac{\partial r}{\partial v}\delta v &= 0 \\ s + \frac{\partial s}{\partial u}\delta u + \frac{\partial s}{\partial v}\delta v &= 0 \end{aligned} \tag{5.29}$$

Solving for δu and δv gives

$$\begin{aligned} \delta u &\simeq \frac{[s\frac{\partial r}{\partial v} - r\frac{\partial s}{\partial v}]}{\det(J)} \\ \delta v &\simeq \frac{[r\frac{\partial s}{\partial u} - s\frac{\partial r}{\partial u}]}{\det(J)} \end{aligned} \tag{5.30}$$

where

$$J = \begin{bmatrix} \frac{\partial r}{\partial u} & \frac{\partial r}{\partial v} \\ \frac{\partial s}{\partial u} & \frac{\partial s}{\partial v} \end{bmatrix} \tag{5.31}$$

J is known as the **Jacobian** matrix of the pair of functions r(u,v) and s(u,v). Equation (5.30) gives δu and δv only approximately. This forms a basis for an iterative scheme to find an exact quadratic factor of f(x). The program that follows implements the above equations. If the Jacobian matrix is singular, then det(J) is zero and problems arise in using this method. The program tests for the possibility that the Jacobian may be singular or that the det(J) becomes very small.

The program to address this problem is Program 5.9.

5.5 Problems

5-1) Show that the function

$$f(x) = x^2 - 6x + 3$$

has roots bracketed by x=(0,1) and x=(5,6).

5-2) Determine the real roots of the function in problem 1 both graphically and using the bisection method. When using bisection compute the estimated error at each step.

5-3) Show graphically how the bisection method is better than the method of false-position when finding the root of the function

$$f(x) = x^7 - 1.$$

5-4) Use the secant method to find the single root of the function

$$f(x) = \ln(x) + x$$

with initial guesses of $x_0 = 0.5$ and $x_1 = 1.0$.

5-5) Show graphically how the secant method searches for a root for the function

$$f(x) = 7x^2 - 5x + 5$$

when one does not exist.

5-6) Repeat problem 4 using Brent's method.

5-7) Determine the highest real root of

$$f(x) = x^3 - 9x^2 + 15x - 1$$

using the Newton-Raphson method with an initial guess of x = 6.

5-8) Write the polynomial

$$12.11x^5 - 323.7x^4 + 0.4474x^3 - 952.78x - 1.430 = 0$$

in the nested multiplication form.

5-9) Determine the single real root of the polynomial

$$x^5 - 3.7x^4 + 7.4x^3 - 10.8x - 6.8 = 0$$

using Newton's method.

5-10) Use Bairstow's method to determine the roots of the polynomial

$$f(x) = x^5 - 3.5x^4 + 2.75x^3 + 2.125x^2 - 3.875x + 1.25$$

with initial guesses of $r = s = -1$. Find both the real and imaginary roots.

Chapter 5 Programs

Program 5.1F Bracketing Routine

```
      SUBROUTINE BRACKET (FCN,N,X,XL,XU)
C______________________________________________________________________

C  THIS PROGRAM COPYRIGHT: (C) 1991 by Marvin N. Zeigler
C  AUTHORS:  Marvin N. Zeigler and Edward Champion
C  TEXT BOOK:  NUMERICAL METHODS FOR ENGINEERING APPLICATIONS
C  SECTION:  5.2.1

C  PURPOSE:  Finds ranges where the function FCN(X) has roots.  The
C   initial range, XL-XU, is divided into N even length subdivisions.
C   The function is evaluated at each end of the subdivisions and
C   checked for a sign change

C  ARGUMENTS:
C   FCN    External function that computes the values of f(x).
C   N      Number of subdivisions.
C   X     Array holding the range pairs in order.  X(1) and X(2) hold
C         the first lower and upper range values, X(3) and x(4) hold
C          the second pair, etc.
C   XL     Lower limit of the initial range.
C   XU     Upper limit of the initial range.

C______________________________________________________________________

      REAL FCN,XL,XU,X(N),XINT,FL,FU
      INTEGER N,K

      XINT = (XU-XL)/REAL(N)
      FL = FCN(XL)
      K = 1

      DO 20 I=1,N
         XU = XL + XINT
         FU = FCN(XU)

C  Check for sign change in the current subdivision
         IF ( FL*FU .LT. 0.0 ) THEN
            X(K) = XL
            X(K+1) = XU
            K = K + 2
         ENDIF

         XL = XU
         FL = FU
20    CONTINUE

      RETURN
      END
```

Program 5.2F Bisection Method

```
      SUBROUTINE BISECT (FCN,X0,X1,X2,XTOL,FTOL,MAXIT)
C_______________________________________________________________________

C  THIS PROGRAM COPYRIGHT: (C) 1991 by Marvin N. Zeigler
C  AUTHORS:  Marvin N. Zeigler and Edward Champion
C  TEXT BOOK:  NUMERICAL METHODS FOR ENGINEERING APPLICATIONS
C  SECTION:  5.2.1

C  PURPOSE:  A root of the function FCN(X) is found in the interval
C   X0-X1 using the bisection method.  Both the X and function
C  tolerances are checked for convergence.  The X location of the root
C   is returned through X2.  If the root is not found in the allowed
C   number of iterations an error message is printed and the routine
C   returns to the calling program.

C  ARGUMENTS:
C   FCN    External function that computes the values of f(x)
C   X0,X1  Limits of the initial range.
C   X2     Value of root returned to calling routine.
C   XTOL   Tolerance value for the error in x.
C   FTOL   Tolerance value for f(x).
C   MAXIT  Maximum number of iterations allowed.

C_______________________________________________________________________

      REAL FCN,X0,X1,X2,XTOL,FTOL,F0,F2,ERROR
      INTEGER MAXIT

      F0 = FCN(X0)

C Test for the function having opposite signs in the initial interval
      IF (F0*FCN(X1) .GT. 0.0) THEN
         WRITE(*,*) ' ILLEGAL FUNCTION RANGE IN BISECT.'
         RETURN
      ENDIF

      ERROR = ABS(X1-X0)

      DO 20 I=1,MAXIT
         X2 = (X0+X1)/2.0
         F2 = FCN(X2)
         ERROR = ERROR/2.0

C  Check if X value has reached tolerance
         IF (ERROR .LE. XTOL) THEN
            WRITE(*,*) ' X TOLERANCE REACHED IN BISECT'
            WRITE(*,*) ' ROOT IS AT X= ',X2,'MAXIMUM ERROR IS ',ERROR
         RETURN
         ENDIF

C  Check if function value has reached tolerance
         IF (ABS(F2) .LE. FTOL) THEN
            WRITE(*,*) ' FUNCTION TOLERANCE REACHED IN BISECT'
            WRITE(*,*) ' ROOT IS AT X= ',X2,'MAXIMUM ERROR IS ',ERROR
         RETURN
         ENDIF
```

```
          IF (F0*F2 .LT. 0.0) THEN
             X1 = X2
          ELSE
             X0 = X2
             F0 = F2
          ENDIF
 20    CONTINUE

       WRITE(*,*) ' MAXIMUM NUMBER OF ITERATIONS REACHED IN BISECT.'
       RETURN
       END
```

Program 5.3F Method of False Position

```
      SUBROUTINE FLSPSTN (FCN,X0,X1,X2,FTOL,MAXIT)
C_____________________________________________________________

C THIS PROGRAM COPYRIGHT: (C) 1991 by Marvin N. Zeigler
C AUTHORS:  Marvin N. Zeigler and Edward Champion
C TEXT BOOK:  NUMERICAL METHODS FOR ENGINEERING APPLICATIONS
C EQUATION:  5.2

C PURPOSE:  A root of the function FCN(X) is found in the interval
C  X0-X1 using the false position method.An approximate error is C
C  not given, nor is a convergence check made based on such an C
C  error, because it is possible for the approximate error to be C
C  somewhat less than the true error if f(X0) is much closer to C
C  zero than f(X1).

C ARGUMENTS:
C  FCN    External function that computes the values of f(x)
C  X0,X1  Limits of the initial range.
C  X2     Value of root returned to calling routine.
C  FTOL   Tolerance value for f(x).
C  MAXIT  Maximum number of iterations allowed.

C_____________________________________________________________

      REAL FCN,X0,X1,X2,FTOL,F0,F1,F2
      INTEGER MAXIT

      F0 = FCN(X0)
      F1 = FCN(X1)

C Test for the function having opposite signs in the initial
C     interval
     IF (F0*F1 .GT. 0.0) THEN
        WRITE(*,*) ' ILLEGAL FUNCTION RANGE IN FLSPSTN.'
        RETURN
     ENDIF

      DO 20 I=1,MAXIT
         X2 = X1 - (X1-X0)*F1/(F1-F0)
         F2 = FCN(X2)

C Check if function value has reached tolerance
         IF (ABS(F2) .LE. FTOL) THEN
            WRITE(*,*) ' ROOT FOUND IN FLSPSTN AT X= ',X2
            RETURN
         ENDIF

         IF (F2 .LT. 0.0) THEN
            X0=X2
            F0=F2
         ELSE
            X1=X2
            F1=F2
         ENDIF

 20   CONTINUE
```

```
      WRITE(*,*) ' MAXIMUM NUMBER OF ITERATIONS REACHED IN
FLSPSTN.'
      RETURN
      END
```

Program 5.4F Secant Method

```
      SUBROUTINE SECANT (FCN,X0,X1,X2,TOL,MAXIT)
C_______________________________________________________________

C  THIS PROGRAM COPYRIGHT: (C) 1991 by Marvin N. Zeigler
C  AUTHORS:  Marvin N. Zeigler and Edward Champion
C  TEXT BOOK:  NUMERICAL METHODS FOR ENGINEERING APPLICATIONS
C  SECTION:  5.2.3

C  PURPOSE: Routine to find a root of the function FCN(X) using the
C   secant method. The root does not have to lie inside the initial
C   interval X0-X1, but it should be near.

C  ARGUMENTS:
C   FCN    External function that computes the values of f(x)
C   X0,X1  Limits of the initial range.
C   X2     Value of root returned to calling routine.
C   TOL   Tolerance value used to determine convergence.
C   MAXIT  Maximum number of iterations allowed.

C_______________________________________________________________

      REAL FCN,X0,X1,TOL,F0,F1,DX
      INTEGER MAXIT
      DATA MAXIT,TOL /50,.0001/

      F0 = FCN(X0)
      DX = X1-X0

      DO 20 I=1,MAXIT
         F1 = FCN(X1)
         DX = F1*DX/(F1-F0)
         X2 = X1 - DX

C  Check if tolerance has been met
         IF (ABS(DX) .LE. TOL) THEN
            WRITE(*,*) ' ROOT FOUND IN SECANT AT X= ',X2
            RETURN
         ENDIF

         DX = X2 - X1
         X0 = X1
         X1 = X2
         F0 = F1
 20   CONTINUE

      WRITE(*,*)' MAXIMUM NUMBER OF ITERATIONS REACHED IN SECANT.'
      RETURN
      END
```

Program 5.5F Brent's Method

```
      PROGRAM CALL
      SUBROUTINE BRENT (FCN,A,B)
C_____________________________________________________________________

C  AUTHORS:  Marvin N. Zeigler and Edward Champion
C  TEXT BOOK:  NUMERICAL METHODS FOR ENGINEERING APPLICATIONS
C  EQUATION:  5.3 - 5.5

C  PURPOSE:  Program using BRENT'S method which takes advantage of
C   inverse quadratic interpolation or bisection depending on the
C   better method for the particular iteration.  A root is found
C   for the function FCN(X) in the interval A-B.

C  ARGUMENTS:
C   FCN    Externally defined function.
C   A      Lower limit of the root bound.
C   B      Upper limit of bound and returned root.
C   XTOL   Tolerance value for convergence.
C   ZTOL   Tolerance value for convergence.
C   MAXIT  Maximum number of iterations allowed.

C_________________________________________________________

      REAL
FCN,A,B,C,DELX,XKEEP,P,Q,R,S,T,EA,EB,EC,ED,XTOL,XMID,ZTOL
      INTEGER MAXIT, IQFLAG
      DATA MAXIT,IQFLAG,XTOL,ZTOL /100,0,1.0E-4,3.0E-8/

      EA = FCN(A)
      EB = FCN(B)

C Test for function having opposite signs in the initial interval
      IF (EA*EB .GT. 0.0) THEN
         WRITE(*,*) ' ILLEGAL FUNCTION RANGE IN BRENT.'
         RETURN
      ENDIF
      EC = EB

      DO 20 I=1,MAXIT

         IF (EB*EC .GT. 0.0) THEN
            C = A
            DELX = B-A
            XKEEP = B-A
            EC = EA
            IQFLAG = 1
         ENDIF
         IF (ABS(EB) .GT. ABS(EC)) THEN
            A = B
            B = C
            C = A
            EA = EB
            EB = EC
            EC = EA
            IQFLAG = 1
         ENDIF
         TTOL = 2.0*ZTOL*ABS(B) + XTOL/2.0
```

```
         XMID = (C-B)/2.0

C Check if tolerance has been met
         IF (ABS(XMID).LE.TTOL .OR. ABS(EB).LT.ZTOL) THEN
            WRITE(*,*) ' ROOT FOUND IN BRENT AT X= ',B
            RETURN
         ENDIF

C Calculate inverse quadratic interpolation values
         S = EB/EA
         IF (IQFLAG .EQ. 1) THEN
            P=2.*XMID*S
            Q=1.-S
            IQFLAG = 0
         ELSE
            ED=EA/EC
            R = EB/EC
            P = S*(2.0*XMID*ED*(ED-R)-(B-A)*(R-1.0))
            Q = (Q-1.0)*(R-1.0)*(S-1.0)
         ENDIF
         IF (P .GT. 0.0) THEN
            Q=-Q
         ELSE
            P=ABS(P)
         ENDIF
         IF (ABS(DELX*Q) .GT. 2.0*P) THEN
C Interpolation values used
            XKEEP = DELX
            DELX = P/Q
         ELSE
C Using bisection instead of interpolation
            DELX = XMID
            XKEEP = XMID
         ENDIF
         A=B
         EA=EB

         IF (ABS(DELX) .GT. TTOL) THEN
C Normal offset used if DELX is not too small
            B = B + DELX
         ELSE
C Offset B by a small amount to give faster convergence
            B=B + SIGN(TTOL,XMID)
         ENDIF

         EB=FCN(B)
20    CONTINUE

      WRITE(*,*) ' MAXIMUM NUMBER OF ITERATIONS REACHED IN BRENT.'
      RETURN
      END
```

Program 5.6F Newton-Raphson Method

```
      SUBROUTINE NEWRAP (FCN,FD,X,XTOL,FTOL,MAXIT)
C______________________________________________________________________
C  THIS PROGRAM COPYRIGHT: (C) 1991 by Marvin N. Zeigler
C  AUTHORS:  Marvin N. Zeigler and Edward Champion
C  TEXT BOOK:  NUMERICAL METHODS FOR ENGINEERING APPLICATIONS
C  EQUATION:  5.9

C  PURPOSE:  Computes a root of the function FCN(X) using the
C   Newton-Raphson method.  The derivative of the function must
C   be provided as FD(X).

C  ARGUMENTS:
C   FCN    External function that computes the values of f(x)
C   FD     External function that computes the values of f'(x)
C   X      Initial guess for root value, estimated to be near it.
C   XTOL   Tolerance value for change in x, i.e. delta x.
C   FTOL   Tolerance value for f(x).
C   MAXIT  Maximum number of iterations allowed.

C______________________________________________________________________
      REAL FCN,FD,FX,DX,X,XTOL,FTOL
      INTEGER MAXIT

      FX = FCN(X)

      DO 20 I=1,MAXIT

C  Compute approximation of root based on Newton-Raphson equation
         DX = FX/FD(X)
         X = X-DX
         FX = FCN(X)

C  Check if function tolerance has been met
         IF (ABS(FX) .LE. FTOL) THEN
            WRITE(*,*) ' FUNCTION TOLERANCE MET IN NEWRAP.'
            WRITE(*,*) ' ROOT IS AT X= ',X
            RETURN
         ENDIF

C  Check if x tolerance has been met
         IF (ABS(DX) .LE. XTOL) THEN
            WRITE(*,*) ' TOLERANCE OF DELTA X MET IN NEWRAP.'
            WRITE(*,*) ' ROOT IS AT X= ',X
            RETURN
         ENDIF
20    CONTINUE

      WRITE(*,*)' MAXIMUM NUMBER OF ITERATIONS REACHED IN NEWRAP.'
      RETURN
      END
```

Program 5.7F Nested Multiplication

```
      PROGRAM RUNNEST
C_______________________________________________________________________
C  THIS PROGRAM COPYRIGHT: (C) 1991 by Marvin N. Zeigler
C  AUTHORS:  Marvin N. Zeigler and Edward Champion
C  TEXT BOOK:  NUMERICAL METHODS FOR ENGINEERING APPLICATIONS
C  EQUATION:  5.11

C  PURPOSE:  Routine to calculate the result of a polynomial for a
C   x value using nested multiplication.  The coefficients of the
C   polynomial must be given in the following order:
C   F(X) = A(1)*X**N + A(2)*X**(N-1) + ... + A(N-1)*X + A(N)

C  ARGUMENTS:
C   N       Number of coefficients in the polynomial.
C   A(N)    Coefficients of the polynomial.
C   X       Value at which polynomial is evaluated.
C   F       Function value returned from routine.
C_______________________________________________________________________

      REAL A(6),F,X
C Coefficients for an example problem
      DATA N,A /6,1.0,-3.7,7.4,-10.8,10.8,-6.8/

      WRITE(*,*) 'INPUT X VALUE TO SOLVE FOR'
      READ*, X

C  THE FOLLOWING STATEMENTS ARE USED VERIFY THAT SUBROUTINE NEST
C  IS WORKING CORRECTLY.  THEY ARE NOT NECESSARY FOR PROGRAM
C  OPERATION.
      F = X**5 - 3.7*X**4 + 7.4*X**3 - 10.8*X**2 + 10.8*X - 6.8
      WRITE(*,*) ' FUNCTION EVALUATED DIRECTLY AT ',X,' IS ',F
C  END VERIFICATION.

      CALL NEST (N,A,X,F)

      END
C_______________________________________________________________________

      SUBROUTINE NEST (N,A,X,F)

      REAL A(N),F,X
      INTEGER N

      F = 0.0
      DO 10 I=1,N-1
         F = (F + A(I))*X
10    CONTINUE
      F = F + A(N)

C     WRITE(*,*) ' FUNCTION VALUE FOUND BY NEST IS ',F
      RETURN
      END
```

Program 5.8F Newton's Method

```
      SUBROUTINE NEWTON (N,A,X,TOL,MAXIT)
C_______________________________________________________________________

C  THIS PROGRAM COPYRIGHT: (C) 1991 by Marvin N. Zeigler
C  AUTHORS:  Marvin N. Zeigler and Edward Champion
C  TEXT BOOK:  NUMERICAL METHODS FOR ENGINEERING APPLICATIONS
C  EQUATION:  5.25

C  PURPOSE:Newton's method is used to find a real zero of the given
C  polynomial.An initial guess of root is necessary to start the
C   routine.The coefficients of the polynomial must be given in the
C   following order:
C   F(X) = A(1)*X**N + A(2)*X**(N-1) + ... + A(N-1)*X + A(N)

C  ARGUMENTS:
C   N      Number of coefficients in the polynomial.
C   A(N)   Coefficients of the polynomial.
C   X      Guess of root read in, calculated root returned.
C   TOL    Tolerance criteria used to check convergence.
C   MAXIT  Maximum number of iterations allowed.

C_______________________________________________________________________

      REAL A(N),B,C,DX,X,TOL
      INTEGER N,MAXIT

      DO 20 I=1,MAXIT
         B = A(1)
         C = B

         DO 10 J=2,N-1
            B = B*X + A(J)
            C = C*X + B
 10      CONTINUE

         B = B*X + A(N)
         DX = B/C
         X = X - DX

C  Check for convergence
         IF (ABS(DX) .LE. TOL) THEN
            WRITE(*,*) ' ROOT FOUND IN NEWTON AT X = ',X
            RETURN
         ENDIF

 20   CONTINUE

      WRITE(*,*)' MAXIMUM NUMBER OF ITERATIONS REACHED IN NEWTON.'
      RETURN
      END
```

Program 5.8P Newton's Method - PASCAL

```
PROGRAM NEWTONCALL; { DRIVER FOR PROCEDURE NEWTON }

USES
  CRT;

CONST
  XTOL = 1E-8; { TOLERANCE FOR DELTA X }
  FTOL = 1E-8; { TOLERANCE FOR FUNCTION F(X) }
  NLIM = 50;   { MAXIMUM ITERATIONS ALLOWED }

VAR
  X : REAL; { INITIAL GUESS AND COMPUTED ROOT VALUE }
  I : INTEGER; { FLAG THAT INDICATES EXIT CONDITION }
               { FROM PROCEDURE NEWTON }


FUNCTION FCN(X:REAL):REAL; { THIS EXAMPLE FUNCTION USED IN }
{ PROCEDURE NEWTON HAS EXACTLY ONE ROOT AT X=0.61546816 }
  BEGIN
    FCN := X-0.2*SIN(X)-0.5
  END;

FUNCTION FDER(X:REAL):REAL; { DERIVATIVE OF EXAMPLE FUNCTION }
  BEGIN
    FDER := 1-0.2*COS(X)
  END;


PROCEDURE NEWTON (VAR X:REAL;VAR I:INTEGER;
                  XTOL,FTOL :REAL; NLIM:INTEGER);
{ A ROOT OF THE FUNCTION F(X)=0 IS FOUND USING NEWTON'S METHOD. }

VAR
  FX,DELX : REAL;
  J : INTEGER;

BEGIN { PROCEDURE NEWTON }
  FX := FCN(X);
  WRITELN('AT ITERATION',0,'  X = ',X:8:6,'  F(X) =',FX:8:6);
  FOR J:=1 TO NLIM DO
    BEGIN { ITERATING TO FIND ROOT }
      DELX := FX / FDER(X);
      X := X - DELX;
      FX := FCN(X);
      WRITELN('AT ITERATION',J,'  X = ',X:8:6,'  F(X) =',FX:8:6);
{ CHECK TO SEE IF EITHER TOLERANCE HAS BEEN MET }
      IF ABS(DELX) <= XTOL THEN
        BEGIN
          I := 1;
          WRITELN('DELTA X TOLERANCE MET IN ',J,' ITERATIONS');
          WRITELN('X = ',X,'  F(X) = ',FX);
          EXIT
        END;
        IF ABS(FX) <= FTOL THEN
          BEGIN
          I := 2;
          WRITELN('FUNCTION TOLERANCE MET IN ',J,' ITERATIONS');
```

```
          WRITELN('X = ',X,'  F(X) = ',FX);
          EXIT
        END
    END; { ITERATING TO FIND ROOT }
{ IF LOOP IS COMPLETED, NLIM IS EXCEEDED AND ROOT WAS NOT FOUND }
  I := -1;
  WRITELN('TOLERANCE NOT MET IN ',J,' ITERATIONS');
  WRITELN('X = ',X,'  F(X) = ',FX);
END; { PROCEDURE NEWTON }

BEGIN { PROGRAM NEWTONCALL }
  CLRSCR;
  X := 0.5;
  NEWTON (X,I,XTOL,FTOL,NLIM);
END. { PROGRAM NEWTONCALL }
```

Program 5.9F Bairstow's Method

```
      PROGRAM BAIRSTOW
C_______________________________________________________________________

C  THIS PROGRAM COPYRIGHT: (C) 1991 by Marvin N. Zeigler
C  AUTHORS:  Marvin N. Zeigler and Edward Champion
C  TEXT BOOK:  NUMERICAL METHODS FOR ENGINEERING APPLICATIONS
C  EQUATION:  5.26-5.31

C  PURPOSE:This program uses Bairstow's method to find both real
C   and complex roots of the polynomial
C   F(X)=X^5-3.5X^4+2.75X^3+2.125X^2-3.875X+1.25.
C   The exact roots are at -1.0, 0.5, 2.0, and 1 +/- 0.5j.
C   The coefficients of a polynomial used in this program must be
C   input in the order:F(X) = A(1)*X^N + A(2)*X^(N-1) + ... + A(N).

C  ARGUMENTS:
C   N      Number of polynomial coefficients.
C   A      Array of polynomial coefficients.
C   R,S    Guesses for roots.
C   ES     Convergence criteria.
C   MAXIT  Maximum number of iterations allowed.

C_______________________________________________________________________

      REAL A(6),B(6),C(6),R,S,DET,DR,DS,EAR,EAS,ES
      INTEGER MAXIT,ITER
      DATA N,A,ES,MAXIT
/6,1.0,-3.5,2.75,2.125,-3.875,1.25,.001,50/

C  Two guesses for roots
      R = -1.0
      S = -1.0

      ITER = 0

      DO 50 WHILE (N .GE. 4 .AND. ITER.LT.MAXIT)
         EAR = 1.1*ES
         EAS = EAR

         DO 30 WHILE (EAR.GT.ES .OR. EAS.GT.ES .AND.
     +      ITER.LT.MAXIT)
            ITER = ITER + 1
            B(1) = A(1)
            B(2) = A(2) + R*B(1)
            C(1) = B(1)
            C(2) = B(2) + R*C(1)

            DO 10 K=1,N-2
               B(K+2) = A(K+2) + R*B(K+1) + S*B(K)
               C(K+2) = B(K+2) + R*C(K+1) + S*C(K)
 10         CONTINUE

            DET = C(N-2)**2 - C(N-3)*C(N-1)

            IF (DET .NE. 0.0) THEN
C  Calculate the jacobian matrix
               DR = (-B(N-1)*C(N-2) + B(N)*C(N-3))/DET
```

```
                DS = (-B(N)*C(N-2) + B(N-1)*C(N-1))/DET
                R = R + DR
                S = S + DS
                EAR = ABS(DR/R)
                EAS = ABS(DS/S)
             ELSE
                R = R + 1.0
                S = S + 1.0
                ITER = 0
             ENDIF

 30       END DO

C  Determine roots of X**2 - R*x - S
          CALL QUAD(X1,X2,R,S)
          N = N-2

          DO 40 J=1,N
             A(J) = B(J)
 40       CONTINUE

 50    END DO

       IF (ITER .LT. MAXIT) THEN
          IF (N .EQ. 3) THEN
             R = A(N-1)/A(N-2)
             S = A(N)/A(N-2)

             CALL QUAD(X1,X2,R,S)
          ELSE
             WRITE(*,*) ' SINGLE ROOT FOUND AT X = ',-A(N)/A(N-1)
          ENDIF
       ELSE
          WRITE(*,*) ' MAXIMUM NUMBER OF ITERATIONS IN BAIRSTOW.'
       ENDIF
       END

C______________________________________________________________________
       SUBROUTINE QUAD (X1,X2,B,C)
C  Routine to solve the quadratic equation
       DSC = B**2 + 4*C

       IF (DSC .LT. 0.0) THEN
          X1 = B/2.0
          X2 = SQRT(-DSC)/2.0
          WRITE(*,*) ' COMPLEX ROOTS FOUND AT X = ',X1,'+/-',X2,'j'
          RETURN
       ELSE
          X1 = (B + SQRT(DSC))/2.0
          X2 = (B - SQRT(DSC))/2.0
          WRITE(*,*) ' REAL ROOTS FOUND AT X = ',X1,'  AND ',X2
       ENDIF

       RETURN
       END
```

6

Eigenvectors and Eigensystems

6.1 Introduction

The determination of the eigenvalues (and hence the eigenvectors) of a matrix is of fundamental importance in understanding the behavior of many physical systems and in understanding and solving many purely mathematical problems. The concept of an eigenvalue is of great importance in pure and applied mathematics. A physical system such as a pendulum or a rotating shaft has associated with it certain numbers characteristic of the system. These are, for example, the period of the pendulum or the critical angular velocities at which the shaft will buckle.

For a given matrix square **A** there are a limited number of values which lead to a non-trivial solution to the following equation.

$$Ay = \lambda y \tag{6.1}$$

The number λ is an *eigenvalue of the matrix A* provided for some nonzero (real or complex) vector **y**. The n-vector **y** is called the eigenvector of **A** belonging to the eigenvalue λ.

The system of equations above is a specific case of the general system of linear equations

$$\begin{aligned} a_{11}x_1 + a_{12}x_2 + \cdots + a_1nx_n &= \lambda x_1 \\ a_{21}x_1 + a_{22}x_2 + \cdots + a_2nx_n &= \lambda x_2 \\ \cdots &= \cdots \\ a_{n1}x_1 + a_{n2}x_2 + \cdots + a_{nn}x_n &= \lambda x_n \end{aligned} \tag{6.2}$$

where λ is a constant. Equation (6.2) will have a non-trivial solution only if

$$f(\lambda) = \det(A - \lambda I) = 0. \tag{6.3}$$

Equation (6.3) is a polynomial equation of degree n in λ which has n roots that can be real or complex. The function $f(\lambda)$ is known as the characteristic polynomial. Any non-zero vector **x** that satisfies Equation (6.3) is known as an eigenvector of the matrix A corresponding to the eigenvalue λ.

Equation (6.3) can be rewritten in the form

$$f(\lambda) = (-1)^n(\lambda^n - \sigma_1\lambda^{n-1} + \dots + (-1)^n\sigma_n) \tag{6.4}$$

where the σ_i are certain combinations of the minors of A. In fact, each σ_i is the sum of the *ith* order principal minor of A. For a square matrix the principal minors are those minors whose diagonal entries are diagonal entries of A. The sum of the diagonal entries of A is the sum of the eigenvalues of A. This value, known as the trace of the matrix A is given as

$$Tr(A) = \sum_{i=1}^{n} \lambda_i = \sum_{i=1}^{n} a_{ii}. \tag{6.5}$$

The product of the eigenvalues is the determinant of A.

$$\prod_{i=1}^{n} \lambda_i = \det(A) \tag{6.6}$$

Before proceeding further into the text for solving for the eigenvalues and eigenvectors, a number of theorems and definitions used later in the text are given without proof in the following sections. There are any number of excellent texts on linear algebra for the proof of the following theorems.

6.1.1 Definitions

A set of vectors $\{y_1,\ldots,y_n\}$ is **orthogonal** if, for all $i \neq j$,

$$y_i^T y_j = 0 \tag{6.7}$$

The set is said to be orthonormal.
An n x n matrix F is orthogonal if

$$F^T F = I \tag{6.8}$$

Two n x n matrices, F and G, are **similar** if there exists a non-singular matrix such that

$$S^{-1} F S = G \tag{6.9}$$

If the matrix A is Hermitian ($A^* = A$), each of the following statements gives a necessary and sufficient condition for A to be positive definite.

1. All of the eigenvalues of A are positive,
2. Coefficients of the characteristic equation alternate in sign,
3. Each leading submatrix of A is positive definite,
4. The product of all eigenvalues is > 0 for $1 \le m \le n$.

A Hermitian matrix A is positive definite if and only if for all $\mathbf{u} \neq 0$, $(\mathbf{u},\mathbf{Au})>0$. The term $(\mathbf{u},\mathbf{Au})$ is the dot product and may be written as

$$(u,Au) = \sum_{i,j=1}^{n} a_{ij}u_i u_j \tag{6.10}$$

If the product is ≥ 0, the product is called positive semidefinite. Similar expressions hold for negative definite and negative semidefinite Hermitian matrices.

6.1.2 Theorems

1. A symmetric n x n matrix with n distinct eigenvalues has n linearly independent eigenvectors.
2. The number of linearly independent eigenvectors associated with an eigenvalue of multiplicity m is m or less.
3. The eigenvalues of a symmetric matrix are real.
4. If A is a symmetric matrix then there exists an orthogonal matrix B such that

$$B^{-1}AB = C \tag{6.11}$$

where C is a diagonal matrix containing the eigenvalues of A.

5. If A is a symmetric matrix, then there exists n eigenvectors of A that

form an orthonormal set.

6. If A and B are similar matrices, then the eigenvalues and associated eigenvectors of A are also the eigenvalues and associated eigenvectors of B.

7. If λ_1 is an eigenvalue of A that is larger in absolute value than any other eigenvalue, it is called the *dominant eigenvalue*. The eigenvector corresponding to that eigenvalue is called the *dominant eigenvector*.

8. An eigenvector is said to be normalized if the coordinate of largest magnitude is equal to unity.

The following sections discuss the methods used to find eigenvalues and eigenvectors.

6.2 The Power Method

An iterative technique for finding the dominant eigenvalue of a matrix A is called the **power method**. The power method is an easy means to compute eigenvalues and it serves as a introduction to the QR iteration method discussed later in this chapter. At each step in the iteration, the vector obtained is immediately normalized. According to the theorems in Section 6.1 the scaling of the integer will not change the relation developed. It is important to (if possible) to choose a starting vector, v_0, close to the eigenvector x_1 so that the method will converge as rapidly as possible. A possible advantage to the power method is that the eigenvector that corresponds to the dominant eigenvalue is generated as the solution converges. A disadvantage is that only one eigenvalue is determined. If we wish to obtain the least eigenvalue and corresponding eigenvector, the power method can be applied to the inverse of the matrix A.

We have included two versions of the power method in this section. The first is the standard power method which is based on the power of the matrix and finds the largest eigenvalue iteratively. The second is the inverse power method. This formulation is based on the inverse power of the matrix and finds the smallest eigenvalue.

If the matrix is of order n, has n linearly independent eigenvectors $\mathbf{u}_i$ with i=1,...,n, and a dominant eigenvalue λ_1, then λ_1 and an associated vector $\mathbf{z}_1$ of unit length can be approximated as given below. First assume that

$$|\lambda_1| > |\lambda_2| \tag{6.12}$$

Define a sequence as follows

$$v_m = \frac{Av_{m-1}}{\|Av_{m-1}\|}, \ for\ m \geq 1. \tag{6.13}$$

Then, taking the limits

$$\begin{aligned} \lim_{m\to\infty} v_m &= z_m \\ \lim_{m\to\infty} \|Av_m - 1\| &= |\lambda_1| \end{aligned} \tag{6.14}$$

Let v_0 be any vector of the form

$$v_0 = c_1x_1 + c_2x_2 + \cdots + c_mx_m \tag{6.15}$$

where the x_i's are eigenvectors.

Since the eigenvectors are linearly independent, any n-component vector can be written as a unique linearly combination of them. If we multiply the matrix A for which we are seeking the largest eigenvalue by Equation (6.15) we have (noting that the x_i eigenvectors have corresponding eigenvalues, λ_i)

$$\begin{aligned} v_1 &= Av_0 = c_1Ax_1 + c_2Ax_2 + \cdots + c_nAx_n \\ v_1 &= c_1\lambda_1x_1 + c_2\lambda_2x_2 + \cdots + c_n\lambda_nx_n \end{aligned} \tag{6.16}$$

Once the dominant eigenvalue is found, the remaining eigenvalues can be found using the following procedure with the assumption that there are no repeated eigenvalues. In order to obtain λ_2, first note that the vector

$$(A - \lambda_1 I)v_0 = (\lambda_1 - \lambda_1)c_1 z_1 + \cdots + (\lambda_n - \lambda_1)c_n z_n \tag{6.17}$$

does not contain the component $\mathbf{z}^1$. Therefore using the same procedure as above with a different starting vector equivalent to Equation (6.17) will give λ_2 and $\mathbf{z}^2$. Once λ_1 and λ_2 are known, a new starting vector given by $(\mathbf{A} - \lambda_1\mathbf{I})(\mathbf{A} - \lambda_2\mathbf{I})$ will give λ_3 and $\mathbf{z}^3$.

Given A and $z^{(0)}$ where $z^{(0)}$ is the starting vector
Repeat for i=1,2,.....
Set $y^{(i)} = Az^{(i-1)}$
μ_i = the element of $y^{(i)}$ with maximum modulus
$z^{(i)} = y^{(i)}/\mu_i$
Answer when converged

The power method is given in Program 6.1 at the end of this chapter.

In order to accelerate the convergence process for cases where the ratio

$$\frac{|\lambda_2|}{|\lambda_1|} \approx 1 \tag{6.18}$$

the following approximation called the Aitken Δ^2 process can be applied. In order to use the approximation, three consecutive approximations to the y^i or z^i vectors should be obtained. The fourth vector based on the preceding three vectors is

$$j_{n+3} = j_{n+2} - \frac{(j_{n+2} - j_{n+1})^2}{j_{n+2} - 2j_{n+1} + j_n} \tag{6.19}$$

The user can substitute either y or z for j. Program 6.2 illustrates the inverse power method.

6.3 Solution Methods for the Complete Eigensystem

In order to find all eigenvalues and eigenvectors of a system, A, most modern methods use similarity transformations to get A in a diagonal form. At this point, the eigenvectors are the columns of the accumulated transformation. If only the eigenvalues are required, then the transformations need to be carried only to the point where A is either upper triangular or lower triangular.
There are two general techniques used to solve the system. The first technique constructs individual transformation matrices for specific tasks handling specific terms in a matrix to zero. An example to be given later is the Jacobi method. For some cases the sequence of transformations cannot completely diagionalize a matrix. When it becomes evident that there can be no farther progress to diagionalize a matrix, then additional techniques are employed to reduce the matrix or one could iterate the finite sequence of transformations again and again until the deviation of the matrix from the diagonal is very small.

6.3.1 The Jacobi Method

The Jacobi method consists of a sequence of orthogonal similarity transformations to transform the matrix A into a diagonal matrix. Each transformation is a plane rotation such that one of the off-diagonal matrix elements is annihilated. The plane rotation matrices R(p,q) are unit matrices except for the elements

$$\begin{aligned} r_{pp} &= \cos\theta \\ r_{pq} &= -\sin\theta \\ r_{qp} &= \sin\theta \\ r_{qq} &= \cos\theta \end{aligned} \tag{6.20}$$

for some chosen value of θ. Since R(p,q) is an orthogonal matrix, a transformation of the form

$$A^* = R^T A R \tag{6.21}$$

preserves the eigenvalues of A.

Therefore, it becomes necessary to choose a value of θ such that the element in position (p,q) reduces to zero. Many texts derive an expression for the necessary values for θ.They are given as follows:

$$\begin{aligned} &\theta = 1/2\tan^{-1}\left(\frac{2a_{pq}}{a_{pp}-a_{qq}}\right) \; for \; a_{pp} \neq a_{qq} \\ &\theta = \pm\frac{\pi}{4} \; for \; a_{pp} = a_{qq} \end{aligned} \tag{6.22}$$

The Jacobi method is an iterative procedure. For each iteration, a decision must be made as to which off-diagonal element should be reduced to zero. If the element of maximum modulus is chosen, this techniques is denoted as the *standard* Jacobi method. The modulus alluded to is $|a_{ij}|$. The value of θ is constrained to lie between [$-\frac{1}{4}\pi, \frac{1}{4}\pi$] and take the sign of a_{pq} of the ith iteration if $a_{pp} = a_{qq}$. The iteration is continued until the modulus of each off diagonal element is within a specified tolerance.

The Jacobi method produces the eigenvalues of the matrix directly. Since the method is iterative, computational time can be extreme for large matrices.

In order to reduce the computational time a modification of the standard Jacobi method is used and it is known as the *cyclic* Jacobi method. In this method the elements are annihilated in strict order. The ordering is usually from left to right and proceeds by row. Should the off-diagonal element be very small or within a given tolerance, the rotation for that particular element is skipped.

The program outline to accomplish this task is given below.

```
Given: A (a matrix), a tolerance value, maximum number of cycles
Repeat though not exceeding maximum number of cycles
      for i=1,2,...,n-1
            for j=i+1,i+2,....,n
                  if |aij|>tolerance value
                        DO Jacobi transformation
                  if tolerance not exceeded then
                        Print out converged
loop until converged
```

The program is given in Program 6.3.

6.3.2 Givens and Householder Methods

For symmetric matrices, the preferred simple form is tridiagonal. The Givens reduction is a modification of the Jacobi method. Instead of trying to reduce the matrix to a diagonal form, the process is stopped when the matrix is tridiagonal. The process is now confined to a finite number of steps instead of iterating until convergence. A symmetric matrix can be transformed into a tridiagonal matrix by the Householder method. The Householder method uses a series of similarity transformations to accomplish this task.

The Householder method is more efficient than the Givens method. Each transformation annihilates the required part of the whole column and the whole corresponding row.

To briefly present the underlying concepts of the Householder method, the following is offered.

Consider a matrix given by

$$A^{(1)} = \begin{bmatrix} a_{11} & \cdots & a_{1n} \\ \vdots & \ddots & \vdots \\ a_{nl} & \cdots & a_{nn} \end{bmatrix} \tag{6.23}$$

The first step is to reduce the matrix $A^{(1)}$ to a tridiagonal matrix to a form denoted by

$$A^{(2)} = \begin{bmatrix} a_{11} & a_{12} & \cdots & 0 & 0 \\ a_{21} & a_{22} & \cdots & a_{2n-1} & a_{2n} \\ \vdots & \vdots & \ddots & \vdots & \vdots \\ 0 & \cdots & \cdots & \cdots & a_{nn} \end{bmatrix} \tag{6.24}$$

The transformation is performed by pre-multiplying and post-multiplying A by a transformation matrix B as

$$A^{(2)} = BA^{(1)}B \tag{6.25}$$

B, in the above equation is defined by

$$B = 1 - \frac{uu^T}{h} \tag{6.26}$$

with

$$u = col(0, a_{2,1} + G, a_{31}, a_{41}, \cdots, a_{nl}) \tag{6.27}$$

where a_{ij} are the matrix elements of $A^{(1)}$ and

and

$$G = [\sum_{i=m+1}^{n} (a_{i,m})^2]^{.5}\ sign(a_{m+1,m}) \tag{6.28}$$

$$h = G^2 + Ga_{m+1,m} \tag{6.29}$$

The term sign(a)=+1 if A>=0 or -1 if sign(a) < 0.

The transformation matrix B has the following properties:

$$\begin{aligned} B &= B^{-1} \\ B^T &= B \\ BB &= I \end{aligned} \tag{6.30}$$

A matrix A of order n is reduced to tridiagonal form by repeating transformations of the above n-2 times. The general form of the transformation is written as

$$A^{m+1} = BA^{(m)}B \tag{6.31}$$

The implementation is given in Program 6.4.

6.3.3 QR Iteration

For a general matrix, the decomposition of that matrix is done by applying Householder transformations to annihilate successive columns of A below the diagonal. Any real matrix can be decomposed in the form

$$A = QR \tag{6.32}$$

where Q is orthogonal and R is upper triangular. If we let A_0 = A, where A is the original matrix in which we wish to determine the eigenvalues, the matrix A_0 is decomposed by

$$A_0 = Q_0 R_0 \tag{6.33}$$

where Q_0 is the orthonormal matrix and R_0 is the upper triangular matrix. The similarity transformation is written as

$$A_1 = Q_0^{-1} A_0 Q_0 \tag{6.34}$$

or

$$A_1 = R_0 Q_0 \tag{6.35}$$

This process is continued until all eigenvalues are found. To improve the rate of convergence, the above equations are modified as follows:

$$A_j - \alpha_j I = Q_j R_j \tag{6.36}$$

and

$$A_{j+1} = R_k Q_k + \alpha_j I \tag{6.37}$$

This modification shifts the eigenvalues of the right side by α_j as well as the eigenvalues of R_jQ_j. Adding $\alpha_j I$ in the second equation shifts the values of A_{j+1} back to the original values. One other step is required. That is to first transform the original matrix to the Hessenberg form. This form of the matrix (upper Hessenberg) has zeros everywhere below the diagonal except for the first subdiagonal row.

The subroutine for this process is given at the end of the chapter in Program 6.5.

The reader is also referred to [66] for a complete discussion of the methods and subroutines for implementation. These subroutines for the matrix reduction TRED1, and eigensolution, TQL1, are perhaps one of the most efficient means for finding all the eigenvalues and eigenvectors of a symmetric, tridiagonal matrix. The name TQL1 stands for "Tridiagonal QL Implicit".

The input parameters are the matrix A, an NxN real symmetric matrix. On output from the routine, the matrix A is overwritten by the orthogonal matrix Q which effects the transformation. D contains the diagonal elements of the tridiagonal matrix and E contains the off-diagonal elements of the tridiagonal elements. These values are imputed directly into the subroutine TQL1. The matrix D is overwritten with the eigenvalues and the corresponding eigenvectors are stored to another matrix.

6.4 Applications of Algorithms for Eigenvectors

To illustrate the application of the above algorithms to the solution of problems requiring eigenvalues and eigenvectors, we offer the following examples.

The first example is an 3 x 3 real matrix given below:

$$\begin{bmatrix} 11 & 6 & -2 \\ -2 & 18 & 1 \\ -12 & 24 & 13 \end{bmatrix} \tag{6.38}$$

The value for the tolerance, η, is 1.0E-9. The maximum number of iterations is 100. Application of the power method results in the following values for the eigenvalues and the eigenvectors.

Eigenvalues

$\lambda 1 = 21.0000$

$\lambda 2 = 14.0000$

$\lambda 3 = 07.0000$

Eigenvectors

0.000000	0.894427	-0.447214
0.316228	0.447214	0.000000
0.948683	0.000000	-0.894427

The inverse power method can be applied to the above problem to determine the dominate eigenvalue.

Another sample data set for use with the Jacobi method is given below. The matrix size is 4 x 4 with the maximum number of iterations at 50 and the tolerance value at 1.0E-10.

$$\begin{bmatrix} 10 & 9 & 7 & 5 \\ 0 & 10 & 8 & 6 \\ 0 & 0 & 10 & 7 \\ 0 & 0 & 0 & 5 \end{bmatrix} \qquad (6.39)$$

For the above matrix the solution is as follows:

Eigenvalues

$\lambda 1 = 30.2886853$

$\lambda 2 = 0.8431071$

$\lambda 3 = 3.8580575$

$\lambda 4 = 0.0101500$

Eigenvectors

0.5209249	0.5676405	-0.6253962	-0.1236975
0.5519547	-0.7603186	-0.2716010	0.2085536
0.5285679	0.3016522	0.6148613	0.5015651
0.3802621	-0.0933051	0.3963056	-0.8304438

6.5 Problems

6-1) Given the matrix

7 2

1 2

Determine the largest eigenvalue and its corresponding eigenvector using the power method.

6-2) Use the power method to find all of the eigenvalues and eigenvectors for

1 3 1

1 2 7

3 4 9

6-3) Determine the dominant eigenvalue of problem 2 using the inverse power method.

6-4) Use a central-difference approximation to find the eigenvalues of the equation

$d^2y/dx^2 + 2dy/dx + q^2y = 0.$

The lowest eigenvalue of the equation is $q = +/- \sqrt{(1+\pi^2)}$.

6-5) Use the inverse power method to find the eigenvalues of the matrix

12 4 10 1

0 11 2 9

14 17 28 12

8 9 4 29

6-6) Find the eigenvalues and eigenvectors for

4 9 3

0 1 6

3 0 9

using the Jacobi method.

6-7) Use the Jacobi method to find the eigenvalues of the matrix given in problem 5.

6-8) Determine the eigenvalues for the matrix given in example 6.38 using the Householder method.

6-9) Find the eigenvalues of the matrix

2 -1 -1

-1 2 -1

-1 -1 2

using the Householder computer program.

6-10) An undamped spring-mass system is represented by the equation

$m\, d^2x/dt^2 = -kx$

where m is the mass and k is the spring constant. Determine the natural frequencies of such a system for $x(0) = 0$ and $x(1) = 0$ using both numerical and non-numerical methods.

Chapter 6 Programs

Program 6.1F Power Method

```
C SUBROUTINE PWRMTH(A,X,C,TOL,NLIMIT,N,NDIM,XWRK)
C THIS SUBROUTINE COMPUTES THE LARGEST EIGENVALUE
C AND THE CORRESPONDING EIGENVECTOR BY THE POWER METHOD.
C
C A      - AN N X N MATRIX WHOSE EIGENVALUES ARE BEING DETERMINED
C X      - ESTIMATE FOR THE EIGENVECTOR USED TO BEGIN ITERATIONS
C          IF NO APPROXIMATIoN TO THIS IS KNOWN, THE USUAL CHOICE
C          IS A VECTOR WITH ALL COMPONENTS EQUAL TO UNITY. X ALSO
C          RETURNS THE FINAL EIGENVECTOR TO THE CALLER.
C C      - RETURNS THE VALUE OF THE EIGENVALUE
C TOL    - TOLERANCE VALUE USED TO DETERMINE CONVERGENCE. ITERATION
C          CONTINUE UNTIL SUCCESSIVE ESTIMATES OF THE EIGENVALUE
C          ARE THE SAME WITHIN TOLERANCE IN VALUE.
C NLIMIT - LIMIT TO THE NUMBER OF ITERATIONS IF NOT CONVERGENT
C N      - SIZE OF THE MATRIX AND THE VECTOR X
C NDIM   - FIRST DIMENSION OF MATRIX A IN THE CALLING PROGRAM
C XWRK   - VECTOR USED TO STORE INTERMEDIATE VALUES. MUST BE
C          DIMENSIONED TO HOLD AT LEAST N ELEMENTS IN THE
C          MAIN PROGRAM.

      REAL A(NDIM,N),X(N),XWRK(N),C,TOL,SAVE
      INTEGER N,NDIM,NLIMIT,IROW,ITER,JCOL

C BEGIN THE ITERATIONS. GET THE PRODUCT OF A AND X, THEN NORMALIZE.
C THE NORMALIZATION FACTORS SHOULD CONVERGE TO THE EIGENVALUE,C, ON
C REPEATED MULTIPLICATIONS. STORE THE CURRENT VALUE OF C FOR
C COMPARISON WITH THE NEXT VALUE TO TEST CONVERGENCE.MAKE SAVE=0.
      SAVE = 0.0
      DO 50 ITER = 1,NLIMIT
            DO 20 IROW = 1,N
                  XWRK(IROW) = 0.0
                  DO 10 JCOL = 1,N
                   XWRK(IROW)=XWRK(IROW)+A(IROW,JCOL)*X(JCOL)
   10        CONTINUE
   20        CONTINUE
C FIND LARGEST ELEMENT OF PRODUCT VECTOR FOR NORMALIZATION
      C = 0.0
      DO 30 IROW = 1,N
            IF ( ABS(C).LT.ABS(XWRK(IROW))) C=XWRK(IROW)
   30 CONTINUE
C NORMALIZE THE PRODUCT VECTOR AND PUT INTO X FOR NEXT ITERATION.
      DO 40 IROW = I,N
            X(IROW) = XWRK(IROW) / C
   40 CONTINUE
C CHECK TOLERANCE. IF NOT, CONTINUE ITERATIONS

      IF( ABS (C - SAVE) . LE. TOL ) RETURN
      SAVE = C
   50 CONTINUE
      RETURN
      END
```

Program 6.1P Power Method Procedure

```
{ procedure PWRMTH                                                           }
{  This procedure computes the largest eigenvalue and the                    }
{  corresponding eigenvector by the power method.                            }
{ A      - An NXN matrix whose eigens are being determined.                  }
{ X      - Estimate for the eigenvector used to being iterations.            }
{          If no approximation is known, the usual choice is a               }
{          vector with all components equal to unity.  X also               }
{          returns the final eigenvector to the caller.                      }
{ C      - Returns the value of the eigenvalue.                              }
{ TOL    - Tolerance value used to determine convergence.Iteration           }
{          continue until successive estimates of the eigenvalue             }
{          are within the tolerance in value.                                }
{ NLIMIT - Limit of the number of iterations if not convergent.              }
{ N      - Size of the matrix and the vector X.                              }
{ NDIM   - First dimension of the matrix.                                    }
{ XWRK   - Vector used to store intermediate values.  Must be                }
{          dimensioned to hold at least N elements in the main               }
{          program.                                                          }

procedure PWRMTH(var A : MAXARRAY;
                 var X : MAXVECTOR;
                 var C : real;
                 TOL   : real;
                 NLIMIT,N,NDIM: integer;
                 var XWRK : MAXVECTOR);

{  Begin the iterations.  Get the product of A and X, then                   }
{  normalize.The normalization factors should converge to the                }
{  eigenvalue C on repeated multiplications. Store the current               }
{  value of Cfor comparison with the next value to test                      }
{  convergence. MAKE SAVE = 0.                                               }

var
  SAVE : real;
  IROW,ITER, JCOL : integer;
begin
   SAVE := 0.0;
   ITER := 0;
   repeat
      ITER := ITER+1;
      for IROW := 1 to N do begin
         XWRK[IROW] := 0.0;
         for JCOL := 1 to N do
            XWRK[IROW] := XWRK[IROW]+A[IROW][JCOL]*X[JCOL];
      end;
{ Find largest element of product vector for normalization.                  }
      C := 0.0;
      for IROW := 1 to N do begin
         if (abs(C) < abs(XWRK[IROW])) then C := XWRK[IROW];
      end;
{ Normalize the product vector and put into X for next iteration.            }
      for IROW := 1 to N do begin
         X[IROW] := XWRK[IROW]/C;
      end;
   until (ABS(C-SAVE)<TOL) or (ITER > NLIMIT);
end;
```

Program 6.2F Inverse Power Method

```
SUBROUTINE INVPWR (SIZE,N,A,P,Z,TOL,MAXITS,
     1      STATUS,LAMBDA,X,NOOFIT,LASTMU,Y,WORKV)
      INTEGER CONVGD,COUNT,I,LIMIT,MAXITS,N,NEWI,
     1      NOOFIT,OLDI,PTINY,SIZE,STATUS
      REAL LAMBDA,LASIMU,NEWMU,OLDMU,P,TOL
      PARAMETER (CONVGD=0,PTINY=1,LIMIT=2)
      REAL A(SIZE,SIZE),WORKV(N),X(N),Y(N),Z(N)
      INTRINSIC ABS
      EXTERNAL LUDECOMP,LUSOLVE
      DO 10, I=1,N
            A(I,I)=A(I,I)-P
   10 CONTINUE
      CALL LUDECOMP (SIZE,N,A,STATUS)
      IF (STATUS.EQ.PTINY) GO TO 999
      NEWMU=0
      NEWI =0
      DO 20, COUNT=1,MAXITS
      CALL LUSOL (SIZE,N,A,Z,STATUS,Y,WORKV)
      IF (STATUS.EQ.PTINY) GO TO 999
      OLDMU=NEWMU
      OLDI=NEWI
      NEWMU=0
      DO 30, I=1,N
            IF (ABS(Y(I)).GT.ABS(NEWMU))
                  NEWMU=Y(I)
                  NEWI=I
            END IF
   30 CONTINUE
      DO 40, I=l,N
            Z(I)=Y(I)/NEWMU
   40 CONTINUE
      IF ( (ABS(NEWMU-OLDMU) .LE. ABS(OLDMU)*TOL
     1.AND. (NEWI.EQ.OLDI) )
            STATUS=CONVGD
            GO TO 21
      END IF
   20 CONTINUE
      STATUS=LIMIT
      DO 50, I=1,N
            X(I)=Z(I)
   50 CONTINUE
      IF(STATUS.EQ.CONVGD) THEN
            LAMBDA=P+1/NEWMU
            NOOFIT=COUNT
      END IF
      IF (STATUS.EQ.LIMIT) THEN
            LASTMU=P+1/NEWMU
      END IF
  999 END

      SUBROUTINE LUDECOMP (SIZE,N,A,STATUS)
      INTEGER FACTRD,I,J,K,N,PTINY,SIZE,STATUS
      READ MULT,PIVOT,TINY
      PARAMETER (FACTRD=0,PTINY=1)
      PARAMETER (TINY=1E-10)
      REAL A(SIZE,SIZE)
      INTRINSIC ABS
```

```
      DO 10, I=1,N-1
            PIVOT=A(I,I)
            IF (ABS(PIVOT).GT.TINY) THEN
                  DO 20, K=I+1,N
                        IF (ABS(A(K,I)).GT.TINY) THEN
                              MULT=A(K,I)/PIVOT
                              A(K,I)=MULT
                              DO 30, J=I+1,N
                                    A(K,J)=A(K,J)-MULT*A(I,J)
                        30    CONTINUE
                        ELSE
                              A(K,I)=0
                        END IF
            20    CONTINUE
            ELSE
                  STATUS=PTINY
                  GO TO 999
            END IF
10    CONTINUE
      STATUS=FACTRD
999   END
      SUBROUTINE LUSOLVE (SIZE,N,A,B,STATUS,X,Y)
      INTEGER I,J,N,PTINY,SIZE,SOLVED,STATUS
      REAL TINY
      DOUBLE PRECISION SUM
      PARAMETER (SOLVED=0,PTINY=1)
      PARAMETER (TINY=1E-10)
      REAL A(SIZE,SIZE),B(N),X(N),Y(N)
      INTRINSIC ABS,DPROD
      IF (ABS(A(N,N).GT.TINY) THEN
            STATUS=SOLVED
            Y(1)=B(1)
            DO 10, I=2,N
                  SUM=B(I)
                  DO 20, J=1,I-1
                        SUM=SUM-DPROD(A(I,J),Y(J))
            20    CONTINUE
                  Y(I)=SUM
      10    CONTINUE
            X(N)=Y(N)/A(N,N)
            DO 30, I=N-1,1,-1
                  SUM=Y(I)
                  DO 40, J=I+1,N
                        SUM=SUM-DPROD(A(I,J),X(J))
            40    CONTINUE
                  X(I)=SUM/A(I,I)
      30    CONTINUE
      ELSE
            STATUS=PTINY
      END IF
      END
```

Program 6.3F Jacobi Method

```
SUBROUTINE JACOBYCY (SIZE,N,A,TOL,MAXCYC,STATUS,DlAG,NOOFlT)
       INTEGER CONVGD,COUNT,I,J,LIMIT,MAXCYC,N,NOOFlT,SIZE,STATUS
       REAL TOL
       LOGICAL TOLEX
       PARAMETER (CONVGD=0,LIMIT=l)
       REAL A(SIZE,SIZE),DIAG(N)
       EXTERNAL JTRANS
       INTRINSIC ABS
       DO 10, COUNT=l,MAXCYC
             TOLEX=.FALSE.
             DO 20, I=l,N-l
             DO 30, J=l+l,N
                   IF (AGS(A(I,J)).GT.TOL) THEN
                         TOLEX=.TRUE.
                         CALL JTRANS (SIZE,N,A,I,J)
                   END IF
   30 CONTINUE
   20 CONTINUE
             IF (.NOT.TOLEX) THEN
                   STATUS=CONVGD
                   GO TO 11
             END IF
   10 CONTINUE
             STATUS=LIMIT
             IF (STATUS.EQ.CONVGD) THEN
                   NOOFIT=COUNT
                   DO 40, I=l,N
                         DIAG(I)=A(I,I)
   40              CONTINUE
             END IF
             END

       SUBROUTINE JTRANS (SIZE,N,A,P,Q)
       INTEGER I,N,P,8,SIZE
       REAL AIP,AIQ,APP,APQ,AQQ,C,CS,CSQ,PI,S,SSQ,TINY,THETA
       PARAMETER (TINY=1E-10)
       REAL A(SIZE,SIZE)
       INTRINSIC ABS,ATAN,COS,SIGN,SIN
       PI=4*ATAN(1.0)
       APP=A(P,P)
       APQ=A(P,Q)
       AQQ=A(Q,Q)
       IF (ABS(APP-AQQ).GT.TINY) THEN
             THETA=ATAN(2*APQ/(APP-A8Q))/2
             IF (ABS(THETA).GT.PI/4) THETA=THETA-SIGN(PI/2,THETA)
       ELSE
             THETA=SIGN(PI/4,THETA)
       END IF
       C=COS(THETA)
       S=SIN(THETA)
       CSQ=C*C
       SSQ=S*S
       CS=C*S
       A(P,P)=CSQ*APP+2*CS*APQ+SSQ*AQQ
       A(Q,Q)=SSQ*APP 2*CS*APQ+CSQ*AQQ
       A(P,Q)=0
       A(Q,P)=0
```

```
      DO 10, I=1,N
            IF (I.NE.P .AND. I.NE.Q) THEN
                  AIP=A(I,P)
                  AIQ=A(I,Q)
                  A(I,P)=C*AIP+S*AIQ
                  A(P,I)=A(I,P)
                  A(I,Q)=-S*AIP+C*AIQ
                  A(Q,I)=A(I,Q)
            END IF
10    CONTINUE
      END
```

Program 6.4F Householder Method

```
      PROGRAM HOUSEHOLDER
C_______________________________________________________________________

C  THIS PROGRAM COPYRIGHT: (C) 1992 by Marvin N. Zeigler
C  AUTHORS:  Marvin N. Zeigler and Edward Champion
C  TEXT BOOK:  NUMERICAL METHODS FOR ENGINEERING APPLICATIONS
C  EQUATION:  6.23-6.31

C  PURPOSE:  This program uses the Householder method to transform
C  a symmetric matrix into a tridiagonal matrix.  Eigenvalues are
C  found from the tridiagonal matrix using a combination of bisection
C  and linear interpolation.

C  ARGUMENTS:
C   H       Symmetric matrix transformed into tridiagonal matrix.
C   XLAM    Resulting eigenvalues.
C   DET     Determinate of current sub-matrix.
C   XTRIAL  Guess value for starting search for eigenvalues.
C   DTOL    Convergence tolerance value.
C   FACTOR  Multiplier for determining XTRIAL.
C   MAXIT   Maximum number of iterations allowed.

C_______________________________________________________________________
      PARAMETER (N=4)
      INTEGER N
      REAL H(N,N+2), XLAM(N,N+1), DET(N+1), XTRIAL
      DATA DTOL, MAXIT, FACTOR /0.001, 100, 5.0/
      WRITE(*,45) 'Householder Method'
      OPEN (1,FILE='DATA')
      READ(1,*) ((H(I,J),J=1,N),I=1,N)
      WRITE(*,45) 'Input matrix'
      DO 10 I=1,N
         WRITE(*,15) (H(I,J),J=1,N)
 10   CONTINUE
      N2 = N-2
      DO 100 K=1,N2
         K1 = K+1
         SUMHB = 0.0
         DO 30 J=K1,N
            SUMHB = SUMHB + H(J,K)*H(J,K)
 30      CONTINUE
         IF (H(K1,K) .LT. 0) THEN
            PLUSMIN = -1.0
         ELSE
            PLUSMIN = 1.0
         ENDIF
         BETA = PLUSMIN * SQRT(SUMHB)
         H(K1,K) = H(K1,K) + BETA
         ALPHA1 = BETA * H(K1,K)
         DO 60 J=1,N
            SUMHH = 0.0
            DO 40 I=K1,N
               SUMHH = SUMHH + H(I,K) * H(J,I)
 40         CONTINUE
            GAMMA = SUMHH / ALPHA1
            DO 50 I=K1,N
               H(J,I) = H(J,I) - GAMMA * H(I,K)
```

```
 50         CONTINUE
 60      CONTINUE

         DO 90 J=K1,N
            SUMHH = 0.0
            DO 70 I=K1,N
               SUMHH = SUMHH + H(I,K) * H(I,J)
 70         CONTINUE
            GAMMA = SUMHH / ALPHA1
            DO 80 I=K1,N
               H(I,J) = H(I,J) - GAMMA * H(I,K)
 80         CONTINUE
 90      CONTINUE
         H(K1,N+1) = H(K1,K)
         H(K1,N+2) = ALPHA1
         H(K1,K) = -BETA
 100  CONTINUE
      WRITE(*,45) 'Tridiagonal Matrix'
      DO 110 I=1,N
         WRITE(*,15) (H(I,J),J=1,N)
 110  CONTINUE

C Find limits for guessing eigenvalues in following routines
      XTRIAL = ABS(H(1,1))
      DO 114 I=1,N
         DO 112 J=1,N
            XTRIAL = MAX(XTRIAL,ABS(H(I,J)))
 112     CONTINUE
 114  CONTINUE
      XTRIAL = XTRIAL * FACTOR

C Solve for the eigenvalues using a bisection/interpolation routine
      XLAM(1,2) = H(1,1)
      DO 140 L=2,N
         L1 = L - 1
         XLAM(L1,1) = -XTRIAL
         XLAM(L1,L+1) = XTRIAL
         DO 130 J=1,L
            XL = XLAM(L1,J)
            XH = XLAM(L1,J+1)
            CALL TRIDET(L, N, DET, H, XL, YL)
            CALL TRIDET(L, N, DET, H, XH, YH)
            DO 120 I=1,MAXIT
               DX = XH-XL
               IF (DX .LT. DTOL) GOTO 125
               IF (DX .GT. 0.5) THEN
C Use bisection scheme
                  XM = (XL+XH)/2.0
                  CALL TRIDET(L, N, DET, H, XM, YM)
               ELSE
C Use linear interpolation
                  XB = XM
                  XM = (XL*YH - XH*YL) / (YH - YL)
                  CALL TRIDET(L, N, DET, H, XM, YM)
                  IF (ABS(XB-XM) .LT. DTOL) GOTO 125
               ENDIF
               IF (YL*YM .LT. 0) THEN
                  XH = XM
                  YH = YM
               ELSE
```

```
                 XL = XM
                 YL = YM
               ENDIF
 120         CONTINUE
      WRITE(*,45) 'Maximum iterations exceeded in HOUSHOLDER'
      STOP
 125     CONTINUE
           XLAM(L,J+1) = XM
 130     CONTINUE
 140  CONTINUE

 150  CONTINUE
      WRITE(*,45) 'Eigenvalues Found'
      DO 160 I=1,N
         WRITE(*,35) 'LAMDA',I,'=', XLAM(N,I+1)
 160  CONTINUE

 15   FORMAT(' ',8(1X,F10.4))
 35   FORMAT(' ',A,I2,A,F11.6)
 45   FORMAT(/,' ',A)
      END

C_____________________________________________________________________
C_____________________________________________________________________
      SUBROUTINE TRIDET(L, N, DET, H, X, Y)
C  PURPOSE:  Calculates the determinate of the current sub-matrix of
C    tridiagonal form.
C_____________________________________________________________________
      REAL DET(N+1), H(N,N+2), X, Y
      DET(1) = 1
      DET(2) = H(1,1) - X
      DO 10 K=2,L
         DET(K+1) = (H(K,K)-X)*DET(K) - H(K,K-1)*H(K,K-1)*DET(K-1)
 10   CONTINUE
      Y = DET(L+1)
      RETURN
      END
```

Program 6.5F QR Iteration

```
      PROGRAM QRITER
C_______________________________________________________________________
C  THIS PROGRAM COPYRIGHT: (C) 1992 by Marvin N. Zeigler
C  AUTHORS:  Marvin N. Zeigler and Edward Champion
C  TEXT BOOK:  NUMERICAL METHODS FOR ENGINEERING APPLICATIONS
C  EQUATION:  6.23-6.31

C   PURPOSE:   This program uses the Householder method to first
transform
C   a symmetric matrix into a tridiagonal matrix.  Then the real and
C   imaginary parts of eigenvalues are found using QR iteration.

C  ARGUMENTS:
C   H        Symmetric matrix transformed into tridiagonal matrix.
C   RE        Vector of the real part of eigenvalues.
C   IM        Vector of the imaginary part of eigenvalues.
C   DTOL      Tolerance value for determining solution accuracy.
C   DATA     File holding the original symmetric matrix of size NXN.

C_______________________________________________________________________
      PARAMETER (N=5)
      INTEGER N
      REAL H(N,N+2),RE(N+2),IM(N+2)
      DATA DTOL /1.0E-13/
      WRITE(*,45) 'Householder Method with QR Iteration'
      OPEN (1,FILE='DATA')
      READ(1,*) ((H(I,J),J=1,N),I=1,N)
      WRITE(*,45) 'Input matrix'
      DO 10 I=1,N
         WRITE(*,15) (H(I,J),J=1,N)
 10   CONTINUE
      N2 = N-2
      DO 100 K=1,N2
         K1 = K+1
         SUMHB = 0.0
         DO 30 J=K1,N
            SUMHB = SUMHB + H(J,K)*H(J,K)
 30      CONTINUE
         IF (H(K1,K) .LT. 0) THEN
            PLUSMIN = -1.0
         ELSE
            PLUSMIN = 1.0
         ENDIF
         BETA = PLUSMIN * SQRT(SUMHB)
         H(K1,K) = H(K1,K) + BETA
         ALPHA1 = BETA * H(K1,K)
         DO 60 J=1,N
            SUMHH = 0.0
            DO 40 I=K1,N
               SUMHH = SUMHH + H(I,K) * H(J,I)
 40         CONTINUE
            GAMMA = SUMHH / ALPHA1
            DO 50 I=K1,N
               H(J,I) = H(J,I) - GAMMA * H(I,K)
 50         CONTINUE
 60      CONTINUE
```

```
         DO 90 J=K1,N
            SUMHH = 0.0
            DO 70 I=K1,N
               SUMHH = SUMHH + H(I,K) * H(I,J)
 70         CONTINUE
            GAMMA = SUMHH / ALPHA1
            DO 80 I=K1,N
               H(I,J) = H(I,J) - GAMMA * H(I,K)
 80         CONTINUE
 90      CONTINUE
         H(K1,N+1) = H(K1,K)
         H(K1,N+2) = ALPHA1
         H(K1,K) = -BETA
 100  CONTINUE

      WRITE(*,45) 'Tridiagonal Matrix'
      DO 110 I=1,N
         WRITE(*,15) (H(I,J),J=1,N)
 110  CONTINUE
      WRITE (*,*) ' '

C Determine real and imaginary parts of eigenvalues using QR
iteration.

      IDC = N
 120  IF (IDC .GT. 0) THEN
         ITER = 0
         IDD = IDC-1
 130     DO 140 IN=IDC,2,-1
            IND = IN-1
            CHECK = DTOL*( ABS(H(IND,IND)) + ABS(H(IN,IN)) )
            IF ( CHECK .GT. ABS(H(IN,IND)) ) GOTO 150
 140     CONTINUE
         IN = 1
 150     CONTINUE
         DI = H(IDC,IDC)

            IF (IN .EQ. IDC) THEN
C Single root found
               RE(IDC) = DI
               IM(IDC) = 0
               IDC = IDC-1
               GOTO 120
            ENDIF

         S = H(IDD,IDD)
         R = H(IDC,IDD)*H(IDD,IDC)

            IF (IDD .EQ. IN) THEN
C Double root found
               A = (S-DI)/2.0
               Q = A*A + R
               S = SQRT(ABS(Q))
               IF (Q .LT. 0) THEN
C Roots are complex
                  RE(IDD) = DI + A
                  RE(IDC) = RE(IDD)
                  IM(IDC-1) = S
                  IM(IDC) = -S
```

```
                 IDC = IDC-2
              ELSE
C Roots are real
                 IF (A .LT. 0) S = -S
                 S = A + S
                 RE(IDC-1) = DI + S
                 RE(IDC) = DI - R/S
                 IM(IDC-1) = 0
                 IM(IDC) = 0
                 IDC = IDC-2
              ENDIF
        GOTO 120
        ENDIF
        IF (ITER .EQ. 30) THEN
           WRITE(*,*) 'ITERATIONS EXCEEDING MAXIMUM IN QRITER'
           STOP
        ENDIF
        IF (ITER.EQ.10 .OR. ITER.EQ.20) THEN
           S=ABS(H(IDC,IDD)) + ABS(H(IDD,IDC-2))
           TR1 = 1.5*S
           S = S*S
        ELSE
           TR1 = DI + S
           S = DI*S - R
        ENDIF
        ITER = ITER + 1
        DO 160 M=IDC-2,IN,-1
           MP1 = M + 1
           MM1 = M - 1
           A = H(M,M)*(H(M,M)-TR1) + S + H(MP1,M)*H(M,MP1)
           Q = H(MP1,M)*(H(M,M)+H(MP1,MP1)-TR1)
           R = H(MP1,M)*H(M+2,MP1)
           TR2 = ABS(A) + ABS(Q) + ABS(R)
           A = A/TR2
           Q = Q/TR2
           R = R/TR2
           IF (M .EQ. IN) GOTO 170
           HH = ABS(H(M,MM1))*(ABS(Q) + ABS(R))
           GG   =   DTOL*ABS(A)*(ABS(H(MM1,MM1))   +
ABS(H(M,M))+ABS(H(MP1,MP1)))
           IF (HH. LT .GG) GOTO 170
 160    CONTINUE
 170    CONTINUE
        DO 180 I=M+2,IDC
           H(I,I-2) = 0.0
 180    CONTINUE
        DO 190 I=M+3,IDC
           H(I,I-3) = 0.0
 190    CONTINUE
        DO 220 L=M,IDD
           LP1 = L + 1
           LP2 = L + 2
           LM1 = L - 1
           IF (L .NE. M) THEN
              R = H(LP2,LM1)
              IF (L .EQ. IDD) R = 0.0
              DI = ABS(H(L,LM1)) + ABS(H(LP1,LM1)) + ABS(R)
              IF (DI .EQ. 0) GOTO 220
              A = H(L,LM1)/DI
              Q = H(LP1,LM1)/DI
```

```
               R = R/DI
            ENDIF
            TR1 = SQRT(A*A + Q*Q + R*R)
            IF (A .LT. 0) TR1=-TR1
            IF (L .NE. M) H(L,LM1) = -TR1*DI
            IF (IN .NE. M) H(L,LM1) = -H(L,LM1)
            A = A + TR1
            DI = A/TR1
            S = Q/TR1
            Z = R/TR1
            Q = Q/A
            R = R/A
            DO 200 I=L,IDC
               A = H(L,I) + Q*H(LP1,I)
               IF (L .NE. IDD) THEN
                  A = A + R*H(LP2,I)
                  H(LP2,I) = H(LP2,I) - A*Z
               ENDIF
               H(LP1,I) = H(LP1,I) - A*S
               H(L,I) = H(L,I) - A*DI
 200        CONTINUE
            I = IDC
            IF (L+3 .LT. IDC) I=L+3
            DO 210 K=IN,I
               A = DI*H(K,L) + S*H(K,LP1)
               IF (IDD .NE. L) THEN
                  A = A + Z*H(K,LP2)
                  H(K,LP2) = H(K,LP2) - A*R
               ENDIF
               H(K,LP1) = H(K,LP1) - A*Q
               H(K,L) = H(K,L) - A
 210        CONTINUE
 220     CONTINUE
         GOTO 130
         WRITE(*,*)  'Maximum  number  of  iterations  exceeded  in
QRITER.'
      ENDIF
C All eigenvalues have been found

      WRITE(*,45) 'Eigenvalues'
      WRITE(*,*) '          Index   Real Part  Imaginary Part'
      DO 250 I=1,N
         WRITE(*,*) I,RE(I),IM(I)
 250  CONTINUE
 15   FORMAT(' ',8(1X,F10.4))
 45   FORMAT(/,' ',A)
      END
```

7

Ordinary Differential Equations

7.1 Introduction

Differential equations occur in many diverse fields of study from medicine to engineering. The equations in general have the form

$$\frac{d^n y}{dx^n} = y^{(n)} = f(x,y,y',y'',\ldots,y^{(n-1)}) \tag{7.1}$$

where y is a function of the single variable x and n is a positive integer. The equation is said to be of order n and corresponds to the order of the highest derivative. The number of independent variables defines the differential equation as an *ordinary differential equation* (a single variable) or *partial differential equation* (two or more variables). A linear differential equation is one of the first degree in the dependent variable and in all of its derivatives. Ordinary differential equations are classified into initial value problems and boundary value problems. Many initial value problems are time-dependent, in which all the conditions for the solution are specified at the initial time. Boundary value problems are those in which the solution is required to satisfy the boundary conditions on the domain.

When we apply a numerical method to a particular problem, we do not solve the differential equation, rather we solve an equation related to it.

For well-posed problems, numerical problems must satisfy the conditions of consistency, stability, and convergence. The definitions of each are given below:

Consistency - If in the limit as h -> 0, the substitute function and the original function are equivalent then the numerical solution is said to be consistent.

Stability - If the errors at one stage of calculations remain finite continually, then the numerical method is said to be stable.

Convergence - If the results of the numerical method approach the true value as h, the solution step size, approaches zero, then the numerical method is said to be convergent.

In this chapter we will look at four methods (three in detail) for solving ordinary differential equations. The methods presented will all follow the above criteria and are discussed below:

The first step in the numerical solution of a differential equation is the division of the interval of question [a,b] into n segments along the axis of the argument by the points designated by

$$i = 0,1,2,3,...,k-1,k,k+1,...,r-1,r$$

such that the spacing between these points is given by

$$h_k = x_k - x_{k-1} \quad (k=1,2,...,r).$$

For equal spacing $h = (b-a)/r$.

The goal of the solution is to find the values of y_i corresponding to the given arguments.

We will begin this chapter with the study of *initial-value* problems followed by *boundary-value* problems. Later in the chapter, we will cover the solution of differential equations by the finite-difference method and by Runge-Kutta methods.

7.2 Initial-Value Problems

7.2.1 Euler and Runge-Kutta Methods

This section will start by considering first order equations of the form

$$\frac{dy}{dx} = f(x,y) \tag{7.2}$$

For this particular equation with no other constraints there are an infinite number of solutions. In order to arrive at a solution, after integrating the function over x, an initial condition that specifies the value of y for a given x is provided. This allows any arbitrary constants arising in the integration process to be determined.

There is no single numerical technique that will be the best solution procedure for all initial-value problems. Several techniques will be given and will be applied to the following initial-value problem.

$$\frac{dy}{dx} = 1/2(x+y) - 1 \tag{7.3}$$

with the initial-value condition

$$y(0) = 1 \tag{7.4}$$

The solution to (7.3) is

$$y = e^{x/2} - 1 \tag{7.5}$$

Equations as those above can be solved by the *series method*. The series method produces an infinite series of powers of x near the initial point. This approach uses Taylor's theorem to expand a function about a point. The function y(x) can be expanded about a point x_0 as follows:

$$y(x) = y(x_0) + (x - x_0)y'(x_0) + \frac{1}{2!}(x - x_0)^2 y''(x_0) + \dots + \frac{1}{n!}(x - x_0)^n y^{(n)}(x_0) \tag{7.6}$$

On the right-hand side of the equation, the first term is given by the initial condition. The other terms are obtained from the differential equation.

This technique is not easily implemented on a computer. For hand calculations with a calculator, arriving at a result is usually easy. The idea behind this technique is to evaluate the function relatively close to the initial value.

Another approach using the above method as a basis, is to use the information at the initial point (x_0, y_0) to obtain an estimate y_1 of $y(x_1)$. The new point (x_1, y_1) will be used to estimate the values y_2 and $y(x_2)$. At each point the value of x is increased by the interval h (the step size) and the corresponding increase in y is estimated. Therefore

$$y_{i+1} = y_i + \delta y_{estimated}. \tag{7.7}$$

The most basic method in this particular process is called the Euler method. This method is not used in practice on simple equations. A major reason for not using the Euler method is that the orders of accuracy are low. If we were to maintain a high accuracy, it would necessitate a small step size. The increase in computational time and increase in round-off errors dictates that this method be used with caution. For more complicated equations (e.g. parabolic partial differential equations and hyperbolic differential equations) the solution techniques are often based on the Euler methods. Euler methods can be divided into three schemes: (1) the forward Euler method, (2) the modified Euler method, and (3) the backward Euler method.

The forward Euler method is obtained by rewriting the forward difference approximation for y' as

$$y_{n+1} = y_n + hf(y_n, t_n) \tag{7.8}$$

Using equation (7.8) recursively results in

$$\begin{aligned} y_1 &= y_0 + hy_0' \\ y_2 &= y_1 + hy_1' \\ \vdots &= \vdots \\ y_n &= y_{n-1} + hy_{n-1}' \end{aligned} \tag{7.9}$$

This subroutine is demonstrated at the end of the chapter in Program 7.1F.

The second example of the Euler method is the modified Euler method. This algorithm is more accurate than the Euler method described above and is more stable. This method is derived by applying the trapezoidal rule to the integration of y' = f(y,x). The result is

$$y_{n+1} = y_n + h/2[f(y_{n+1},t_{n+1}) + f(y_n t_n)] \tag{7.10}$$

If f is a nonlinear function of y, equation (7.10) becomes a nonlinear function of y_{n+1}. A commonly used method to solve nonlinear equations is to use the method of successive substitution. Later in the chapter, the reader will notice the similarity of this algorithm with the Runge-Kutta method. The modified Euler algorithm is illustrated in the subroutine at the end of the chapter in Program 7.2F.

Similar to the forward difference algorithm, the backward Euler method is based on the backward difference approximation and is written as

$$y_{n+1} = y_n + hf(y_{n+1},t_{n+1}) \tag{7.11}$$

The backward difference algorithm is stable. The algorithm is illustrated in the subroutine at the end of the chapter in Program 7.3F.

One of the more widely used methods is the Runge-Kutta method. This method is a one-step procedure that involves only first-order derivative evaluations. These results will produce results of equivalent accuracy to the higher order Taylor formulas. One-step methods do not show instability problems if the step sizes for integration are small. For example, the Taylor's series method can be adjusted to the differential equation and to the accuracy required. Higher derivatives can be calculated numerically or can be determined from various tables.

The Runge-Kutta methods are perhaps the more widely used numerical methods for calculating the solution to ordinary differential equations. The advantages of the Runge-Kutta methods are the ease of computer implementation, stability, and the ability to change step sizes with minimal hassle. However one must necessarily expect an increase in computational time and consider the lack of ability to estimate accurately the error associated with the method.

All Runge-Kutta methods have algorithms of the form

$$y_{i+1} = y_i + h\phi(x_i, y_i, h). \tag{7.12}$$

The ϕ variable is called the increment function and is a suitably chosen approximation to f(x,y) on the interval $x_i \leq x \leq x_{i+1}$. Without derivation, the second, third, and fourth order Runge-Kutta algorithms are given below, respectively:

$$y_{i+1} = y_i + \frac{h}{2}[f(x_i, y_i) + f(x_{i+1}, \bar{y}_{i+1})] \tag{7.13}$$

where

$$\bar{y}_{i+1} = y_i + hf(x_i, y_i) \tag{7.14}$$

The third order method is given as

$$y_{i+1} = y_i + \frac{h}{6}(k_1 + 4k_2 + k_3) \tag{7.15}$$

where

$$\begin{aligned} k_1 &= f(x_i, y_i) \\ k_2 &= f(x_i + \frac{h}{2}, y_i + \frac{h}{2}k_1) \\ k_3 &= f(x_i + h, y_i + 2hk_2 - hk_1). \end{aligned} \tag{7.16}$$

The fourth order method ascribed to Kutta is given by

$$y_{i+1} = y_i + \frac{h}{8}(k_1 + 3k_2 + 3k_3 + k_4) \tag{7.17}$$

where

$$\begin{aligned}
k_1 &= f(x_i, y_i) \\
k_2 &= f(x_i + \frac{h}{3}, y_i + \frac{h}{3} k_1) \\
k_3 &= f(x_i + \frac{2h}{3}, y_i - \frac{h}{3} k_1 + hk_2) \\
k_4 &= f(x_i + h, y_i + hk_1 - hk_2 + hk_3).
\end{aligned} \tag{7.18}$$

One of the most widely used fourth-order method is one attributed to Gill is given by

$$y_{i+1} = y_1 + \frac{h}{6}[k_1 + 2(1 - \frac{1}{\sqrt{2}})k_2 + 2(1 + \frac{1}{\sqrt{2}})k_3 + k_4] \tag{7.19}$$

where

$$\begin{aligned}
k_1 &= f(x_i, y_i) \\
k_2 &= f(x_i + \frac{h}{2}, y_i + hk_1) \\
k_3 &= f(x_i + \frac{h}{2}, y_i + (-\frac{1}{2} + \frac{1}{\sqrt{2}})hk_1 + (1 - \frac{1}{\sqrt{2}})hk_2) \\
k_4 &= f(x_i + h, y_i - \frac{1}{\sqrt{2}} hk_2 + (1 + \frac{1}{\sqrt{2}})hk_3).
\end{aligned} \tag{7.20}$$

The subroutine in Program 7.4F at the end of the chapter solves a pair of ordinary differential equations using the Runge-Kutta method. The user can substitute equations as required in the subroutine VALDER at the end of the program.

7.2.2 Minimizing Errors

To produce an accurate numerical solution to an equation or a set of equations it is necessary to evaluate the local truncation error and assure that the error value never exceeds a predefined tolerance. It is necessary to calculate at each stage of the solution the error. If this value is larger than the predetermined value then the solution should return to the previous step and recalculate the value using a smaller step size.

For the fourth order Runge-Kutta method the truncation error is given by

$$e_t = \frac{16}{15}(y_{n+1,2} - y_{n+1,1}) \tag{7.21}$$

Using the above estimate of the error involves significant computational effort if the function on the right-hand side of the differential equation is complicated. For the fourth-order Runge-Kutta method a total of 11 (3x4 - 1) function evaluations would be required at each stage of the numerical solution. A perhaps better method is called the Runge-Kutta-Fehlberg method that employs fourth and fifth order estimates to arrive at an estimate of the truncation value. These are given below.

$$y_{i+1} = y_i + (\frac{25}{216}k_1 + \frac{1408}{2565}k_3 + \frac{2197}{4104}k_4 - \frac{1}{5}k_5) \tag{7.22}$$

The values for k_i are as follows:

$$
\begin{aligned}
k_1 &= hf(x_i,y_i)\\
k_2 &= hf(x_i+\frac{h}{4},y_i+\frac{k_1}{4})\\
k_3 &= hf(x_i+\frac{3h}{8},y_i+\frac{3k_1}{32}+\frac{9k_2}{32})\\
k_4 &= hf(x_i+\frac{12h}{13},y_i+\frac{1932k_1}{2197}-\frac{7200k_2}{2197}+\frac{7296k_3}{2197})\\
k_5 &= hf(x_i+h,y_i+\frac{439k_1}{216}-8k_2+\frac{3680k_3}{513}-\frac{845k_4}{4104})\\
k_6 &= hf(x_i+\frac{h}{2},y_i-\frac{8k_1}{27}+2k_2-\frac{3544k_3}{2565}+\frac{1859k_4}{4104}-\frac{11k_5}{40})
\end{aligned}
\tag{7.23}
$$

The estimate of the local truncation error is given by

$$
E = |y^*_{i+1}\ y_{i+1}| = \left|\frac{k_1}{360}-\frac{128k_3}{4275}-\frac{2197k_4}{75240}+\frac{k_5}{50}+\frac{2k_6}{55}\right| \tag{7.24}
$$

7.3 Multistep Methods

Multistep methods, whether explicit or implicit, are very stable for a reasonably small step size, h, and only require one derivative calculation. As with the one-step methods the multistep methods such as the Adams-Bashforth-Moulton method can be adjusted to the differential equation and to the desired solution accuracy. These methods are known as *predictor-corrector methods*. A predictor-corrector method consists of a predictor step and a corrector step in each interval. The predictor estimates the solution for the new point and the corrector improves its accuracy. Predictor-corrector methods use the solutions for previous points instead of using intermediate values in each interval.

In solving first order initial-value problems, the single-step methods use information from the previous calculated data. The multistep methods use information from several points. It is impossible to initially start using the multistep method until several points are calculated in the solution field. These points can be determined from a single-step method or a series method.

There are several ways to construct a multistep method. For example, consider the differential equation in 7.2 integrated between the limits of x_j and x_{i+1} with $j<=i$.

$$\frac{dy}{dx} = f(x,y) \tag{7.25}$$

Integrating the equation gives

$$\int_{x_j}^{x_{i+1}} \frac{dy}{dx}dx = \int_{x_j}^{x_{i+1}} f(x,y)dx \tag{7.26}$$

from which we get

$$y(x_{i+1}) - y(x_j) = \int_{x_j}^{x_{i+1}} f(x,y)dx \tag{7.27}$$

Numerical integration is used to obtain the estimate of the right hand side of the equation. If we set $i=j$ in the above equation and approximate the function $f(x,y)$ by a third degree polynomial passing through the points $(x_i,f_i),.....,(x_{i-3},f_{i-3})$ a four step explicit formula is obtained.

$$y_{i+1} = y_i + \frac{h}{24}(55f_1 - 59f_{i-1} + 37f_{i-2} - 9f_{i-3}) \tag{7.28}$$

If we approximate f(x,y) by a third-degree polynomial passing through the points (i+1) to (i-2), a three-step implicit formula is obtained.

$$y_{i+1} = y_i + \frac{h}{24}(9f_{i+1} + 19f_1 - 5f_{i-1} + f_{i-2}) \qquad \textbf{(7.29)}$$

The above two equations are a fourth-order predictor-corrector pair that are often used with the fourth-order Runge-Kutta method.

It can be shown that the error in the corrected estimate of $y(x_{i+1})$ is given by

$$y(x_{i+1}) - y^c_{i+1} \simeq \frac{19}{270}(y^p_{i+1} - y^c_{i+1}). \qquad \textbf{(7.30)}$$

The superscripts p and c stand for predictor and corrector, respectively.

Those methods of the explicit form are known as Adams-Bashforth formula and those of the implicit form are known as Adams-Moulton formula.

A routine for the Adams-Brashforth-Moulton method follows.

```
Given: f(x,y), (xi,yi) where i=0,1,2,3, h,n, tolerance
Set:  x=x3
      yc=y3
      f1=f(x0,y0)
      f2=f(x1,y1)
      f3=f(x2,y2)
      f4=f(x3,y3)

for i=3,4,5,..., n-1
      Set x=x+h
          y3=yc
          f0=f1
          f1=f2
          f2=f3
          f3=f4

          yp=y3+h(55f3-59f2+37f1-9f0)/24
          f4=f(x,yp)
          yc=y3+h(9f4+19f3-5f2+f1)/24
          f4=f(x,yc)
          error=abs(19(yp-yc)/270)/h
         if error>tolerance then

          h is too large
else

          output x,yc.
```

The Program 7.5F implements the Adams-Brashforth-Moulton method in FORTRAN is given at the end of the chapter.

7.4 Systems of Equations

There will be times when one equation does not describe the physical situation. You require a system of differential equations of the form

$$\frac{dy_i}{dx} = f_i(x,y_1,y_2,\cdots,y_m) \text{ for } i=1,\ldots,m. \tag{7.31}$$

These equations are to be solved for the functions $y_i(x)$ over an interval [a,b] subject to the conditions

$$y_i(a) = c_i \tag{7.32}$$

where the c_i are constants.

The methods described earlier can be applied to a series of equations. The following algorithm is the Adams-Bashforth-Moulton method applied to a pair of equations.

```
Given:                  f(x,y,z), g(x,y,z), (xi,yi,zi), h,n
For i=0,1,2,3
        Set fi=f(xi,yi,zi)
            gi=g(xi,yi,zi)
For i=3,4,...,n-1
        Set xi+1=xi+h
            yi+1=yi+h(55fi-59fi-1+37fi-2-9fi-3)/24
            zi+1=zi+h(55gi-59gi-1+37gi-2-9gi-3)/24
            fi+1=f(xi+1,yi+1,zi+1)
            gi+1=g(xi+1,yi+1,zi+1)
            yi+1=yi+h(9fi+1+19fi-5fi-1+fi-2)/24
            zi+1=zi+h(9gi+1+19gi-5gi-1+gi-2)/24
            fi+1=f(xi+1,yi+1,zi+1)
            gi+1=g(xi+1,yi+1,zi+1)
```

The subroutine is given at the end of the chapter in Program 7.6F.

7.5 Stiff Ordinary Differential Equations

In reference to a set of "Stiff" Ordinary Differential Equations, one considers the equation to have a very short time constant. Theses types of problems pose a problem for most of the usual numerical methods. A system is considered stiff if the general solution is a sum of two or more particular solutions which vary with respect to time at very different rates. An example of this condition would be a system whose general solution is

$$y(t) = \alpha e^{50t} + \beta e^{-10t} \tag{7.33}$$

where for

$$\begin{aligned} y(0) &= 1 \\ y'(0) &= -10 \end{aligned} \tag{7.34}$$

the exact solution is

$$y(t) = e^{-10t} \tag{7.35}$$

Most numerical methods will return an approximate solution of the form

$$y(t) = e^{-10t} + \epsilon e^{50t} \tag{7.36}$$

For very small values of ε, the e^{50t} term becomes very large for positive t.

7.6 Higher Order Initial Value Problems

In order to solve a general nth order differential equation with n conditions the equation can be replaced by a series of transformations that result in a system of ordinary first order differential equations. This system of equations can be solved as discussed in the previous section.

7.7 Boundary-Value Problems

Any differential equation of any order must be accompanied by the necessary conditions in order to solve the equation. In discussing the initial-value problems, all conditions necessary to solve the problem were defined at a point x. Equations of order greater than one usually have conditions that specify values of y and/or its derivatives at more than one value of x. Conditions given at the endpoints of the interval over which to solve the problem put these types of equations into the class called boundary-value problems.

A boundary-value problem may have a unique solution, an infinite number of solutions, or no solution. There are theorems that allow you to determine if a solution exists and if that solution is unique. For example, consider a two point boundary-value problem. The following theorem is applicable.

Let $f(x,y,y')$ be defined and continuous for all (x,y,y') in a region D defined by $a<=x<=b$, $-\infty<y<\infty$, $-\infty<y'<\infty$ and let $\partial f/\partial y$ and $\partial f/\partial y'$ be continuous in D. The two-point boundary-value problems have a unique solution if for all (x,y,y') in D:

$$\frac{\partial f}{\partial y} > 0 \; , \; \left|\frac{\partial f}{\partial y'}\right| \leq |K| \tag{7.37}$$

Numerical solutions to boundary-value problems are approached from two directions. The first technique uses the methods of the initial value problems discussed previously. The second technique replaces the boundary-value problems by a system of algebraic equations.

7.7.1 The Shooting Method

The shooting method involves estimating a value of y'(a) (a being one of the end points on the interval [a,b]) in order to produce an initial-value problem. There are methods to accelerate the estimates for the derivative of the function.

Consider the following equation:

$$\begin{aligned} \frac{d^2y}{dx^2} &= f(x,y,y') \ with \\ y(a) &= A \\ y(b) &= B \end{aligned} \tag{7.38}$$

If the value of y'(a) were known in addition to the given conditions, the differential equation could be solved over the interval [a,b] using the conditions of y(a)=A and y'(a)=C. Note that y_n is a function of C. This is like solving the problem

$$y_n(C) - D = 0 \tag{7.39}$$

The above equation is a non-linear equation and could be solved using for example, any of the methods discussed earlier such as the secant method. A disadvantage of using this method though is that the trial and error procedure could make the solution become unstable.

The following algorithm is for the shooting method.

```
Given:      f(x,y,y'),a,A,b,B,C1,C2
Set h=(b-a)/n
Repeat (for i=1,2,3...)
      For j=1,2,...,n
      Set x=a+ih
      Evaluate y_i,z_i by solving the system y'=z
                                             z'=f(x,y,z)
                                             y(a)=A
                                             z(a)=C_j
Set   B_i=y_n
      If i>=2 then
      Set C_{i+1}=((B_i-B)C_{i-1} - (B_{i-1} - B)C_i)/(C_i - C_{i-1})

Do until convergence
```

The Program 7.7F at the end of the chapter illustrates the shooting method.

7.8 The Finite-Difference Method

A differential equation can be replaced by a finite-difference approximation to that equation. The finite-difference method replaces the interval of integration [a,b] by a selected set of points and approximates the differential equation and its initial-value equations or its boundary-value equations by the respective difference equations related to these selected set of points. This allows the system of resulting algebraic equations to be solved by conventional matrix methods. The finite difference approximations have been discussed earlier in chapters 2 and 4. An advantage to this method id that there is no instability problems and no trial and error method for linear problems. However, the program may have to be developed for each problem under consideration.

7.9 An Example of Application of Solution to ODE's to a Medical Problem

The problem discussed in this section presents a one-dimensional inviscid solution for flow through a model of a compliant tube with a stenosis. Flow through the stenosis causes the pressure inside the tube to decrease, and the tube may collapse if this internal pressure becomes less than the external pressure. The model was used to represent an artery with an atherosclerotic plaque and to investigate the possibility that the plaque creates favorable conditions for arterial collapse. The general theory for flow through collapsible tubes is based on a "tube law", which relates the mechanics of the flow to the mechanics of the tube. This theory was used to develop the equations governing flow through the stenotic artery model.

A FORTRAN program, which uses a finite difference method, was written to solve the equations.

7.9.1 Medical Relevance

Atherosclerosis is responsible for approximately half of the deaths in the United States per year and is by far the leading cause of death in the Western World. The most frequent occurrences of the disease in humans is in the infrarenal abdominal aorta, the vessels supplying the lower extremities, the carotid arteries and the coronary arteries. Because of the latter two locations, myocardial infarcts (heart attacks) and cerebral infarcts (strokes) are two major consequences of the disease.

Atherosclerosis is a disease in which lesions are formed within the intimal layer of arteries. The lesion consists of a raised focal plaque having a core of lipid (mainly cholesterol), connective tissue, and smooth muscle cells and a covering fibrous cap. As the plaque grows it encroaches on the lumen and makes the artery stenotic. One of the major complications associated with

1advanced atherosclerosis is rupture of the fibrous cap with subsequent thrombosis or distal embolization. Although the mechanism causing plaque fracture is not well understood, one possibility is that the plaque creates favorable hemodynamic conditions for arterial collapse (ie. the collapse is caused by sharp drops in pressure resulting from increased fluid velocity through the stenosis). This collapse may produce sufficient stress and fatigue on the arterial wall to cause plaque fracture.

When the lesion first begins to form the affected artery expands to maintain its lumen diameter. As the lesion continues to grow it may eventually exceed the rate of enlargement of the artery and the lumen area will begin to decrease. Clinical symptoms occur when the vessel lumen is occluded enough to limit flow to supplied tissues, or when the fractured plaque causes distal embolization. Since the flow rate through a stenosis does not decrease significantly until the flow area is reduced by more than 90%, a fully developed plaque may be present before any symptoms caused by flow reduction may occur. Arterial collapse may account for the symptoms that occur before there is significant flow reduction.

7.9.2 Governing Equations

Flow through an artery with a stenotic lesion is complicated with unsteady, three dimensional, viscous, and turbulent characteristics. However, for analysis of flow limitation and collapse, it is convenient and relevant to use a simplified one-dimensional model to describe the first order flow behavior. The governing equations for collapsible tube flow were developed by Shapiro [74] and Elad et al. [35] based on steady, incompressible, and inviscid flow.

The continuity equation is given by

$$Q = uA \tag{7.40}$$

The conservation of mass is given by

$$\frac{\partial(Au)}{\partial x} = 0 \tag{7.41}$$

The momentum equation is

$$u\frac{\partial u}{\partial x} = -\frac{1}{\rho}\frac{\partial p}{\partial x} \tag{7.42}$$

and is the same for compressible and open channel flows.

Flow mechanics through compliant tubes that are partially collapsed because of a negative transmural pressure difference (internal pressure minus external pressure, p-p_e) are closely related to the structural mechanics of the tube. For a positive transmural pressure the tube is inflated. For a negative transmural pressure the tube is partially collapsed and the pressure difference is balanced by the bending stiffness of the tube wall. The relationship between transmural pressure and cross-sectional area for a compliant tube may be modeled by

$$P = \frac{p-p_e}{K_p} = \alpha^{n_1} - \alpha^{-n_2} \tag{7.43}$$

is defined as here P is the dimensionless transmural pressure difference and the area ratio, α, is defined as

$$\alpha = \frac{A}{A_0} \tag{7.44}$$

where n_1 and n_2 are numerical constants.

A_0 is the nominal tube cross-sectional area. The length of the tube and the stenosis length can be varied to model different sizes of atherosclerotic plaques. The shape of the stenosis is given by the formula

$$A_0 = A_{0i} - \lambda_A A_{0i} \sin^2(\pi(x-x_1)/(x_2-x_1)) \tag{7.45}$$

where A_{oi} is the nominal inlet tube area,$\lambda_1 A_o$ is the area reduction of the stenosis, x is the length coordinate, and the end-points of the stenosis are x_1 and x_2. A function of the form $\sin^2 x$ was chosen to give the stenosis a smooth continuous shape.

The stiffness of the tube wall is given by the value of K_p. For a linear elastic material, K_p can be estimated from an analysis of the loads on a circular tube subjected to external pressure before buckling. K_p is proportional to

$$K_p \propto \frac{Et^3}{12R_0^3(1-\nu^2)} \tag{7.46}$$

where E is Young's modulus, t is the tube wall thickness, R_0 is the nominal tube radius and ν is Poisson's ratio. The stiffness of the stenosis can either increase or decrease according to the formula

$$K_p = K_{pi} - \lambda_{K_p} K_{pi} \sin^2(\pi(x-x_1)/(x_2-x_1)) \tag{7.47}$$

where K_{pi} is the inlet stiffness, and $\lambda_{Kp}K_{pi}$ is the amplitude of the stiffness perturbation. If $\lambda_{Kp}$is positive the stenosis is more stiff than the inlet tube and if λ_{K_p} is negative the stenosis is less stiff.

The local wave speed of a small disturbance c, is given by

$$c^2 = \frac{A}{\rho}\frac{\partial(p-p_e)}{\partial A} = \frac{k_p}{\rho}(n_1\alpha^{n_1} + n_2\alpha^{-n_2}) \tag{7.48}$$

Using the above relation for the dimensionless transmural pressure difference, the wave speed can be found as a function of the area ratio, α. An important dimensionless parameter for flow through compliant tubes is the speed index,

$$S = \frac{u}{c} \tag{7.49}$$

After rearranging the governing differential equations involving area and stiffness changes only, the system of equations becomes

$$\frac{d\alpha}{dx} = \frac{\alpha}{1-S^2}\left[\frac{S^2}{A_0}\frac{dA_0}{dx} - \frac{P}{\rho c^2}\frac{dK_p}{dx}\right] \tag{7.50}$$

$$\frac{dS^2}{dx} = \frac{S^2}{1-S^2}\left[\frac{[-2+(2-M)S^2]}{A_0}\frac{dA_0}{dx} + \left(\frac{MP}{\rho c^2} - \frac{1-S^2}{K_p}\right)\frac{dK_p}{dx}\right] \tag{7.51}$$

where M is defined as

$$M = 3 + \alpha\frac{\dfrac{d^2P}{d\alpha^2}}{\dfrac{dP}{d\alpha}} \tag{7.52}$$

Once α(x) and S(x) have been determined, the values for u(x), c(x), p(x), etc.,

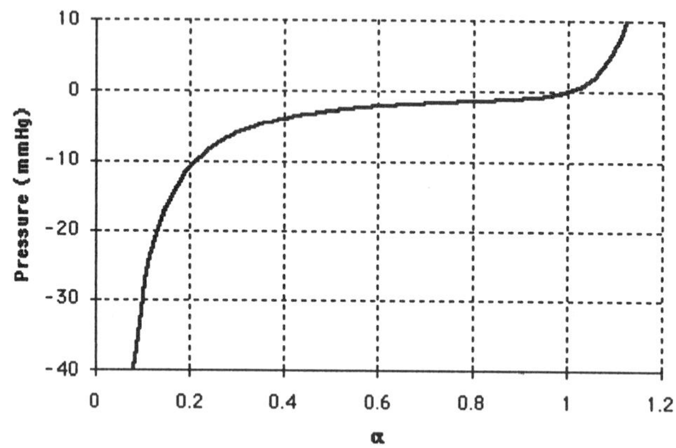

Figure 7-1: Tube Law for an Artery; Kp=125, n1=20, n2=1.5

can be easily found. These differential equations are coupled and must be solved numerically.

As the flow approaches the stenosis center the fluid velocity, u, will increase and the local wave speed, c, will decrease, driving the speed index, S, toward unity. At the same time the values of dA_0/dx and dK_p/dx become zero at the stenosis center. When S=1 at the stenosis center the conditions $d\alpha/dx = 0/0$ and $dS^2/dx = 0/0$ are present. The condition $dS^2/dx = 0/0$ provides the only possible method of having a smooth transition from subcritical to supercritical flow and vice versa. Flow limitation or "choking" is related to the states in which S=1 because there is a maximum steady flow rate that the conducting system can transmit independent of difference between upstream and downstream pressures.

Another important characteristic of collapsible tube flow is the possibility of elastic jumps, in which there is a rapid transition from supercritical to subcritical flow. The occurrence of supercritical flows and elastic jumps depends on the conditions at the choke point and the boundary conditions at the downstream end. For the inviscid solution presented here, the elastic jump is the only method for any significant energy dissipation.

Oates[59] and Cowley[25] used the tube law relationship with conservation of mass and momentum to develop a relationship between flow properties just before and just after an elastic jump. The quantity $\phi(\alpha)$ given by

$$\phi(\alpha)=\frac{\tilde{Q}^2}{\alpha}+\frac{K_p}{\rho}\left[\frac{n_1}{n_1+1}\alpha^{n_1+1}+\frac{n_2}{1-n_2}\alpha^{1-n_2}\right] \tag{7.53}$$

is conserved across an elastic jump; where

$$\tilde{Q}^2 = u_1^2\alpha_1^2 = u_2^2\alpha_2^2 \tag{7.54}$$

Knowing u and α upstream of the jump allows the calculation of the value of α downstream of the jump.

Since this study is investigating tube collapse distal to the stenosis center, a transition from upstream subcritical flow to downstream supercritical flow is desired. This presents a numerical problem of solving the governing differential equations at the stenosis center. This problem was resolved by expanding the equations using a Taylor series method and solving for a small distance from the stenosis center.

The program model is a one dimensional compliant tube with a stenosis at its center. The simplifications imposed for the model are:

1. The flow is steady.
2. The fluid is incompressible and inviscid.
3. The flow is one-dimensional.
4. A universally approximate "tube law"

connecting tube cross-sectional area to transmural pressure is applicable.

7.9.3 Computer Program

A FORTRAN program was written to solve the two governing differential equations. The program uses a fourth-order Runge-Kutta finite difference method to solve for the values of S^2 and α for points that are dx apart along the tube. The input data for the program is:

1. Initial inlet fluid speed (cm/sec)
2. Pressure at tube inlet (mmHg)
3. Pressure at tube outlet (mmHg)
4. Inlet tube-wall stiffness (Pa)
5. Nominal tube diameter (mm)
6. Stenosis area amplification factor
7. Stenosis stiffness amplification factor
8. Step size along tube length
9. Small step size of x from center of stenosis
10. End-points of stenosis
11. Two trial values for the elastic jump position

Equation (7.43) is solved by a Newton's Method subroutine to get the inlet value of α, which is fixed by the inlet pressure and stiffness. Since the inlet α is fixed, the inlet wave speed is also fixed (equation 7.48). The program uses this value for c and an initial value for u to get the inlet speed index (equation 7.49). A fourth-order Runge-Kutta subroutine is used to solve for the values of S^2 and α, from equations (7.50) and (7.51), along the tube up to the stenosis center.

If S becomes equal to one before the stenosis center, an error will occur because the denominator in equations (7.50) and (7.51) is zero. The only way for a smooth transition from subcritical (S<1) to supercritical (S>1) flow to occur is for S to equal one at the stenosis center where $dA_0/dx=0.0$ and/or $dK_p/dx=0.0$. If the error does occur, a lower inlet velocity must be used so that S remains less than one up to the stenosis center. If S is less than one at the stenosis center, the inlet velocity is increased by an amount based on the value of S.

$$u_{i+1} = u_i + e^{10.0*S} \tag{7.55}$$

The above formula was found by trial and error to be the quickest approach to finding the critical inlet velocity. The calculations of S^2 and α up to the stenosis center are repeated with a new inlet velocity until $S \approx 1.0$.

The value of S is then set to exactly 1.0 and α is left unchanged. Since c is fixed by α, the critical velocity and flow rate can now be calculated from equations (7.49) and (7.40). A Taylor's series expansion is used to calculate S^2 and α at a small distance, Δx, on both sides of the stenosis center. The values of S and a upstream of the stenosis center are then calculated backward from the center to the tube inlet.

The flow will be subcritical (S<1) upstream of the stenosis center and supercritical (S>1) downstream. In order for the flow to become subcritical at the tube exit an elastic jump must occur. The program calculates the value of a after an elastic jump by solving equation (7.48) using Newton's Method. This value of α and equations (7.49) and (7.41) are used to get the value for S after the elastic jump. The program then solves for S^2 and α up to the tube outlet.

The position of the elastic jump is governed by the tube outlet pressure. The program uses two trial jump positions to get two values for the outlet pressure. Then the program sends these two values to a Lagrangian Interpolation subroutine to estimate a new jump position. The values of S^2 and downstream of the jump are calculated and a new outlet pressure is found. Now the three outlet pressures are used to estimate a new jump position. This procedure is repeated until the new estimate for the jump position is the same as the previous estimate.

The program output consists of the critical inlet velocity (cm/sec), the critical flow rate (cm^3/sec), and the following values for each x- position:

1. Speed index
2. Area ratio
3. Fluid velocity (cm/sec)
4. Wave speed (cm/sec)
5. Transmural pressure (mmHg)
6. Tube stiffness (Pa)
7. Cross-sectional area (mm^2)

```
      PROGRAM COLTUBE

      REAL WM,ALMDA,ELMDA,EKP,YO(2),XO,YEND(2),YWRK(4,4),F(2)
      REAL SST(1003),AST(1003),XJMP(100),POUT(100)
      INTEGER N,NLIM,LEN
      COMMON/DERIV/ AO,AOI,EKP,EKPI,X1,X2,ALMDA,ELMDA,WM,PDP,FTL
      DATA N,DEN,PI /2,995.0,3.141592654/
      DATA NLIM,YTOL,FTOL /50,.0001,.0001/

C     READ AND PRINT INPUT DATA

      CALL READAT (UI,PIN,EKPI,DO,ALMDA,ELMDA,HX,SMHX,HLI,X1,X2
     +             ,XEND,FT,PDN,XJMP)

C     INITIALIZE VARIABLES

      UI = UI/100.
      AOI = PI/4.0 * DO**2
      FTL = 2.*XEND*FT
      PIN = PIN*133.3
      PDN = PDN*133.3
      XCNT = (X2+X1)/2.0
      IF (FT .EQ. 0.0) THEN
      IT = INT((X2-X1)/2.0/HX)
      XSTR = X1
      XSTP = X2
```

```
      ELSE
      IT = INT(XCNT/HX)
      XSTR = 0.0
      XSTP = XEND
      ENDIF

      PALIN = PIN/EKPI

C     INITIAL GUESS FOR THE VALUE OF ALPHA AT TUBE INLET

      ALIN = 2.0

C     SOLVE FOR ALPHA AT TUBE INLET

      CALL ALINET (ALIN,YTOL,FTOL,NLIM,PALIN)
      CI = SQRT(EKPI/DEN*(20.*ALIN**20 + 1.5*ALIN**(-1.5)))
      WRITE (3,*)
      WRITE (3,*) '. INLET AREA RATIO =',ALIN
      WRITE (3,*) '. INLET WAVE SPEED =',CI*100.,'(cm/sec)'
      WRITE (*,15)
C ---------------------------------------------------------
C     FIND THE CRITICAL VELOCITY
 10   XO = XSTR
      YO(1) = (UI/CI)**2
      YO(2) = ALIN
C     SOLVE THE SYSTEM OF DIFFERENTIAL EQUATIONS
      DO 20 I=1,IT
       CALL SLVDIF(XO,HX,YO,YEND,YWRK,F,N)
       XO = XO + HX
       YO(1) = YEND(1)
       YO(2) = YEND(2)
 20   CONTINUE
      S = SQRT(YEND(1))
      PS = EKP*(YEND(2)**20 - YEND(2)**(-1.5))
      C = SQRT(EKP/DEN*(20.*YEND(2)**20 + 1.5*YEND(2)**(-1.5)))
      WRITE(*,35) UI*100.,S,YEND(2),S*C*100.,C*100.,PS/133.3,EKP,AO

C     WHEN THE VALUE OF S GETS CLOSE TO 1, THE STEP SIZE FOR
C     LAMBDA IS DECREASED
      HL = EXP(-HLI*S)

C     THE FLUID SPEED IS INCREASED UNTIL THE WAVE SPEED REACHES THE
C     CRITICAL VALUE, S = 1
      IF (S .LT. 0.99) THEN
      UI = UI + HL
      GOTO 10
      ENDIF
      WRITE(3,15)
      WRITE(3,35) UI*100.,S,YEND(2),S*C*100.,C*100.,PS/133.3,EKP,AO
      WRITE(3,*)
C ---------------------------------------------------------
C     SET S=1 AT CENTER STENOSIS
      U = C
      QI = U*YEND(2)*AO
      WRITE(*,*) ' CRITICAL FLOW RATE (cc/sec) =',QI
      WRITE(3,*) '. CRITICAL FLOW RATE (cc/sec) =',QI
      SST(IT+2) = 1.0
      AST(IT+2) = YEND(2)

C     STEP FORWARD FROM STENOSIS CENTER BY DX
```

```
      DAOO = AOI*2.*ALMDA*(PI/(X2-X1))**2
      DEKPO = EKPI*2.*ELMDA*(PI/(X2-X1))**2
      TERM = SQRT(DAOO/AO - PDP/EKP*DEKPO)
      SST(IT+3) = 1.0 + TERM*SQRT(WM) * SMHX
      AST(IT+3) = YEND(2) + TERM/SQRT(WM) * SMHX

C     STEP BACKWARD FROM STENOSIS CENTER BY DX
      SST(IT+1) = 1.0 + TERM*SQRT(WM) * (-SMHX)
      AST(IT+1) = YEND(2) + TERM/SQRT(WM) * (-SMHX)

C ---------------------------------------------------------
C     CALCULATE VALUES ALONG TUBE FOR THE CRITICAL % STENOSIS
      HXTMP = HX - SMHX
      XO = XCNT -SMHX
      YO(1) = SST(IT+1)
      YO(2) = AST(IT+1)
       CALL SLVDIF(XO,-HXTMP,YO,YEND,YWRK,F,N)
        XO = XO - HXTMP
        YO(1) = YEND(1)
        YO(2) = YEND(2)
        SST(IT) = YEND(1)
        AST(IT) = YEND(2)
      DO 30 I=IT-1,1,-1
       CALL SLVDIF(XO,-HX,YO,YEND,YWRK,F,N)
       XO = XO - HX
       YO(1) = YEND(1)
       YO(2) = YEND(2)
       SST(I) = YEND(1)
       AST(I) = YEND(2)
 30   CONTINUE

C     PRINT VALUES ALONG TUBE FOR THE CRITICAL % STENOSIS UP TO X+DX
      XO = XSTR
      WRITE (3,25)
      WRITE (3,*) '9999'
      WRITE (3,*) 2*IT+4
       S = SQRT(SST(1))
       C = SQRT(EKPI/DEN*(20.*AST(1)**20 + 1.5*AST(1)**(-1.5)))
       PS = EKPI*(AST(1)**20 - AST(1)**(-1.5))
      WRITE(*,*)' VALUES AT TUBE INLET'
      WRITE(*,35) XO,S,AST(1),S*C*100.,C*100.,PS/133.3,EKPI,AOI
      DO 32 I=1,IT
        IF (XO .GT. X1 .AND. XO .LT. X2) THEN
         AO = AOI - ALMDA*AOI * (SIN( PI * (XO-X1)/(X2-X1) ))**2
         EKP = EKPI + ELMDA*EKPI * (SIN( PI * (XO-X1)/(X2-X1) ))**2
        ELSE
         AO = AOI
         EKP = EKPI
        ENDIF
       S = SQRT(SST(I))
       C = SQRT(EKP/DEN*(20.*AST(I)**20 + 1.5*AST(I)**(-1.5)))
       PS = EKP*(AST(I)**20 - AST(I)**(-1.5))
       WRITE(3,35) XO,S,AST(I),S*C*100.,C*100.,PS/133.3,EKP,AO
       XO = XO + HX
 32   CONTINUE

       XO = XCNT - SMHX
      DO 34 I = IT+1,IT+3
       AO = AOI - ALMDA*AOI * (SIN( PI * (XO-X1)/(X2-X1) ))**2
       EKP = EKPI + ELMDA*EKPI * (SIN( PI * (XO-X1)/(X2-X1) ))**2
```

```
      S = SQRT(SST(I))
      C = SQRT(EKP/DEN*(20.*AST(I)**20 + 1.5*AST(I)**(-1.5)))
      PS = EKP*(AST(I)**20 - AST(I)**(-1.5))
      WRITE(3,35) XO,S,AST(I),S*C*100.,C*100.,PS/133.3,EKP,AO
      XO = XO + SMHX
 34   CONTINUE

C ----------------------------------------------------------
C     GENERATE DATA FROM STENOSIS CENTER TO TUBE OUTLET
      HXTMP = HX - SMHX
      YO(1) = SST(IT+3)
      YO(2) = AST(IT+3)
       CALL SLVDIF(XO,HXTMP,YO,YEND,YWRK,F,N)
       XO = XO + HXTMP
       YO(1) = YEND(1)
       YO(2) = YEND(2)
       SST(1) = YEND(1)
       AST(1) = YEND(2)
       S = SQRT(YEND(1))
       C = SQRT(EKP/DEN*(20.*YEND(2)**20 + 1.5*YEND(2)**(-1.5)))
       PS = EKP*(YEND(2)**20 - YEND(2)**(-1.5))
       WRITE(3,35) XO,S,YEND(2),S*C*100.,C*100.,PS/133.3,EKP,AO

C     SOLVE THE SYSTEM OF DIFFERENTIAL EQUATIONS
      DO 40 I=2,IT
       CALL SLVDIF(XO,HX,YO,YEND,YWRK,F,N)
      IF (YO(2) .LT. 2.0E-2) GOTO 42
       XO = XO + HX
       YO(1) = YEND(1)
       YO(2) = YEND(2)
       SST(I) = YEND(1)
       AST(I) = YEND(2)
 40   CONTINUE
C ------------------------------------------------------
C     FIND JUMP POSITION
 42   WRITE(*,55)
      L = NINT((XJMP(1)-XCNT)/HX)
      DO 60 J=1,2
      XO = XJMP(J)
      YO(2) = AST(L)
      WRITE(*,*) ' JUMP POSITION =',XJMP(J)
C     FIND VALUES AFTER ELASTIC JUMP
      CALL JUMP (QI,XO,YO,DEN,YTOL,FTOL,NLIM)
        DO 50 I=1,NINT((XSTP-XJMP(J))/HX)
         CALL SLVDIF(XO,HX,YO,YEND,YWRK,F,N)
         XO = XO + HX
         YO(1) = YEND(1)
         YO(2) = YEND(2)
 50     CONTINUE
        S = SQRT(YEND(1))
        C = SQRT(EKP/DEN*(20.*YEND(2)**20 + 1.5*YEND(2)**(-1.5)))
        PS = EKP*(YEND(2)**20 - YEND(2)**(-1.5))
        POUT(J) = PS
        WRITE(*,35) XO,S,YEND(2),S*C*100.,C*100.,PS/133.3,EKP,AO
       L = L + NINT((XJMP(J+1) - XJMP(J))/HX)
 60   CONTINUE

      J = J - 1
 80   CALL ESTIMT (POUT,XJMP,J,PDN,XJN)
      J = J + 1
```

```
      XJMP(J) = NINT(XJN/HX)*HX
      I = NINT((XJN-XCNT)/HX)
      WRITE(*,*)' JUMP POSITION =',XJN,'    ROUNDED = ',XJMP(J)
       IF (XJN .LT. XCNT .OR. XJN .GT. XSTP) THEN
        WRITE (*,*) ' JUMP POSITION OUT OF RANGE'
        XJMP(J) = XSTP
        GOTO 72
       ENDIF
      IF (ABS(XJMP(J)-XJMP(J-1)) .LT. 1E-6) GOTO 72
      XO = XJMP(J)
      YO(2) = AST(I)
C     FIND VALUES AFTER ELASTIC JUMP
      CALL JUMP (QI,XO,YO,DEN,YTOL,FTOL,NLIM)
      DO 70 I=1,NINT((XSTP-XJMP(J))/HX)
       CALL SLVDIF(XO,HX,YO,YEND,YWRK,F,N)
       XO = XO + HX
       YO(1) = YEND(1)
       YO(2) = YEND(2)
 70   CONTINUE
       S = SQRT(YEND(1))
       C = SQRT(EKP/DEN*(20.*YEND(2)**20 + 1.5*YEND(2)**(-1.5)))
       PS = EKP*(YEND(2)**20 - YEND(2)**(-1.5))
       POUT(J) = PS
       WRITE(*,35) XO,S,YEND(2),S*C*100.,C*100.,PS/133.3,EKP,AO
     GOTO 80
 72  WRITE(*,*)' FINAL JUMP POSITION =',XJMP(J)

C ------------------------------------------------
C    PRINT VALUES DOWNSTREAM OF STENOSIS CENTER
     XO = XCNT + HX
     DO 100 I=1,NINT((XJMP(J)-XO)/HX)
       IF (XO .GT. X1 .AND. XO .LT. X2) THEN
        AO = AOI - ALMDA*AOI * (SIN( PI * (XO-X1)/(X2-X1) ))**2
        EKP = EKPI + ELMDA*EKPI * (SIN( PI * (XO-X1)/(X2-X1) ))**2
       ELSE
        AO = AOI
        EKP = EKPI
       ENDIF
      S = SQRT(SST(I))
      C = SQRT(EKP/DEN*(20.*AST(I)**20 + 1.5*AST(I)**(-1.5)))
      PS = EKP*(AST(I)**20 - AST(I)**(-1.5))
      WRITE(3,35) XO,S,AST(I),S*C*100.,C*100.,PS/133.3,EKP,AO
      XO = XO + HX
 100 CONTINUE
     YO(2) = AST(I)
CJ    FIND VALUES AFTER ELASTIC JUMP
     CALL JUMP (QI,XO,YO,DEN,YTOL,FTOL,NLIM)
      S = SQRT(YO(1))
      C = SQRT(EKP/DEN * (20.*YO(2)**20 + 1.5*YO(2)**(-1.5)))
      PS = EKP*(YO(2)**20 - YO(2)**(-1.5))
      WRITE(3,35) XO,S,YO(2),S*C*100.,C*100.,PS/133.3,EKP,AO
     DO 90 I=1,NINT((XSTP-XJMP(J))/HX)
      CALL SLVDIF(XO,HX,YO,YEND,YWRK,F,N)
      XO = XO + HX
      YO(1) = YEND(1)
      YO(2) = YEND(2)
      S = SQRT(YEND(1))
      C = SQRT(EKP/DEN*(20.*YEND(2)**20 + 1.5*YEND(2)**(-1.5)))
      PS = EKP*(YEND(2)**20 - YEND(2)**(-1.5))
      WRITE(3,35) XO,S,YEND(2),S*C*100.,C*100.,PS/133.3,EKP,AO
```

```
 90   CONTINUE
C -----------------------------------------------------
15   FORMAT (/2X,'. VALUES AT CENTER STENOSIS'/' . UI(cm/s)    S',
     +'        A RATIO    U (cm/s)   C (cm/s)     P (mmHg)',
     +'  KP(Pa) AO(smm)')
25   FORMAT(/2X,'. VALUES ALONG TUBE AT CRITICAL STENOSIS'/2X,'. X',
     +8X,'S',' 	A RATIO','    U (cm/s)  C (cm/s)    P (mmHg)',

     +'   KP(Pa) AO(smm)')
35 FORMAT(1X,F6.3,2(1X,F10.7),2(1X,F10.4),1X,F12.6,1X,F7.1,1X,F6.2)
55   FORMAT (4X,'VALUES AT TUBE EXIT'/4X,'X',8X,
     +'S      A RATIO    U (cm/s)    C (cm/s)    P (mmHg)   KP(Pa)
      AO(smm)')
      END
C_____________________________________________________
      SUBROUTINE VALDER (Y,X,F,N)
C     THIS SUBROUTINE COMPUTES THE DERIVATIVE VALUES OF S^2 AND ALPHA
C     PARAMETERS -
C     X       - NORMALIZED DISTANCE ALONG TUBE
C     Y(1)    - SQUARE OF SPEED INDEX, S^2
C     Y(2)    - NONDIMENSIONAL AREA RATIO, ALPHA
C     F(1)    - DERIVATIVE OF S^2 WITH RESPECT TO X, dS^2/dx
C     F(2)    - dALPHA/dx
C     AO      - CROSS SECTIONAL AREA OF TUBE
C     DAO     - dAO/dx
C     WM      - PROPERTY OF THE TUBE LAW
C
WM=3+(N1*(N1-1.)-N2*(N2+1.)*Y(2)**(-N1-N2))/(N1+N2*Y(2)**(-N1-N2))

      REAL WM,N1,N2,ALMDA,ELMDA,EKP,X,Y(N),F(N)
      COMMON/DERIV/ AO,AOI,EKP,EKPI,X1,X2,ALMDA,ELMDA,WM,PDP,FTL
      DATA N1,N2,PI /20.,1.5,3.141592654/

      IF (X .GT. X1 .AND. X .LT. X2) THEN
C     AREA PERTURBATION FUNCTIONS, AREA REDUCTION
C      AO = AOI - ALMDA*AOI * SIN( PI * (X-X1)/(X2-X1) )
C      DAO = -ALMDA*AOI* PI/(X2-X1) * COS(PI * (X-X1)/(X2-X1))
      AO = AOI - ALMDA*AOI * (SIN( PI * (X-X1)/(X2-X1) ))**2
      DAO = -ALMDA*AOI* PI/(X2-X1) * SIN(2.*PI*(X-X1)/(X2-X1))

C     STIFFNESS PERTURBATION FUNCTIONS.  MINUS SIGN BEFORE ELMDA
C     INDICATES
C     A DECREASE IN STIFFNESS, PLUS SIGN INDICATES AN INCREASE IN
C     STIFFNESS
C      EKP = EKPI + ELMDA*EKPI * SIN( PI * (X-X1)/(X2-X1) )
C      DKP = +ELMDA*EKPI* PI/(X2-X1) * COS(PI * (X-X1)/(X2-X1))
      EKP = EKPI + ELMDA*EKPI * (SIN( PI * (X-X1)/(X2-X1) ))**2
      DKP = +ELMDA*EKPI* PI/(X2-X1) * SIN(2.*PI*(X-X1)/(X2-X1))

      ELSE
      AO = AOI
      EKP = EKPI
      DAO = 0.0
      DKP = 0.0
      ENDIF

      DAOAO = DAO/AO
      DKPKP = DKP/EKP
      ALN = Y(2)**(-21.5)
      ONSQ = 1.0-Y(1)
```

```
C       DIO = 1.12838*SQRT(AO)
C       FRIC = FTL * Y(1)/(Y(2)*DIO)

      WM = 3. + ( (380. - 3.75*ALN) / (20. + 1.5*ALN) )
      PDP =  (Y(2)**N1 - Y(2)**(-N2)) / (N1*Y(2)**N1 + N2*Y(2)**(-N2))

     F(1) =Y(1)/ONSQ * ((-2.+(2.-WM)*Y(1))*DAOAO +(WM*PDP-ONSQ)*DKPKP)
     F(2) = Y(2)/ONSQ * ( Y(1)*DAOAO - PDP*DKPKP )

C       F(1) = Y(1)/ONSQ * ( (-2.+(2.-WM)*Y(1))*DAOAO +
C     +         (WM*PDP-ONSQ)*DKPKP + WM*FRIC)
C       F(2) = Y(2)/ONSQ * ( Y(1)*DAOAO - PDP*DKPKP - FRIC )

      RETURN
      END
C_____________________________________________________________________
      SUBROUTINE SLVDIF(XO,H,YO,YEND,YWRK,F,N)
C     THIS SUBROUTINE SOLVES A SYSTEM OF N FIRST ORDER DIFFERENTIAL
C    EQUATIONS OF THE FORM  DY1/DX = F1(Y,X) , DY2/DX = F2(Y,X) ETC.
C     USING THE FOURTH ORDER RUNGE-KUTTA METHOD.
C     PARAMETERS -
C    VALDER - A SUBROUTINE THAT COMPUTES VALUES OF THE N DERIVATIVES.
C     XO      - THE INITIAL VALUE OF THE INDEPENDENT VARIABLE
C     H       - THE INCREMENT TO X, THE STEP SIZE
C       YO        - THE ARRAY THAT HOLDS THE INITIAL VALUES OF THE
FUNCTIONS
C    YEND   - AN ARRAY THAT RETURNS THE FINAL VALUES OF THE FUNCTIONS
C     YWRK    - AN ARRAY USED TO HOLD INTERMEDIATE VALUES
C     N       - THE NUMBER OF EQUATIONS
C     F       - AN ARRAY THAT HOLDS THE VALUES OF THE DERIVATIVES

      REAL YO(N),YEND(N),YWRK(4,N),F(N)
      CALL VALDER (YO,XO,F,N)
      DO 10 I=1,N
       YWRK (1,I) = H * F(I)
       YEND (I) = YO(I) + YWRK(1,I)/2.0
 10   CONTINUE

      CALL VALDER (YEND,XO+H/2.0,F,N)
      DO 20 I=1,N
       YWRK (2,I) = H * F(I)
       YEND (I) = YO(I) + YWRK(2,I)/2.0
 20   CONTINUE

      CALL VALDER (YEND,XO+H/2.0,F,N)
      DO 30 I=1,N
       YWRK (3,I) = H * F(I)
       YEND (I) = YO(I) + YWRK(3,I)
 30   CONTINUE

      CALL VALDER (YEND,XO+H,F,N)
      DO 40 I=1,N
       YWRK (4,I) = H * F(I)
 40   CONTINUE

      DO 50 I=1,N
       YEND(I) = YO(I) + ( YWRK(1,I) + 2.0*YWRK(2,I) +
     +           2.0*YWRK(3,I) + YWRK(4,I) )/6.0
 50   CONTINUE
```

```
      RETURN
      END
C_______________________________________________________________
      SUBROUTINE ESTIMT (X,Y,N,XINT,YOUT)
C   THIS SUBROUTINE PERFORMS LAGRANGIAN INTERPOLATION WITHIN A SET OF
C   (X,Y) PAIRS TO GIVE THE Y VALUE CORRESPONDING TO XINT. THE DEGREE
C     OF THE INTERPOLATION POLYNOMIAL IS ONE LESS THAN THE NUMBER OF
C     POINTS SUPPLIED.
C     PARAMETERS:
C     X       - ARRAY OF VALUES OF THE INDEPENDENT VARIABLE
C     Y       - ARRAY OF FUNCTION VALUES CORRESPONDING TO X
C     N       - NUMBER OF POINTS
C     XINT    - THE X VALUE FOR WHICH ESTIMATE OF Y IS DESIRED
C     YOUT    - THE Y VALUE RETURNED
      REAL X(N),Y(N),XINT,YOUT,TERM
      INTEGER N,I,J

       YOUT = 0.0
       DO 20 I=1,N
        TERM = Y(I)
        DO 10 J=1,N
         IF (I .NE. J) THEN
         TERM = TERM * (XINT - X(J))/(X(I) - X(J))
         ENDIF
10      CONTINUE
        YOUT = YOUT + TERM
20     CONTINUE
      RETURN
      END
C_______________________________________________________________
      SUBROUTINE JUMP (QI,XO,YO,DEN,YTOL,FTOL,NLIM)
      REAL YO(2)
      COMMON/DERIV/ AO,AOI,EKP,EKPI,X1,X2,ALMDA,ELMDA,WM,PDP,FTL
      DATA PI/3.141592654/
      AO = AOI - ALMDA*AOI * (SIN( PI * (XO-X1)/(X2-X1) ))**2
      EKP = EKPI + ELMDA*EKPI * (SIN( PI * (XO-X1)/(X2-X1) ))**2
      U = QI / (AO*YO(2))
      W = (U*YO(2))**2
     PHI1=U**2*YO(2)+EKP/DEN*(20./21.*YO(2)**21.-3.*YO(2)**(-0.5))
C     INITIAL GUESS FOR THE VALUE OF ALPHA AFTER ELASTIC JUMP
      YO(2) = 2.0
C     SOLVE FOR ALPHA AFTER ELASIC JUMP
      CALL AFTJUMP (YO(2),YTOL,FTOL,NLIM,W,EKP/DEN,PHI1)
      U = QI / (AO*YO(2))
      C = SQRT(EKP/DEN * (20.*YO(2)**20 + 1.5*YO(2)**(-1.5)))
      YO(1) = (U/C)**2
CD      PS = EKP*(YO(2)**20 - YO(2)**(-1.5))
C     WRITE(3,35) XO,(U/C)*100.,U*100.,C*100.,YO(2),EKP,PS/133.3
C 35  FORMAT (1X,F9.4,2X,F10.7,2(1X,F10.3),2X,F9.7,2X,F8.3,1X,F12.7)
      RETURN
      END
C_______________________________________________________________
      SUBROUTINE AFTJUMP (X,XTOL,FTOL,NLIM,W,EKPR,PHI1)
C     THIS SUBROUTINE FINDS THE ROOT OF A FUNCTION, F(X)=0, BY
C     NEWTON'S METHOD.
C     PARAMETERS -
C     FCN       - FUNCTION THAT COMPUTES THE VALUES FOR F(X).
C     FDER      - FUNCTION THAT COMPUTES THE DERIVATIVE OF F.
C     X         - INITIAL VALUE OF X, SHOULD BE NEAR THE ROOT. ALSO
C                 RETURNS THE VALUE OF THE ROOT.
```

```
C     XTOL,FTOL- TOLERANCE VALUES FOR X, F(X) TO TERMINATE EXECUTION.
C      NLIM     - LIMIT TO NUMBER OF ITERATIONS.

       REAL FCN,FDER,X,FX,DELX,XTOL,FTOL,W,EKPR,PHI1
       INTEGER J,NLIM
       FCN(X,W,EKPR,PHI1)=W/X+EKPR*(20./21.*X**21-3.*X**(-0.5))-PHI1
       FDER(X,W,EKPR) = -W/X**2 + EKPR*(20.*X**20 + 1.5*X**(-1.5))

C      COMPUTE SEQUENCE OF POINTS CONVERGING TO THE ROOT
       FX = FCN(X,W,EKPR,PHI1)
       DO 20 J = 1,NLIM
       DELX = FX / FDER(X,W,EKPR)
       X = X - DELX
       FX = FCN(X,W,EKPR,PHI1)

C      CHECK TO SEE IF EITHER TOLERANCE MET.
       IF (ABS(DELX) .LE. XTOL) THEN
        RETURN
       ENDIF
       IF (ABS(FX) .LE. FTOL) THEN
        RETURN
       ENDIF

   20  CONTINUE

C      IF LOOP IS COMPLETED, NLIM IS EXCEEDED AND ROOT WAS NOT FOUND.
       WRITE(*,205) NLIM,X,FX
       RETURN

  205  FORMAT(/' VALUE OF ALPHA AFTER ELASTIC JUMP NOT FOUND AFTER ',
     + I4,' ITERATIONS:   ALPHA =',E12.5,3X,' PHI2 =',E12.5)
       END
C______________________________________________________________
       SUBROUTINE ALINET (X,XTOL,FTOL,NLIM,PALIN)
C      THIS SUBROUTINE WORKS THE SAME AS SUBROUTINE AFTJUMP
       REAL AFCN,AFDER,X,FX,DELX,XTOL,FTOL
       INTEGER J,NLIM
       AFCN(X,PALIN) = X**20 - X**(-1.5) - PALIN
       AFDER(X) = 20.*X**19 + 1.5*X**(-2.5)
       FX = AFCN(X,PALIN)
       DO 20 J = 1,NLIM
       DELX = FX / AFDER(X)
       X = X - DELX
       FX = AFCN(X,PALIN)
       IF (ABS(DELX) .LE. XTOL) THEN
        RETURN
       ENDIF
       IF (ABS(FX) .LE. FTOL) THEN
        RETURN
       ENDIF
   20  CONTINUE
       WRITE(*,205) NLIM,X,FX
       RETURN
  205  FORMAT(/' VALUE OF ALPHA AT INLET NOT FOUND AFTER ',
     + I4,' ITERATIONS:   ALPHA =',E12.5,3X,' PHI2 =',E12.5)
       END
C______________________________________________________________
       SUBROUTINE READAT (UI,PIN,EKPI,DO,ALMDA,ELMDA,HX,SMHX,HLI,X1,X2
     +                  ,XEND,FT,PDN,XJMP)
C      PROGRAM TO ANALYZE STEADY FLOW IN COLLAPSIBLE TUBES
```

```
C     THE PROGRAM FINDS A CRITICAL % STENOSIS FOR THE GIVEN INLET
C     CONDITIONS BY INCREASING THE INLET VELOCITY UNTIL S = 1 AT THE
C     CENTER OF THE STENOSIS.
C     THE INDEPENDENT VARIABLE IS THE DISTANCE ALONG THE TUBE, X, IN
C     DIAMETERS.  THE DEPENDANT VARIABLES ARE THE SPEED INDEX, S,
C     AND THE AREA RATIO, ALPHA.

C     PARAMETERS -
C     X       - DISTANCE ALONG TUBE IN DIAMETERS
C     Y(1)    - SQUARE OF SPEED INDEX, S^2
C     Y(2)    - AREA RATIO, ALPHA
C     ALMDA    - % STENOSIS (AREA REDUCTION)
C     ELMDA   - STIFFNESS PERTURBATION AMPLITUDE
C     UI      - FLUID VELOCITY AT TUBE INLET
C     EKPI    - TUBE WALL STIFFNESS, KP
C     HX      - STEP SIZE FOR X
C     SMHX    - SMALL STEP SIZE FOR X FROM STENOSIS CENTER
C     PIN     - PRESSURE AT TUBE INLET

      REAL XJMP(100)
C     READ IN INITIAL CONDITIONS
      OPEN (2,FILE='COLTUBE.DAT')
      READ(2,*) UI
      READ(2,*) PIN,PDN
      READ(2,*) EKPI
      READ(2,*) DO
      READ(2,*) ALMDA
      READ(2,*) ELMDA
      READ(2,*) HX
      READ(2,*) SMHX
      READ(2,*) HLI
      READ(2,*) X1,X2
      READ(2,*) XEND,FT
      READ(2,*) XJMP(1),XJMP(2)
      OPEN (3,FILE='COLTUBE.PRN')
      WRITE(3,105) 'INPUT DATA FILE'
      WRITE(3,115) 'INLET FLUID VELOCITY (cm/sec)',UI
      WRITE(3,115) 'INLET & OUTLET PRESSURE (mmHg)',PIN,PDN
      WRITE(3,115) 'STIFFNESS FACTOR Kp',EKPI
      WRITE(3,115) 'TUBE DIAMETER (mm)',DO
      WRITE(3,115) 'INITIAL STENOSIS AMPLITUDE',ALMDA
      WRITE(3,115) 'STIFFNESS PERTURBATION AMPLITUDE',ELMDA
      WRITE(3,115) 'INITIAL X STEP SIZE',HX
      WRITE(3,115) 'SMALL X STEP SIZE FROM CENTER',SMHX
      WRITE(3,115) 'VELOCITY INCREASE FACTOR',HLI
      WRITE(3,115) 'END POINTS OF STENOSIS',X1,X2
      WRITE(3,115) 'TUBE LENGTH & FRICTION FACTOR',XEND,FT
      WRITE(3,115) 'JUMP POSITION GUESSES',XJMP(1),XJMP(2)
  105 FORMAT (/' . ',A25/)
  115 FORMAT (1X,'. ',A33,' =',3F9.3)
      RETURN
      END
```

DATA FOR ABOVE PROBLEM

```
40.000          FLUID VELOCITY AT TUBE INLET (cm/sec)
100.0 30.0      INLET & OUTLET PRESSURE (mmHg)
125.0           INLET STIFFNESS FACTOR (Kp)
6.0             NOMINAL TUBE DIAMETER (mm)
0.900           STENOSIS AREA AMPLITUDE
```

```
0.0            STIFFNESS PURTURBATION AMPLITUDE
0.010          STEP SIZE FOR X ALONG TUBE
0.00001        SMALL STEP FOR X FROM CENTER STENOSIS
9.00           STEP SIZE FACTOR FOR VELOCITY INCREASE
0.3 1.7        ENDPOINTS OF PURTURBATION
2.0 0.0        TUBE LENGTH & FRICTION COEFFICIENT
1.30 1.50      GUESSES FOR JUMP POSITION
```

7.9.4 Results

Figures 7-2 and 7-3 give the results for specific parameters that are considered to be physiologically relevant to study of flow through a stenotic artery.

Figure 7-3 shows the variation of the speed index, S, along the tube. The flow is subcritical (S<1) at the tube inlet and becomes supercritical (S>1) after passing through the area reduction.

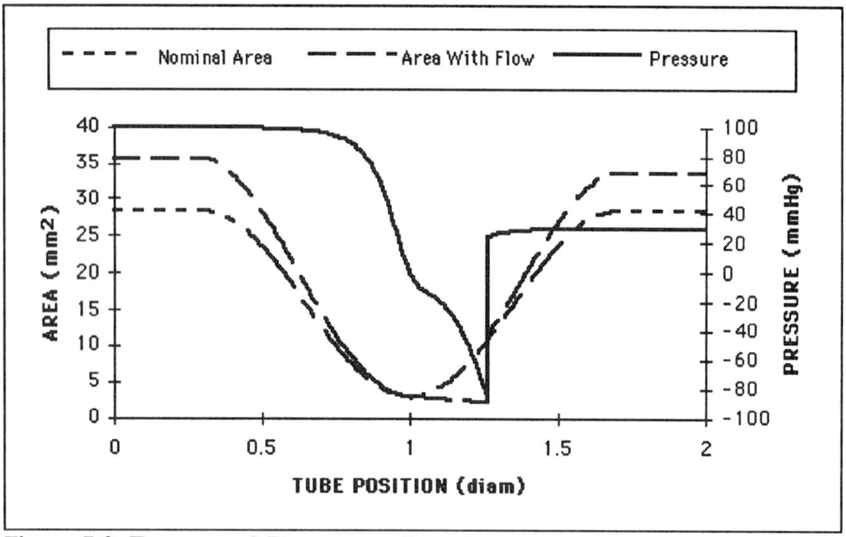

Figure 7-2: Transmural Pressure vs. Length

The tube length (x) is in multiples of the tube diameter, flow is from left to right and the area reduction is 90% by equation (7.41). The following parameters are used: the stenosis runs from x=0.3 to x=1.7, with the center at x=1.0; the inlet pressure is 100 mmHg and the outlet pressure is 30 mmHg;

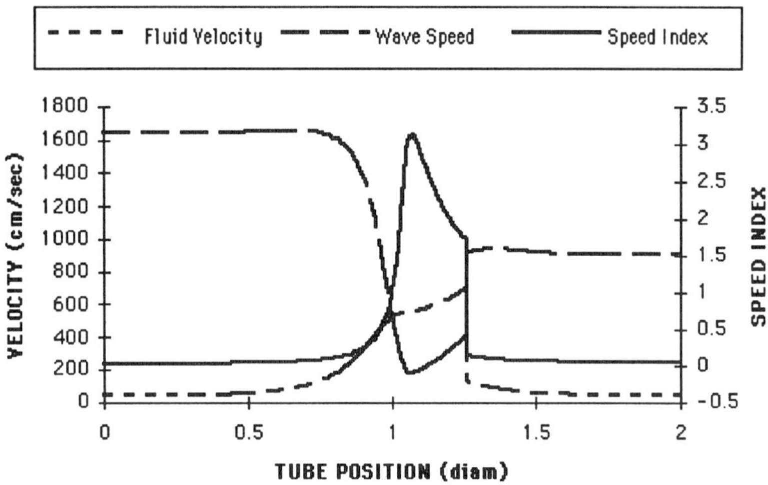

Figure 7-3: Velocity, Wave Speed, Speed Index vs Tube Length for 90% Stenosis

the inlet diameter is 6 mm; and the nominal tube-wall stiffness (K_{po}) is 125 Pa. The stiffness increases by a factor of 10 from the edge of the stenosis to the center according to equation (7.47). The stenosis length used in these figures was chosen to represent an atherosclerotic lesion of approximately one diameter in length.

S=1 at the stenosis center. The value of S reaches a maximum when the wave speed reaches its lowest point at x=1.07. Downstream of the stenosis center, at $x=x_{jump}=1.26$, an elastic jump occurs and the flow becomes subcritical again.

Figure 7-3 also shows the variation of the fluid velocity, u, and the wave speed, c, along the tube. The value of u is governed by equation (7.40) and since the flow-rate is constant, u is controlled by the change in actual area, A. Therefore, u increases as A decreases up to x_{jump}, and then decreases as A increases. The value of c is governed by equation (7.44) and since K_p and ρ are constant, it is controlled by the area ratio, α. When α is close to 1, the first term in equation (7.48) is dominant; and when α is small, the last term is dominant.

Figure 7-2 shows the cross-sectional area along the tube for resting and flow conditions. The tube is inflated upstream of the stenosis center and at the tube outlet because of the positive inlet and outlet pressure. The tube becomes collapsed until the elastic jump occurs, when the actual area expands to greater than the rest area.

Figure 7-2 also shows the transmural pressure along the tube, which is governed by the tube law, equation (7.43). When α is close to 1 the first term in equation (7.43) is dominant, and as a gets smaller the second term becomes dominant. The pressure is not 0 until $\alpha=1$ and reaches a minimum of -11 mmhg at the elastic jump position, $x=1.26$.

In conclusion, this study has shown quantitative flow conditions that cause collapse in a model of a stenotic artery. Based on the theory of collapsible tube flow, a stenotic artery can collapse under realistic physiologic conditions, and this collapse may be a cause of atherosclerotic plaque fracture with subsequent embolization.

7.10 Problems

7-1) Solve the initial value problem

$dy/dx = xy + 2x$

$y(0) = 1$

using a Taylor series expansion.

7-2) Use the second order Runge-Kutta algorithm to numerically integrate

$f(x,y) = 2x^3 + 5.7x^2 - 20x + 8$

$y(0) = 1$

from x=0 to x=2 using a step size of 0.5.

7-3) Use the third order Runge-Kutta algorithm to integrate

$f(x,y) = e^{0.1x} - 0.2y$

with $y(0) = 1$ from x=0 to x=1 with a step size of 0.25.

7-4) Repeat problem 3 using the fourth order Runge-Kutta computer program at the end of the chapter. Compare the results of the two methods.

7-5) Modify the computer program to incorporate the Runge-Kutta-Fehlberg error estimate. The program should decrease the step size h until the error is below a prescribed tolerance.

7-6) Use the Adams-Brashford-Moulton (predictor corrector) method to solve the differential equation

$dy/dx = x + y$

from x=0 to x=1 with the initial value y(0)=0. Compare the results to the exact solution

$y = e^x - 1 - x.$

7-7) Use the Adams-Brashforth-Moulton method to solve the pair of differential equations

$$dy/dx = -0.5y$$

$$dz/dx = 4 - 0.3z - 0.1y$$

from x=0 to x=2 with the initial values y(0)=4 and z(0)=6.

7-8) The deflection of a cantilever beam with a force distributed along its length is represented by the differential equation

$$d^2y/dx^2 - F_l/2EI\ (L-x)^2 = 0$$

where F_l is the distributed load, E is the modulus of elasticity, I is the moment of interia, and L is the total length of the beam. Find the deflection of the end of the beam using fourth order Runge-Kutta computer program with the following parameters; $F_l = 150$ N/m, $E = 7.2 \times 10^9$ kg/m^2, $I = 0.13 \times 10^{-4}$ m^4, and L = 10 m.

7-9) Repeat problem 6 with an initial value of y(1) = 0.

7-10) The temperature distribution from along a fin that is exposed to still air with fixed end temperatures is given by the differential equation

$$d^2T/dx^2 - h_f[T(x)-T_\infty] = 0$$

where T(x) is the temperature along the fin, T_∞ is the ambient air temperature, and h_f is a net heat-transfer coefficient (m^{-2}). With the end temperatures of the rod fixed at

$$T(0) = 50°C \qquad T(10) = 90°C$$

use the shooting method to find the temperature along the rod. Use $h_f = 0.006$ and $T_\infty = 20°C$.

Chapter 7 Programs

Program 7.1F Euler Forward Difference

```
 PROGRAM EULERFOR

C______________________________________________________________________

C  THIS PROGRAM COPYRIGHT: (C) 1992 by Marvin N. Zeigler
C  AUTHORS:  Marvin N. Zeigler and Edward Champion
C  TEXT BOOK:  NUMERICAL METHODS FOR ENGINEERING APPLICATIONS
C  EQUATION:  7.8-7.9

C  PURPOSE:  This program uses the forward Euler method to solve an
C   initial value problem of the form dy/dt = f(y,t).  The function
is
C   evaluated over the range XL to XH with an initial value YL.

C  ARGUMENTS:
C   F       Differential being evaluated. y'=f(y,t).
C   Y       Dependent variable of equation.
C   T       Independent variable of equation.
C   TL,TH   Range over which function is evaluated.
C   NSTEPS  The range is divided into NSTEPS.
C   H       Step size along the range.

C______________________________________________________________________

      REAL T,Y,H,TL,TH,YL,F
      INTEGER NSTEPS

      F(Y,T) = 8.0*T*T*T - 12.0*T*T - 2.0*T  + 3.0

C Define initial data
      TL = 0.0
      TH = 2.0
      YL = 3.0
      NSTEPS  = 40

      WRITE(*,*) '   T        Y-CALC       Y-SOL     ERROR'
      YSOL = 2.0*TL**4 - 4.0*TL*TL*TL - 1.0*TL*TL + 3.0*TL + 3.0
      ERR = (YO-YSOL)/YSOL
      WRITE(*,*) TL, YL, YSOL , ERR

C Perform the integration
      H = (TH-TL)/NSTEPS
      T = TL
      Y = YL
      DO 20 I=1,NSTEPS
         Y = Y + H * F(Y,T)
         T = T + H
         YSOL = 2.0*T**4 - 4.0*T*T*T - 1.0*T*T + 3.0*T + 3.0
         ERR = (Y-YSOL)/YSOL
         WRITE(*,*) T,Y,YSOL,ERR
 20   CONTINUE

      END
```

Program 7.2F Modified Euler Method

```
PROGRAM EULERMOD
C_______________________________________________________________________

C  THIS PROGRAM COPYRIGHT: (C) 1992 by Marvin N. Zeigler
C  AUTHORS:  Marvin N. Zeigler and Edward Champion
C  TEXT BOOK:  NUMERICAL METHODS FOR ENGINEERING APPLICATIONS
C  EQUATION:  7.10

C  PURPOSE:  This program solves the differential equation given by
C   F(Y,T) using the modified Euler method.  This algorithm is more
C   accurate than either the forward Euler or backward Euler method,
C   even with fewer divisions.

C  ARGUMENTS:
C   F      Differential being evaluated. y'=f(y,t).
C   Y      Dependent variable of equation.
C   T      Independent variable of equation.
C   TL,TH  Range over which function is evaluated.
C   NSTEPS The range is divided into NSTEPS.
C   H      Step size along the range.

C_______________________________________________________________________

      REAL T,Y,H,TL,TH,YL,F,YI,FI
      INTEGER NSTEPS

      F(Y,T) = 8.0*T*T*T - 12.0*T*T - 2.0*T  + 3.0

C Define initial data
      TL = 0.0
      TH = 2.0
      YL = 3.0
      NSTEPS  = 20

      WRITE(*,*) '    T        Y-CALC        Y-SOL     ERROR'
      YSOL = 2.0*TL**4 - 4.0*TL*TL*TL - 1.0*TL*TL + 3.0*TL + 3.0
      ERR = (YO-YSOL)/YSOL
      WRITE(*,*) TL, YL, YSOL , ERR

C Perform the integration
      H = (TH-TL)/NSTEPS
      T = TL
      Y = YL
      DO 20 I=1,NSTEPS
         YI = Y
         FI = F(Y,T)
         Y = YI + H*FI
         T = T + H
         Y = YI + H/2.0 * (F(Y,T) + FI)
         YSOL = 2.0*T**4 - 4.0*T*T*T - 1.0*T*T + 3.0*T + 3.0
         ERR = (Y-YSOL)/YSOL
         WRITE(*,*) T,Y,YSOL,ERR
 20   CONTINUE

      END
```

Program 7.3F Euler Backward Difference

```
      PROGRAM EULERBAC
C_______________________________________________________________________

C  THIS PROGRAM COPYRIGHT: (C) 1992 by Marvin N. Zeigler
C  AUTHORS:  Marvin N. Zeigler and Edward Champion
C  TEXT BOOK:  NUMERICAL METHODS FOR ENGINEERING APPLICATIONS
C  EQUATION:  7.11

C  PURPOSE:  This program solves the differential equation given by
C   F(Y,T) using the backward Euler method.  This algorithm is stable.

C  ARGUMENTS:
C   F      Differential being evaluated. y'=f(y,t).
C   Y      Dependent variable of equation.
C   T      Independent variable of equation.
C   TL,TH  Range over which function is evaluated.
C   NSTEPS The range is divided into NSTEPS.
C   H      Step size along the range.

C_______________________________________________________________________

      REAL T,Y,H,TL,TH,YL,F,YI
      INTEGER NSTEPS

      F(Y,T) = 8.0*T*T*T - 12.0*T*T - 2.0*T  + 3.0

C Define initial data
      TL = 0.0
      TH = 2.0
      YL = 3.0
      NSTEPS  = 40

      WRITE(*,*) '   T        Y-CALC        Y-SOL     ERROR'
      YSOL = 2.0*TL**4 - 4.0*TL*TL*TL - 1.0*TL*TL + 3.0*TL + 3.0
      ERR = (YO-YSOL)/YSOL
      WRITE(*,*) TL, YL, YSOL , ERR

C Perform the integration
      H = (TH-TL)/NSTEPS
      T = TL
      Y = YL
      DO 20 I=1,NSTEPS
         YI = Y
         Y = YI + H*F(Y,T)
         T = T + H
         Y = YI + H*F(Y,T)
         YSOL = 2.0*T**4 - 4.0*T*T*T - 1.0*T*T + 3.0*T + 3.0
         ERR = (Y-YSOL)/YSOL
         WRITE(*,*) T,Y,YSOL,ERR
 20   CONTINUE

      END
```

Program 7.4F Runge-Kutta Solver

```
      PROGRAM RKSOLV
C     A FOURTH ORDER RUNGE-KUTTA METHOD IS USED TO SOLVE THE PAIR OF
C      DIFFERENTIAL EQUATIONS DY1/DX=-0.5*Y1, DY2/DX=4-0.3*Y2-0.1*Y1
C      FROM X=0 TO X=2 WITH INITIAL CONDITIONS Y1(0)=4 AND Y2(0)=6.

C      N      NUMBER OF DIFFERENTIAL EQUATIONS TO SOLVE
C      H      STEP SIZE ALONG X
C      XL,XU  LIMITS OF THE INDEPENDENT VARIABLE X
C      YO     ARRAY OF INITIAL VALUES FOR THE DEPENDANT VARIABLES

      INTEGER N
      PARAMETER (N=2)
      REAL X,H,YO(N),YEND(N),YWRK(4,N),F(N)
      DATA H,XL,XU,YO /0.25,0.0,2.0,4.0,6.0/

      X = XL
      ITER = NINT((XU-XL)/H)
      WRITE(*,*) '       X             Y1              Y2'
      WRITE(*,*) X,YO(1),YO(2)

      DO 10 I=1,ITER
         X = X + H
         CALL SLVDIF (X,H,YO,YEND,YWRK,F,N)
         WRITE(*,*) X,YEND(1),YEND(2)
         YO(1) = YEND(1)
         YO(2) = YEND(2)
 10   CONTINUE
      END
C_____________________________________________________________________
      SUBROUTINE SLVDIF(XO,H,YO,YEND,YWRK,F,N)
C     THIS SUBROUTINE SOLVES A SYSTEM OF N FIRST ORDER DIFFERENTIAL
C     EQUATIONS OF THE FORM  DY1/DX = F1(Y,X) , DY2/DX = F2(Y,X) ETC.
C     USING THE FOURTH ORDER RUNGE-KUTTA METHOD.

C     VALDER - A SUBROUTINE THAT COMPUTES VALUES OF THE N DERIVATIVES.
C      XO      - THE INITIAL VALUE OF THE INDEPENDENT VARIABLE
C      H       - THE INCREMENT TO X, THE STEP SIZE
C     YO      - THE ARRAY THAT HOLDS THE INITIAL VALUES OF THE FUNCTIONS
C     YEND    - AN ARRAY THAT RETURNS THE FINAL VALUES OF THE FUNCTIONS
C      YWRK    - AN ARRAY USED TO HOLD INTERMEDIATE VALUES
C      N       - THE NUMBER OF EQUATIONS
C      F       - AN ARRAY THAT HOLDS THE VALUES OF THE DERIVATIVES

      REAL YO(N),YEND(N),YWRK(4,N),F(N)
      CALL VALDER (YO,XO,F,N)
      DO 10 I=1,N
       YWRK (1,I) = H * F(I)
       YEND (I) = YO(I) + YWRK(1,I)/2.0
 10   CONTINUE

      CALL VALDER (YEND,XO+H/2.0,F,N)
      DO 20 I=1,N
       YWRK (2,I) = H * F(I)
       YEND (I) = YO(I) + YWRK(2,I)/2.0
 20   CONTINUE
```

```
      CALL VALDER (YEND,XO+H/2.0,F,N)
      DO 30 I=1,N
       YWRK (3,I) = H * F(I)
       YEND (I) = YO(I) + YWRK(3,I)
 30   CONTINUE

      CALL VALDER (YEND,XO+H,F,N)
      DO 40 I=1,N
       YWRK (4,I) = H * F(I)
 40   CONTINUE

      DO 50 I=1,N
       YEND(I) = YO(I) + ( YWRK(1,I) + 2.0*YWRK(2,I) +
     +           2.0*YWRK(3,I) + YWRK(4,I) )/6.0
 50   CONTINUE

      RETURN
      END
C_____________________________________________________________________
      SUBROUTINE VALDER (Y,X,F,N)
C   ROUTINE TO COMPUTE VALUES OF THE DERIVATIVES

C     X      INDEPENDENT VARIABLE
C     Y(N)   ARRAY OF DEPENDANT VARIABLES
C     F(N)   VALUES OF THE DERIVATIVES
C     N      NUMBER OF EQUATIONS
      REAL X,Y(N),F(N)
      INTEGER N

      F(1) = -0.5*Y(1)
      F(2) = 4.0 - 0.3*Y(2) - 0.1*Y(1)

      RETURN
      END
```

Program 7.5F Adams-Brashforth-Moulton/One Equation

```
      PROGRAM ONEEQABM
C_______________________________________________________________________

C  THIS PROGRAM COPYRIGHT: (C) 1991 by Marvin N. Zeigler
C  AUTHORS:  Marvin N. Zeigler and Edward Champion
C  TEXT BOOK:  NUMERICAL METHODS FOR ENGINEERING APPLICATIONS
C  EQUATIONS:  7.25, 7.26, 7.27

C  PURPOSE:  The Adams-Brashford-Moulton (Predictor Corrector) method
C   is used to solve the single differential equation dy/dx = x+y from

C   x=0 to x=1 with the initial value y=0.  The exact solution of the
C   differential equation is y = e^x - 1 - x.  A fourth order Runge-
C   Kutta routine is used to determine the first four function values
C   needed to start the ABM method.

C  ARGUMENTS:
C   F       Function that computes the derivative value.
C   H       Step size for the independent variable x.
C   XL,XU   Limits of the x interval.
C   YL      Initial value of the dependant variable y.
C   TOL     Tolerance for determining if step size needs reducing.
C_______________________________________________________________________

       EXTERNAL F
       REAL F,H,XL,YL,TOL
       DATA H,TOL,XL,XU,YL /0.01,0.001,0.0,1.0,0.0/

 1     WRITE(*,*)'         X              Y              ERROR'
       ITER = NINT((XU-XL)/H)

C  Calculate the first four values using Runge-Kutta routine.
       X = XL
       Y = YL
       F1 = F(X,Y)
       WRITE(*,15) X,Y

       CALL SLVDIF1(X,H,Y,YNEW,F)
       X = X + H
       F2 = F(X,YNEW)
       Y = YNEW
       WRITE(*,15) X,YNEW

       CALL SLVDIF1(X,H,Y,YNEW,F)
       X = X + H
       F3 = F(X,YNEW)
       Y = YNEW
       WRITE(*,15) X,YNEW

       CALL SLVDIF1(X,H,Y,YNEW,F)
       X = X + H
       F4 = F(X,YNEW)
       WRITE(*,15) X,YNEW

C  Begin predictor corrector scheme.
```

```
      YC = YNEW
      DO 20 I=4,ITER
         X = X + H
         Y3 = YC

C  Use the predictor formula first.
         YP = Y3+H*(55.*F4-59.*F3+37.*F2-9.*F1)/24.
         F5 = F(X,YP)

C  Now use corrector formula.
         YC = Y3+H*(9.*F5+19.*F4-5.*F3+F2)/24.

C  Calculate the error in the corrected estimate.
         ERR = ABS(19.*(YP-YC)/270.)

C  Check to see if the step size needs to be reduced.
         IF (ERR .GT. TOL) THEN
            WRITE(*,*) 'STEP SIZE BEING REDUCED BY 1/2'
            H = H/2.0
            GOTO 1
         ELSE
            WRITE(*,15) X,YC,ERR
         ENDIF

         F1 = F2
         F2 = F3
         F3 = F4
         F4 = F(X,YC)

 20   CONTINUE
 15   FORMAT(2X,F9.4,2X,F9.4,2X,E11.6)

      END
C______________________________________________________________________
C______________________________________________________________________

      SUBROUTINE SLVDIF1(XI,H,YI,YNEW,F)
C______________________________________________________________________

C  PURPOSE:  A single first order differential equation is solved
C  using the fourth order Runge-Kutta method.  The equation is of the
C  form dy/dx=f(x,y).

C  ARGUMENTS
C   XI      The initial value of the independent variable.
C   H       The incremental step size for x.
C   YI      Initial value of the function.
C   YNEW    Returns the calculated value of the function.
C   F       External function that computes the derivative value.
C   K       Intermediate Kutta values.
C______________________________________________________________________

      REAL F,XI,H,H13,YI,YNEW,K1,K2,K3,K4

      H13 = H/3.0
      K1 = F(XI,  YI)
      K2 = F(XI + H13,  YI + H13*K1)
      K3 = F(XI+ 2.0*H13,  YI - H13*K1 + H*K2)
      K4 = F(XI + H,  YI + H*(K1 - K2 + K3))
      YNEW = YI + H/8.0*( K1 + 3.0*K2 + 3.0*K3 + K4 )
```

```
      RETURN
      END
C_______________________________________________________________________
C_______________________________________________________________________

      FUNCTION F(X,Y)
C_______________________________________________________________________
C PURPOSE:  Routine to compute the value of the function derivative.

C ARGUMENTS:
C  X      independent variable.
C  Y      Array of dependent variables.
C  F      Values of the derivatives.
C_______________________________________________________________________

      REAL X,Y

      F = X+Y

      RETURN
      END
```

Program 7.6F Adams-Brashforth-Moulton Method

```
      PROGRAM ONEEQABM
C_______________________________________________________________________

C  THIS PROGRAM COPYRIGHT: (C) 1991 by Marvin N. Zeigler
C  AUTHORS:  Marvin N. Zeigler and Edward Champion
C  TEXT BOOK:  NUMERICAL METHODS FOR ENGINEERING APPLICATIONS
C  EQUATIONS:  7.25, 7.26, 7.27

C PURPOSE:  The Adams-Brashford-Moulton (Predictor Corrector) method
C  is used to solve the single differential equation dy/dx = x+y from

C  x=0 to x=1 with the initial value y=0.  The exact solution of the
C  differential equation is y = e^x - 1 - x.  A fourth order Runge-
C  Kutta routine is used to determine the first four function values
C  needed to start the ABM method.

C  ARGUMENTS:
C   F      Function that computes the derivative value.
C   H      Step size for the independent variable x.
C   XL,XU  Limits of the x interval.
C   YL     Initial value of the dependant variable y.
C   TOL    Tolerance for determining if step size needs reducing.
C_______________________________________________________________________

      EXTERNAL F
      REAL F,H,XL,YL,TOL
      DATA H,TOL,XL,XU,YL /0.01,0.001,0.0,1.0,0.0/

 1    WRITE(*,*)'         X              Y             ERROR'
      ITER = NINT((XU-XL)/H)

C  Calcualte the first four values using Runge-Kutta routine.
      X = XL
      Y = YL
      F1 = F(X,Y)
      WRITE(*,15) X,Y

      CALL SLVDIF1(X,H,Y,YNEW,F)
      X = X + H
      F2 = F(X,YNEW)
      Y = YNEW
      WRITE(*,15) X,YNEW

      CALL SLVDIF1(X,H,Y,YNEW,F)
      X = X + H
      F3 = F(X,YNEW)
      Y = YNEW
      WRITE(*,15) X,YNEW

      CALL SLVDIF1(X,H,Y,YNEW,F)
      X = X + H
      F4 = F(X,YNEW)
      WRITE(*,15) X,YNEW

C  Begin predictor corrector scheme.
```

```
      YC = YNEW
      DO 20 I=4,ITER
         X = X + H
         Y3 = YC

C  Use the predictor formula first.
         YP = Y3+H*(55.*F4-59.*F3+37.*F2-9.*F1)/24.
         F5 = F(X,YP)

C  Now use corrector formula.
         YC = Y3+H*(9.*F5+19.*F4-5.*F3+F2)/24.

C  Calculate the error in the corrected estimate.
         ERR = ABS(19.*(YP-YC)/270.)

C  Check to see if the step size needs to be reduced.
         IF (ERR .GT. TOL) THEN
            WRITE(*,*) 'STEP SIZE BEING REDUCED BY 1/2'
            H = H/2.0
            GOTO 1
         ELSE
            WRITE(*,15) X,YC,ERR
         ENDIF

         F1 = F2
         F2 = F3
         F3 = F4
         F4 = F(X,YC)

 20   CONTINUE
 15   FORMAT(2X,F9.4,2X,F9.4,2X,E11.6)

      END
C_____________________________________________________________________
C_____________________________________________________________________

      SUBROUTINE SLVDIF1(XI,H,YI,YNEW,F)
C_____________________________________________________________________

C  PURPOSE:  A single first order differential equation is solved
C  using the fourth order Runge-Kutta method.  The equation is of the
C  form dy/dx=f(x,y).

C  ARGUMENTS
C   XI     The initial value of the independent variable.
C   H      The incremental step size for x.
C   YI     Initial value of the function.
C   YNEW   Returns the calculated value of the function.
C   F      External function that computes the derivative value.
C   K      Intermediate Kutta values.
C_____________________________________________________________________

      REAL F,XI,H,H13,YI,YNEW,K1,K2,K3,K4

      H13 = H/3.0
      K1 = F(XI,  YI)
      K2 = F(XI + H13,  YI + H13*K1)
      K3 = F(XI+ 2.0*H13,  YI - H13*K1 + H*K2)
      K4 = F(XI + H,  YI + H*(K1 - K2 + K3))
      YNEW = YI + H/8.0*( K1 + 3.0*K2 + 3.0*K3 + K4 )
```

```
      RETURN
      END
C______________________________________________________________________
C______________________________________________________________________

      FUNCTION F(X,Y)
C______________________________________________________________________
C  PURPOSE:  Routine to compute the value of the function derivative.

C  ARGUMENTS:
C   X      independent variable.
C   Y      Array of dependent variables.
C   F      Values of the derivatives.
C______________________________________________________________________

      REAL X,Y

      F = X+Y

      RETURN
      END
```

Program 7.7F Shooting Method

```
      PROGRAM SHOOT
C_____________________________________________________________________

C THIS PROGRAM COPYRIGHT: (C) 1991 by Marvin N. Zeigler
C AUTHORS:  Marvin N. Zeigler and Edward Champion
C TEXT BOOK:  NUMERICAL METHODS FOR ENGINEERING APPLICATIONS
C EQUATIONS:  7.31

C  PURPOSE:  Program that uses the shooting method to solve the
boundary
C   value problem T''+R(Ta-T)=0.  This second order equation is
C  expressed as two first order equations: dT/dx=Z and dZ/dx=R(T-Ta).
C  A fourth order Runge-Kutta routine is used to solve the equations
C   along the range of x for each initial slope value.

C ARGUMENTS:
C   AA,BB  Limits of dependent variable Y.
C   C1,C2  Initial guesses of the slope and subsequently calculated
C          approximations of the slope.
C   XL,XU  Limits of the x range for evaluation.
C   NS     Number of steps used in the interval XL-XU.
C   H      Step size for x.
C   MAXIT  Maximum number of iterations allowed.
C   TOL    Tolerance value used to determine convergence.
C   R,TA   Constants used in the differential equations.
C_____________________________________________________________________

      REAL AA,BB,XL,XU,R,TA,H,X,Y(2),T(2),C1,C2,F(2),TOL
      INTEGER NS,MAXIT
      COMMON R,TA
      DATA NS,C1,C2,AA,BB,XL,XU /5,10.,25.,40.,200.,0.,10./
      DATA R,TA,MAXIT,TOL /.02,20.,10,0.01/

C  Input values for initial slope and number of interval steps.
      WRITE(*,*) 'ENTER INITIAL SLOPE AND NUMBER OF INTERVALS.'
      READ*, C1,NS

      H = (XU-XL)/NS

C  Use the first slope guess to solve the differential equations.
      X = XL
      Y(1) = AA
      Y(2) = C1

      DO 10 J = 1,NS
C  Solve the set of equations using the Runge-Kutta routine.
         CALL SLVDIF (X,H,Y,T,F,2)
         X = X+H
         Y(1) = T(1)
         Y(2) = T(2)
 10   CONTINUE
      WRITE(*,*)' APPROXIMATION OF T =',Y(1)
      BL = T(1)

C  Now use the second slope guess to solve the differential equations.
      X = XL
      Y(1) = AA
```

```
      Y(2) = C2

      DO 20 J = 1,NS
         CALL SLVDIF (X,H,Y,T,F,2)
         X = X+H
         Y(1) = T(1)
         Y(2) = T(2)
 20   CONTINUE
      WRITE(*,*)' APPROXIMATION OF T =',Y(1)
      BU = T(1)

C  Iterate to improve the approximation for T.
      DO 40 I=1,MAXIT
         X = XL
         Y(1) = AA
C  Approximate the initial slope from the calculated value using
C  linear interpolation.
         Y(2) = C1 + (C2-C1)/(BU-BL)*(BB-BL)
         C1 = C2
         C2 = Y(2)

         DO 30 J = 1,NS
            CALL SLVDIF (X,H,Y,T,F,2)
            X = X+H
            Y(1) = T(1)
            Y(2) = T(2)
 30      CONTINUE
      WRITE(*,*)' APPROXIMATION OF T =',Y(1)

         IF (ABS(T(1)-BB) .LT. TOL) THEN
            WRITE(*,*)' TOLERANCE REACHED IN PROGRAM SHOOT.'
            STOP
         ENDIF

            BL = BU
            BU = T(1)
 40   CONTINUE
      WRITE(*,*)' MAXIMUM ITERATIONS REACHED IN PROGRAM SHOOT.'
      END

C_______________________________________________________________________
C_______________________________________________________________________

      SUBROUTINE SLVDIF(XI,H,YI,YNEW,F,N)
C_______________________________________________________________________

C  PURPOSE:A system of N first order differential equations are solved
C  using the fourth order Runge-Kutta method. The equations are of the
C  form dy1/dx=f1(x,y), dy2/dx=f2(x,y), ..., dyN/dx=fN(x,y).

C  ARGUMENTS
C   XI     The initial value of the independent variable.
C   H      The incremental step size for x.
C   YI     Array of the initial values of the functions.
C   YNEW   Array that returns the calculated values of the functions.
C   F      Array of the function derivatives as shown above.
C   N      Number of equations being solved.
C   DYDXF  A subroutine that computes the values of the N derivatives.
C   K      Arrays of the intermediate Kutta values.
C_______________________________________________________________________
```

```
      REAL XI,H,H13,YI(N),YNEW(N),F(N)
       REAL K1(10),K2(10),K3(10),K4(10)

       H13 = H/3.0

      CALL DYDXF (XI,YI,F)
      DO 10 I=1,N
       K1(I) = F(I)
       YNEW (I) = YI(I) + H13*K1(I)
 10   CONTINUE

      CALL DYDXF (XI+H13,YNEW,F)
      DO 20 I=1,N
       K2(I) = F(I)
       YNEW (I) = YI(I) - H13*K1(I) + H*K2(I)
 20   CONTINUE

      CALL DYDXF (XI+2.0*H13,YNEW,F)
      DO 30 I=1,N
       K3(I) = F(I)
       YNEW (I) = YI(I) + H * (K1(I) - K2(I) + K3(I))
 30   CONTINUE

      CALL DYDXF (XI+H,YNEW,F)
      DO 40 I=1,N
       K4(I) = F(I)
 40   CONTINUE

      DO 50 I=1,N
       YNEW(I) = YI(I) + H/8.0*( K1(I) + 3.0*K2(I) +
     +           3.0*K3(I) + K4(I) )
 50   CONTINUE
      RETURN
      END

C______________________________________________________________________
C______________________________________________________________________

       SUBROUTINE DYDXF (X,Y,F)
C______________________________________________________________________
C    PURPOSE:    Routine to compute the values of the function
derivatives.

C  ARGUMENTS:
C   X       independent variable.
C   Y       Array of dependent variables.
C   F       Values of the derivatives.
C______________________________________________________________________

       REAL X,Y(2),F(2)
       COMMON R,TA
      DATA HP,TA /0.01,20.0/

      F(1) = Y(2)
      F(2) = HP*(Y(1)-TA)

       RETURN
       END
```

8

Numerical Solutions of Partial Differential Equations

8.1 Introduction

A partial differential equation of order n is a functional equation that contains at least one partial derivative of the unknown function of two or more variables. The order of a partial differential equation is the order of the highest derivative.

A linear partial differential equation is one in which the coefficients of the function and derivatives of the function do not depend on the function or any of its derivatives. The general form of the equation is given as follows with the subscripts of the function representing the partial derivative.

$$A\Phi_{xx}+B\Phi_{xy}+C\Phi_{yy}+D\Phi_{x}+E\Phi_{y}+F\Phi=G \tag{8.1}$$

The equations are classified further as

1. Elliptic if $B^2 - 4AC < 0$
2. Hyperbolic if $B^2 - 4AC > 0$
3. Parabolic if $B^2 - 4AC = 0$.

Elliptic equations are of a steady-state type with conditions that require phi and/or its normal derivatives to have specific values on the boundary and the boundary specifies the region of application. Parabolic and hyperbolic equations are of the non-steady state type. Their solutions are allowed to propagate in the open region of interest.

These types of equations can be solved in many ways such as separation of variables, the Laplace transform method, the complex variable method, finite-difference methods, and finite element methods. Since this text is concerned with the numerical solution of equations and numerical methods only the finite-difference and the finite element methods will be discussed.

Some of the basic techniques needed in the formulation of finite-difference representations are introduced. In the finite-difference approach, the problem domain is discretized so that the dependent variables are considered to exist only at discrete points. Derivatives are approximated by differences resulting in an algebraic representation of the partial differential equation (PDE). The nature of the resulting system of algebraic equations depends on the character of the problem posed by the original PDE (or system of PDEs). Partial differential equations in two independent variables can be classified as elliptic, parabolic, or hyperbolic. Each class has distinguishing features. The Laplace and Poisson equations, which govern many important physical processes are examples of elliptic PDEs. More specifically, Laplace's equation governs the steady-state temperature distribution in a homogeneous solid. The unsteady heat conduction (or "diffusion") equation is an example of a parabolic PDE, and the wave equation is an example of a hyperbolic PDE. Each of these equations will be discussed in the material to follow. Elliptic PDEs are said to govern "equilibrium" problems, and both parabolic and hyperbolic equations govern "marching" problems.

Equilibrium problems occur on a closed domain, and the solution must meet prescribed conditions on all the boundaries. A single solution is sought that satisfies both the PDE and the boundary conditions. In contrast, marching problems arise on open domains; at least, the solution cannot be forced to meet specific conditions at more than one value of the timelike variable. Initial conditions must be specified for marching problems, but the other end of the time interval is open. The solution is " marched" in time from the initial conditions. For problems occurring in a region (physical space) of finite extent, boundary conditions must be specified. For some problems for which the extent of the region and boundary conditions are fixed, a time-asymptotic (steady) solution is approached at large values of the timelike variable.

The distinction as to elliptic, parabolic, or hyperbolic for second-order PDEs depends on the coefficients of the second derivative terms as discussed previously.

In some cases it is possible to classify systems of equations. Procedures for doing this are discussed in [52]. Some systems are mixed in classification, exhibiting behavior of more than one type. Examples of such mixed systems appear in subsequent chapters of this handbook. Simple examples of hyperbolic, parabolic, and elliptic PDEs occurring in heat transfer applications follow.

The finite-difference method replaces the region of interest by a specific grid and approximates the governing equations by difference relations in terms of the unknown variable. The difference equations are related to the grid. From this a set of algebraic equations is developed and the solution of the set of equations lends itself to techniques developed earlier in the book.

The finite element method draws upon many of the ideas presented earlier in the text as far as numerical techniques are concerned. Only a brief description of the finite element method will be given in the text that follows. Chapter 9 illustrates the coding of a finite element application.

8.2 Partial Differential Equations Encountered in Practice

The following differential equations are given to illustrate the types of equations often encountered by scientists and engineers. Though not exhaustive, the equations do give examples of the three basic types of equations.

8.2.1 Unsteady Heat Conduction

One dimensional heat conduction in a rod and variable in time is given by the equation

$$\frac{\partial\left(k\frac{\partial T}{\partial x}\right)}{\partial x} = \rho c_p \frac{\partial T}{\partial t} \tag{8.2}$$

where T is the temperature, k is the material thermal conductivity, ρ is the material density, and c_p is the material specific heat.

8.2.2 Vorticity Transport Equation

For an incompressible fluid in two-dimensional motion, the vorticity behaves according to the following equation:

$$\frac{\partial \omega}{\partial t} + u\frac{\partial \omega}{\partial x} + v\frac{\partial \omega}{\partial y} = \nu \nabla^2 \omega \tag{8.3}$$

In this equation u and v are the x and y component velocities and ν is the kinematic velocity.

8.2.3 Poisson's Equation

A standard notation for Poisson's equation is

$$\nabla^2 \psi = -\omega \tag{8.4}$$

This equation is a field equation and is found in areas such as fluid dynamics. Here ψ is the stream function and ω is the vorticity. For electrostatics, ψ is the electrical potential and ω is the ratio of charge density to the dielectric constant. In the study of elasticity, with $\omega=2$, ψ is a function from which the angle of twist of a cylinder in torsion can be calculated.

8.2.4 Laminar Flow Heat Exchanger Equation

For steady laminar flow in a cylindrical heat exchanger the following equation is applicable:

$$\frac{\partial^2 T}{\partial r^2} + \frac{1}{r}\frac{\partial T}{\partial r} + \frac{\partial^2 T}{\partial z^2} = \frac{\rho u c_p}{k}\frac{\partial T}{\partial z} \tag{8.5}$$

For the above equation, ρ, k, and c_p are the fluid density, fluid thermal conductivity, and fluid specific heat, respectively. The quantity u is the axial velocity and is a function of the radius, r.

8.2.5 Telephone Equation

To predict the voltage variation along a transmission cable, the following equation is used:

$$\frac{\partial^2 V}{\partial x^2} = LC\frac{\partial^2 V}{\partial t^2} + (RC + GL)\frac{\partial V}{\partial t} + RGV \tag{8.6}$$

In this equation, R, L, C, and G represent the resistance, inductance, capacitance, and conductance to ground per unit length of cable.

8.2.6 Wave Equation

The angle of twist, ϕ, at any section of a circular shaft undergoing torsional vibrations is governed by the following equation:

$$\frac{\partial^2 \phi}{\partial t^2} = \frac{G}{\rho} \frac{\partial^2 \phi}{\partial x^2} \tag{8.7}$$

G is the modulus of rigidity of the shaft and ρ is the shaft material density.

8.2.7 Vibrating Beam Equation

For a vibrating beam, the transverse deflection y can be determined by the following equation:

$$\frac{\partial^2 y}{\partial t^2} + \frac{EI}{A\rho} \frac{\partial^4 y}{\partial x^4} = 0 \tag{8.8}$$

In the above equation E, I, A, and ρ are the modulus of elasticity, cross-sectional moment of inertia, cross-sectional area, and density, respectively.

The following subsections of this chapter deal with initial value problems and boundary value problems. It is not a complete discourse on the topic and it is intended to give to the reader general information for solutions to the problems.

8.3 Derivative Approximations by Finite Differences

As a recap for derivative approximations given in the earlier chapters, consider the grid layout given in Figure 8-1.

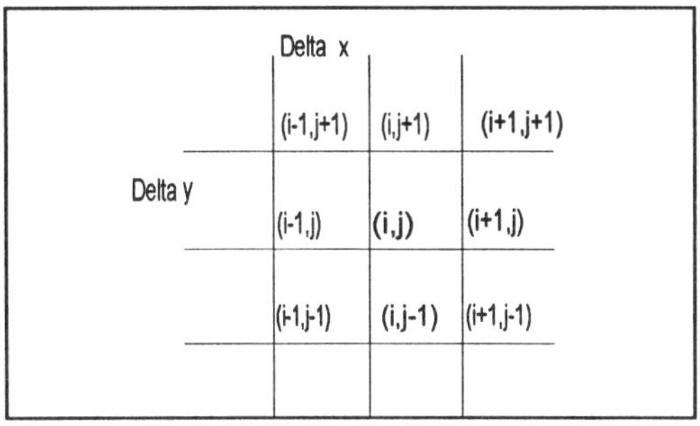

Figure 8-1: Grid Point Definition for Finite Differences

Expanding in a Taylor's series for the points $u_{i-1,j}$ and $u_{i+1,j}$ about the central point $u_{i,j}$ and manipulating the equations by addition or subtraction gives the following relations:

$$\frac{\partial u}{\partial x} = \frac{u_{i+1,j} - u_{i,j}}{\Delta x} \tag{8.9}$$

$$\frac{\partial u}{\partial x} = \frac{u_{i,j} - u_{i-1,j}}{\Delta x} \tag{8.10}$$

$$\frac{\partial u}{\partial x} = \frac{u_{i+1} - u_{i-1}}{2\Delta x} \tag{8.11}$$

The above formulas are known as the forward, backward, and central difference formulas, respectively. Further derivation leads to the following formula:

$$\frac{\partial^2 u}{\partial x^2} = \frac{u_{i-1,j} - 2u_{ij} + u_{i+1,j}}{(\Delta x^2)} \quad (8.12)$$

$$\frac{\partial^2 u}{\partial x \partial y} = \frac{u_{i+1,j+1} - u_{i-1,j+1} - u_{i+1,j-1} + u_{i-1,j-1}}{4\Delta x \Delta y} \quad (8.13)$$

There may be notation where derivatives will be represented by $u_x \equiv \partial u/\partial x$, etc.

8.4 Solution of a Simple Parabolic Differential Equation–Explicit Method

To illustrate the concepts in the solution of the types of differential equations given above, examples from the engineering field will be given. For the parabolic differential equation solution, consider an insulated bar with an initial temperature distribution at time t=0 having ends that are subsequently maintained at temperatures that may be functions of time. Because the parabolic partial differential equations represent time-dependent diffusion processes, we usually let t and x be the independent variables in which t is time and x is the one-dimensional space coordinate.

The equation describing this problem is given by

$$\frac{\partial u}{\partial t} = \frac{\partial^2 u}{\partial x^2} \quad (8.14)$$

for all x between 0 and 1 and for all t between 0 and T.

The initial and boundary conditions are given by

$$\begin{aligned} u(x,0) &= f(x) \\ u(0,t) &= g_0(t). \\ u(1,t) &= g_1(t) \end{aligned} \quad (8.15)$$

To approximate the solution of (8.15), a network of grid points is superimposed over the region of interest for x and t. For any grid point that is not on the boundaries of x or at t=0, the partial derivatives are replaced by the finite difference approximations:

$$\frac{u_{i,n+1}-u_{i,n}}{\Delta t} = \frac{u_{i-1,n}-2u_{i,n}+u_{i+1,n}}{(\Delta x)^2}. \tag{8.16}$$

The values of $u_{i,n}$ are known at any time level t_n. This means that the above equation allows $u_{i,n+1}$ to be calculated from the previously calculated values. This is an *explicit* method of solving the equation along with the appropriate boundary conditions.

By letting $\lambda=(\Delta t)/(\Delta x)^2$ it can be shown that to insure convergence, the value of λ has to be between 0 and ½ and can be equal to ½. The error for this approximation is on the order of (Δx^2).

The subroutine is given at the end of the chapter in Program 8.1.

8.5 Solution of Parabolic Equations Implicitly

The explicit method depends only upon previously calculated values. This fact makes this method easy to implement. However, the restriction that λ have a value between 0 and ½ places restrictions on the time increment. For problems needing to be solved over large values of time, computational effort is intense. The implicit method masters these problems while sacrificing simplicity in implementation. This method uses the representation of the u_{xx} term evaluated at an advanced time, t_{n+1} instead of at t_n. The governing equation becomes

$$\frac{u_{i,n+1}-u_{i,n}}{\Delta t}=\frac{u_{i-1,n+1}-2u_{i,n+1}+u_{i+1,n+1}}{(\Delta x)^2}. \tag{8.17}$$

The boundary conditions of the explicit method hold. At any given time level, the above equation will be written for each grid point. A system of M-1 simultaneous equations is developed for the M-1 unknowns. The implicit method for solving the partial differential equation converges to a solution regardless of the value of the ratio $\Delta t/(\Delta x)^2$.

Let $\lambda=\Delta t/(\Delta x)^2$. At each time step there are M-1 linear equations to be solved. These are of the form

$$(1+2\lambda)u_{1,n+1}-\lambda u_{2,n+1}=u_{1,n}+\lambda g_0(t_{n+1}) \tag{8.18}$$

$$-\lambda u_{i-1,n+1}+(1+2\lambda)u_{i,n+1}-\lambda u_{i+1,n+1} = u_{i,n} \quad \textit{for } 2\le i\le M-2 \tag{8.19}$$

$$-\lambda u_{M-2,n+1}+(1+2\lambda)u_{M-1,n+1}=u_{M-1,n}+\lambda g_1(t_{n+1}) \tag{8.20}$$

Letting N=M-1, dropping the subscripts (n+1) and letting the right-hand side of the above equations (all known values) be represented by d_i, the above equations can be replaced by the following equation:

$$
\begin{aligned}
b_1u_1 + c_1u_2 &= d_1 \\
a_iu_{i-1} + b_iu_i + c_iu_{i+1} &= d_i \\
a_Nu_{N-1} + b_Nu_N &= d_N
\end{aligned}
\tag{8.21}
$$

The matrix of coefficients a,b, and c is called a tridiagonal matrix and can be solved by any number of ways such as Gaussian elimination or by the Gauss-Seidel method. Without derivation, a recursive solution of the form can be developed for computer implementation. This is summarized in the following equations.

$$u_N = \frac{d_N - a_N\gamma_{N-1}}{b_N - \dfrac{a_Nc_{N-1}}{\beta_{N-1}}} = \gamma_N. \tag{8.22}$$

$$u_i=\gamma_i-\frac{c_iu_{i+1}}{\beta_i} \ for \ i=N-1,N-2,...,1 \tag{8.23}$$

and the β's and γ's are determined from the following:

$$
\begin{aligned}
\beta_1 &= b_1 \\
\gamma_1 &= \frac{d_1}{\beta_1} \\
\beta_i &= b_i - \frac{a_i c_{i-1}}{\beta_{i-1}} \\
\gamma_i &= \frac{d_i - a_i \gamma_{i-1}}{\beta_i}
\end{aligned}
\tag{8.24}
$$

for i=2,3,....,N.

The program is given as Program 8.2 at the end of the chapter.

8.6 Stability of a Finite-Difference Equation

Given certain conditions, the finite-difference forms of equations are convergent. This means that the approximation of the equation will converge to the exact solution as the space and time grids on which the equations are solved tend to zero. If a finite-difference equation algorithm is convergent, then the solution must be consistent and stable. For a solution to be consistent means that the solution is for that particular equation and not the solution to some other equation.

The stability of a solution procedure is a property of the finite-difference equation used as the time increment is made vanishingly small. There is an upper limit to the extent to which any variable or boundary condition can be amplified in the calculations due to any error in the calculation. Stability implies that there is a bound to the solution.

The method generally preferred and most often used for stability analysis of finite-difference solution schemes is called the von Neumann method. This analysis is considered local in that the coefficients of the difference equations are varying slowly enough to be considered constant in space and in time. The independent solutions of the resulting difference equations are of the form

$$u_j^n = \xi^n e^{ijk\Delta x} \tag{8.25}$$

In the above equation, k is a real spatial wave number and $\xi=\xi(k)$ is a complex number that depends on k. It is noted that the time dependence of a single solution value is the successive integer powers of the complex number ξ. Therefore the difference equations are unstable (ie have exponentially growing modes) is

$$|\xi(k)|>1 \tag{8.26}$$

for some k. The number ξ is called the amplification factor. $\xi(k)$ is determined by substituting (8.26) into any of the difference equations. For example, to obtain the equation for ξ for a forward time-center spaced algorithm, the result would be

$$\xi(k) = 1 - i\frac{v\Delta t}{\Delta x}\sin\, k\Delta x. \tag{8.27}$$

This scheme is unconditionally unstable since the value of ξ is always greater than 1 for all k.

All algorithms for any partial differential should be investigated for the most important concept of stability.

For marching problems, the errors will not grow if the solution scheme is stable. A stable scheme is thus defined as one for which errors from any source are not permitted to grow in the sequence of numerical procedures as the calculation proceeds from one marching step to the next. Thus, it is seen that an acceptable finite-difference representation for a marching problem must meet the conditions of consistency and stability. A scheme meeting these requirements for a marching problem is generally found to be convergent. Convergence here means that the solution to the finite-difference equation approaches the true solution to the PDE having the same initial and boundary conditions as the mesh is refined. Perhaps a more descriptive terminology for this type of convergence is "truncation convergence," since the word convergence is used in other contexts, as in "iteration convergence." It has been shown and proven that given a properly posed initial value problem (governed by a linear PDE) and a finite-difference approximation to it that satisfies the consistency condition, stability is the necessary and sufficient condition for convergence. This important result is known as Lax's equivalence theorem. Although the theorem has been proved only for linear PDEs, computational work generally proceeds as though it were applicable also to nonlinear PDEs. Stability will be considered in more detail in the following sections. Consistency is also important for equilibrium problems. The solution to equilibrium problems (elliptic PDEs) leads to a system of simultaneous algebraic equations that has to be solved only once rather than in a marching manner. Thus, the concept of stability developed previously is not directly applicable. To achieve truncation convergence for equilibrium problems, it would seem that it is only necessary to devise a consistent solution scheme in which the error in solving the simultaneous algebraic equations can be controlled as the mesh size is refined without limit. When the algebraic equations are solved in an iterative fashion for equilibrium problems, there is some resemblance between the sequence of iterative solutions and the time-dependent sequence of solutions obtained in marching problems. Because of this resemblance, ideas from the stability analysis for marching problems are sometimes used to establish the convergence properties of schemes for equilibrium problems.

Although it is important to use a convergent algorithm in solving a PDE numerically, it is never possible to refine the mesh size to zero. Thus, calculations must be made on a finite grid, and there are errors in the numerical solution. These errors in the solution are referred to as discretization errors. More specifically, discretization error is the error in the solution to the PDE caused by replacing the continuous problem by a discrete one and is defined as the difference between the exact solution of the PDE (roundoff free) and the exact solution of the finite-difference equations (roundoff free) Thus, the difference between the exact solution of the PDE and the computer solution is equal to the sum of the discretization error and the roundoff error associated with the finite-difference calculations. It can also be observed that the discretization error is the error in the solution caused by the truncation error in the difference representation of the PDE plus any error induced by treatment of boundary conditions.

There are several common methods for solutions of the PDE's that are stable and convergent. When the fully explicit and implicit representations of the heat diffusion term are averaged, the scheme suggested by Crank and Nicolson results in

$$\frac{u^{t+1}_i - u^t_i}{\Delta t} = \alpha \frac{\delta^2_x u^t_i + \delta^2_x u^{t+1}_i}{2(\Delta x)^2} \tag{8.28}$$

where the operator notation used earlier is used. The averaging results in an algorithm whose truncation error is on the order of the sum of $(\Delta t)^2$ and $(\Delta x)^2$. This implicit scheme is also unconditionally stable and can be solved by the tridiagonal matrix algorithm.

A combined explicit implicit representation of which the Crank-Nicolson method [26] is a special case is given by the following with θ a weighting factor with values between 0 and 1.

$$\frac{u^{t+1}_i - u^t_i}{\Delta t} + \alpha \frac{\Theta\delta^2_x u^{t+1}_i + (1-\Theta)\delta^2_x u^t_i}{(\Delta x)^2} \tag{8.29}$$

The use of a central time representation in the simple explicit scheme as suggested by Richardson gives an unstable algorithm. The unstable Richardson method can be made stable by replacing the term $(u^t)_i$ in the diffusion term with the time averaged expression $((u^{t+1})_i + (u^{t-1})_i)/2$. The resulting explicit, three-time level expression is

$$\frac{u^{t+1}_i - u^{t-1}_i}{2\Delta t} = \alpha \frac{u^t_{i+1} - u^{t+1}_i - u^{t-1}_i + u^t_{i-1}}{(\Delta x)^2} \tag{8.30}$$

and was first proposed by DuFort and Frankel [34]. The von Neumann analysis applied to this scheme reveals that it is unconditionally stable.

The scheme proposed by Barakat and Clark [7] is representative of a class of methods known as Alternating Direction Explicit (ADE) methods. The two-step scheme involves the simultaneous marching of two approximate solutions $(p^{t+1})_i$ and $(q^{t+1})_i$. The two solutions are advanced separately and at each time level the solution to the difference equation is considered to be the average of p and q. For the heat diffusion equation, the algorithm is given by

$$\begin{aligned} \frac{p^{t+1}_i - p^t_i}{\Delta t} &= \alpha \frac{p^{t+1}_{i-1} - p^{t+1}_i - p^t_i + p^t_{i+1}}{(\Delta x)^2} \\ \frac{q^{t+1}_i - q^t_i}{\Delta t} &= \alpha \frac{q^t_{n-1} - q^t_i - q^{t+1}_i + q^{t+1}_{i+1}}{(\Delta x)^2} \\ u^{t+1}_i &= 1/2(p^{t+1}_i + q^{t+1}_i \end{aligned} \tag{8.31}$$

The method is unconditionally stable for the heat diffusion equation.

8.7 Multidimensional Parabolic Equations

The extension of the above discussions into two or three dimensions is quite straightforward, especially if you are using a rectangular region. Consider the equation for non-steady heat flow in two dimensions:

$$\frac{\partial u}{\partial t} = \frac{k}{cp}\left(\frac{\partial^2 u}{\partial x^2} + \frac{\partial^2 u}{\partial y^2}\right) \tag{8.32}$$

By letting $\Delta x = \Delta y$ and letting $\beta = k\Delta t/cp(\Delta x)^2$, the explicit algorithm results in the following:

$$\begin{aligned} u^{t+1}_{i,j} &= \beta(u^{t}_{i+1,j} + u^{t}_{i-1,j} + u^{t}_{i,j+1} + u^{t}_{i,j-1}) + \\ &\cdots + (1-4\beta)u^{t}_{i,j} \end{aligned} \tag{8.33}$$

The maximum value for β given the constant end conditions is ¼ and for a more general case of $\Delta x \neq \Delta y$,

$$\frac{k\Delta t}{cp[(\Delta x)^2 + (\Delta y)^2]} \leq 1/8 \tag{8.34}$$

For three dimensions the above quantity is $\leq 1/6$.

In using the explicit scheme, the restriction of Δt can make the calculation procedure long. It is best to use an implicit scheme to minimize the restrictions on Δt. An implicit scheme can be developed by using the averages of central difference approximations to give $\partial^2 u/\partial x^2$ and $\partial^2 u/\partial y^2$ at the time averaged position. For the above equation, this is

$$u_{i,j}^{t+1} - u_{i,j}^{t} = \frac{\beta}{2}[u_{i+1,j}^{t+1} - 2u_{i,j}^{t+1} + u_{i-1,j}^{t+1} + u_{i+1,j}^{t} - 2u_{i,j}^{t} + u_{i-1,j}^{t} + u_{i,j+1}^{t+1} - 2u_{i,j}^{t+1} + u_{i,j-1}^{t+1} + u_{i,j+1}^{t} - 2u_{i,j}^{t} + u_{i,j-1}^{t}] \tag{8.35}$$

The β is proportional to the ratio of Δt and the sum of Δx^2 and Δy^2.

The implicit alternating-direction method (ADI) as presented by Peaceman and Rachford [63] and Douglas [33] avoids the cumbersome calculational requirements of the above equation. The algorithm employs two difference equations which are used in turn over successive time steps each of duration Δt/2. The first equation is implicit only in the x-direction and the second equation is implicit only in the y direction. The advantage of a tridiagonal matrix is retained in the ADI method. Let u^* be the intermediate value at the end of the first time step and let the central difference operator be denoted as given earlier in the book by δ, the equations for the ADI method can be written as

$$\frac{u^{*}_{i,j} - u^{t}_{i,j}}{\Delta t/2} = \delta_x^2 u^{*}_{i,j} + \delta_y^2 u^{t}_{i,j} \tag{8.36}$$

and

$$\frac{u^{t+1}_{i,j} - u_{i,j}^{*}}{\Delta t/2} = \delta_x^2 u^{*}_{i,j} + \delta_y^2 u^{t+1}_{i,j}. \tag{8.37}$$

The stability of the procedure is investigated using the technique as described above. Without proof, the ADI method given above is unconditionally stable for any value of Δt.

The example program is given at the end of the chapter in Program 8.3.

8.8 Boundary Conditions

The formulation of a problem requiring the solution of a partial differential equation also requires the specification of appropriate boundary or initial conditions. The term "initial" refers to the fact that in many physical problems governed by hyperbolic or parabolic PDEs, the "marching" coordinate is time. Specification of the dependent variable or its time derivative at time zero is referred to as an initial condition. For physical problems, the term "boundary" literally means on the physical boundary of the region in space in which the solution is sought.

The three most common boundary conditions are:

1. Value of the dependent variable specified on the boundary. This boundary condition is known as a Dirichlet condition or a boundary condition of the first kind.

2. Normal derivative specified on the boundary. This condition is known as a Neumann condition or a boundary condition of the second kind.

3. A linear combination of conditions 1 and 2; for example, au + b(du/dn) specified on the boundary. Convective boundary conditions in heat transfer are of this type. This condition is known as a Robbins condition, a mixed condition, or a boundary condition of the third kind.

Other boundary conditions may be encountered in heat transfer such as a radiation flux condition or an interface condition in a conduction problem involving materials of different thermal conductivities. However, the problems that occur most frequently involve the three conditions listed above.

A more complete discussion of the mathematical requirements of "appropriate" boundary conditions can be found in any of the references on partial differential equations. Generally, however, for elliptic equations, either Dirichlet or Neumann conditions are specified on all the boundaries. For parabolic equations, Dirichlet or Neumann conditions are specified on the space boundaries, and a distribution of the dependent variable is given for an "initial" value of the marching coordinate. For second-order hyperbolic equations on a finite domain, Dirichlet or Neumann conditions are usually specified on the space boundaries, and a distribution of the dependent variable and its first derivative with respect to the marching (time or timelike) coordinate is given for an "initial" value of the marching coordinate.

8.9 Elliptic Partial Differential Equations

All of the examples that normally are presented in discussing elliptic partial differential equations illustrate the fact that coefficient matrix is sparse when the finite-difference method is used to solve the equations. Especially in a three-dimensional case, the number of nonzero coefficients is a small fraction of the total.

The relative sparseness increases as the number of equations increases. For example a 21 x 21 coefficient matrix has 81% of its positions filled with zeros. If symmetry were taken into account, there would have been a 14 x 14 coefficient matrix with 71% zeros. Decreasing the value of h (the spacing) so as to have 105 points within the region gives rise to a coefficient matrix with 96% zeros. For a 30 x 30 x 30 three-dimensional case, there would be 90,176 nonzero coefficients, but this is only 0.012% of the 729E6 values in the coefficient matrix!

Earlier in the book it was shown that iterative methods are usually preferred for sparse matrices unless they have a tridiagonal structure. The reason for this is that elimination does not preserve the sparseness (unless the matrix is tridiagonal), so that one cannot work only with nonzero terms. Iterative methods such as the relaxation methods (SOR) are simple to program and it is easy to optimize the iteration parameter for optimum convergence. However, as the grid size grows, the convergence rate becomes slower. The other type of iterative method, the ADI method, requires more programming expertise and for some problems it may be slower than SOR. It does however have a speed advantage over the SOR method for a large class of problems. The successive relaxation methods include the Jacobi-iterative method, the Gauss-Seidel method, and the SOR methods. The SOR methods are well-known iterative methods.

Frequently the coefficient matrix has a band structure. A band structure is a goal to strive for. The ordering of the points must be carefully done to attain this.

The best of the band structures is tridiagonal, with corresponding economy of storage and speed of solution, as discussed earlier. A method for steady-state heat equation called the alternating-direction-implicit (A D I) method results in tridiagonal matrices and is of growing popularity. It was initially developed for unsteady-state problems as discussed previously.

The relations are implicit but, by properly ordering the equations, we obtain tridiagonal coefficient matrices. This requires that we proceed down each column of points in turn, then processing the equations row-wise. (This is the reason that the method is called the alternating direction implicit method).

The iterations begin with an initial estimate of the u vector. The iterations are terminated when successive calculations agree within a given tolerance. Since there is a bias in the calculated results in the first, third, fifth,... iterates, one usually ignores these and records the results of only the even-numbered iterations.

Except for the additional work of recalculating right hand sides, this method has a minimum of computational effort, equivalent to S.O.R.(successive over relaxation), but converging faster. The coefficient matrices are always the same, so the reduction step need be made only once, if one uses a L decomposition procedure or equivalent.

Since Laplace's equation is a special case of Poisson's equation, two computer programs (one a subroutine to be used with previous subroutines) are given for two dimensions. They can be applied Laplace's equation by defining f(x, y) = 0.

The first is a subroutine that forms the coefficient matrix for Laplacian on a rectangular region. The parameters passed to the subroutine COEFFMTX include the matrix A into which the coefficients are placed, NDIM, first dimension of matrix A in the calling program, and N and M, the number of intervals in the x- and y-directions. Only nonzero values are inserted, so the calling program must first set all the values of A to zero. The subroutine returns arrays.

This routine is used in connection with a linear equation solver such as the subroutines given earlier in the book; the main program must establish the right hand side values based on the boundary conditions and f(x, y) values, combining them with the matrix produced by COEFFMTX and sending them to the linear equation solver to obtain the solution vector, which will then be the calculated steady-state potential values at each node of the grid. The difficulty with this technique is large memory requirements that may be needed to store the values of matrix A. In general, one will need to do an L decomposition of the coefficient matrix, then use a method such as Gauss elimination with partial pivoting. The coefficient matrix will be in the form of a tridiagonal matrix and can be solved by the subroutine TRIDG given previously.

Program 8.4 at the end of the chapter illustrates a program to form the coefficient matrix for the Laplacian on a rectangular region.

The Program 8.5 illustrates how to solve Poisson's equation by the finite difference method and the program will be written in BASIC. Without going into the derivation the equation for steady-state, two-dimensional heat conduction with constant thermal conductivity is given as

$$\frac{\partial^2\theta}{\partial\bar{x}^2} + \frac{\partial^2\theta}{\partial\bar{y}^2} + \frac{q'''l^2}{k(T_0-T_1)} = 0 \tag{8.38}$$

where the "bar" indicates a dimensionless variable, l is a reference length. When the heat generation term is zero, equation (8.38) reduces to the Laplace equation.

When the nodes are located on the surface, the equations must account for the energy crossing the surface as well as the conduction to the surface from within the solid. Boundary conditions at the surface may specify constant temperature, insulation, convection, radiation, or any combination thereof. The solution to steady-state, two-dimensional heat transfer requires the determination of nodal temperatures in a defined finite difference grid. The temperature of each node must satisfy the appropriate nodal equation. For interior nodes an equation of the following form is used:

$$\theta_{i,j} = \frac{\theta_{i,j-1}+\theta_{i-1,j}+\theta_{i,j+1}+\theta_{i+1,j}+q'''\delta\bar{x}^2/k}{4} \tag{8.39}$$

For this problem it is assumed that there exists no convection, radiation, and that the walls have adiabatic boundary conditions. The program assumes a 10 x 5 grid with the rows and columns denoted by i and j, respectively. A temperature gradient is applied across the region.

The program first determines if a node is a surface node or an interior node. If J is not equal to one, left-surface nodal equations are not required and the program jumps from Statement 440 to 500. Then at 500 if J is not equal to five, right surface nodal equations are not required, and the program jumps to Statement 560 where the non-dimensional interior nodal equation is applied to the node. Throughout the calculations, X(I, J) represents the current calculated values in the nodal matrix, T(I,J).

Statement 570 defines Y(I, J), which is the difference between the current nodal value and the previously calculated nodal value. With the exception of the isothermal surface nodes, the values of all nodes are recalculated during each successive iteration.

The value of Y(I, J) indicates the convergence of the calculations. As the calculations proceed, the difference between the newly calculated nodal values and the previously calculated values becomes smaller, and the correct temperature distribution is approached. The allowable difference is a function of the accuracy required in the calculations. This allowable difference is determined by the programmer and is read into the program as variable E1 in Statement 210. Statement 580 determines whether the convergence criterion has been met for each node. If not, the value of U is increased in increments of one for each node that does not satisfactorily meet the convergence criterion. Then the calculated nodal values of X(I, J) are defined as the previously calculated nodal values T(T, J) in Statement 600 for use in the next iteration.

When a surface node is encountered, a different logical path is followed. If J=I at Statement 440, then Statement 450 determines if the node falls on an adiabatic surface or on an isothermal surface. If the number of adiabatic nodes on the left surface denoted by I1 is greater than or equal to the row value I, then the non-dimensional adiabatic nodal equation is used in Statement 480. If I1 is less than I, the node is assigned the isothermal value Tl in Statement 460. A similar determination is made for the nodes on the right face (J = 5) in Statements 500 to 550. The nondimensional nodal equation for the right vertical surface appears in Statement 540.

After one complete iteration of all 40 nodes, the value of N is increased by an increment in Statement 630 to account for the number of iterations completed. If convergence has not been achieved after a maximum number of iterations (M=50), provision is made to terminate the iteration procedure in Statement 640. The user defines the desired value in Statement 200. If a solution has been obtained, the iteration is terminated by Statement 640 which is satisfied when all nodes fall within the convergence criterion set in Statement 580. Statement 652 is controlled by the print variable P. The value of P is initially zero as defined in Statement 202 which allows the print Statements 654- to 656 to be accomplished after the first iteration only. The program concludes by printing out the final temperature matrix along with parameters I1, I2, and N1.

8.10 Hyperbolic Differential Equations

The third classification of partial differential equations, hyperbolic differential equations, includes the "wave equation" which is fundamental to the study of vibrating systems. They are also involved in transport problems (diffusion of matter,. neutron diffusion, and radiation transfer), wave mechanics, gas dynamics, supersonic flow, and other important areas. The technology in most of these applications is so complex that their study is beyond our scope of this book. This section will concentrate on the simple situation, that of vibrations of a string, held at fixed ends.

Consider an elastic string stretched between two fixed endpoints and set to vibrating by plucking it with a finger. A number of simplifying assumptions are made: the string is perfectly elastic, neglect gravitational forces, so that the only force is the tension force in the direction of the string; the string is uniform in density and thickness and has a weight per unit length of w lb/ft; the lateral displacement of the string is so small the tension can be considered to be a constant value of T_{lb}; and the slope of string is small enough that sin α = tan α, where α is the angle of inclination.

We take x = O at the left end. L is the total length of the string. We are interested in how y, the lateral displacement varies with time t and with distance x along the string.

The forces acting in the y-direction are the vertical components of the two tensions. Take the upward direction as positive and use Newton's law to equate the force to mass times the acceleration results in the one-dimensional wave equation

$$\frac{\partial^2 y}{\partial t^2} = \frac{Tg}{w}\frac{\partial^2 y}{\partial x^2} \tag{8.40}$$

The conditions imposed on the solution are the end conditions $[ay+b(\partial y/\partial x) = f(t)]$ at x = O and x = L the initial conditions at t = O. Initial conditions specifying both y = f(x) and the velocity $\partial y/\partial t$ = g(x) are usual for this problem. The problem of solving the one-dimensional wave equation using the finite-difference approach is by replacing derivatives by difference quotients. We use superscripts to denote time, and subscripts for position:

$$\frac{y^{t+1}_i - 2y^t_i + y^{t-1}_i}{(\Delta t)^2} = \frac{Tg}{w}\frac{y^t_{i+1} - 2y^t_i + y^t_{i-1}}{(\Delta x)^2} \qquad (8.41)$$

Note that selecting the ratio $Tg(\Delta t)^2/w(\Delta x)^2 = 1$ gives some simplification. Equation (8.41) is the usual way that the one-dimensional wave equation is solved numerically. We are, of course, interested in the validity of choosing the ratio at unity. the finite-difference replacements to the derivatives do not have mixed error terms, as they did in the explicit method for the heat equation. If the ratio is greater than one, we cannot be sure of convergence. Stability also sets a limit of unity to the ratio. It is surprising to find that if one uses a value of the ratio of less than one, the results are less accurate, while with the ratio equal to one we can get exact analytical answers.

Program 8.6 illustrates the Wave Equation solution.

8.11 Problems

8-1) Find the temperature distribution over time in a 10cm insulated aluminum rod with an initial temperature of 25°C from t=0 to t=10 sec. The end at x=0 is held at 25°C and the end at x=20 is held at 110°C. Use the explicit method, dividing the rod into 40 segments. What is the largest time step that can be used?

8-2) Solve problem 1 for a copper rod.

8-3) Solve problem 1 using the implicit method.

8-4) Determine the temperatures throughout a rectangular steel plate, 8cm wide and 16cm long, that is initially at 0°C from t=0 to t=10 sec. The left and top edges are held at 60°C and the other two adjacent edges are held at 95°C. Use the alternating-direction-implicit scheme.

8-5) If the steel plate in problem 5 were initially at -30° C how long does it take for the center of the plate to reach 40°C?

8-6) Use program PSOLV to find the steady-state temperature distribution in a rectangular solid 20cm by 10cm. The top and bottom surfaces are at 75°C and 25°C respectively and the left and right sides are adiabatic. There is no heat generation.

8-7) Rework problem 6 with unit heat generation and steel material.

8-8) Modify program PSOLV to use variable rectangular geometry.

8-9) Use the Wave Program to solve the one dimensional wave equation with initial displacements given by F(X)=X*SIN((X-10)/2) and initial velocities given by G(X)=X for the first half of the string length and G(X)=10-X for the remainder of the length.

8-10) Solve the wave equation for initial displacements given by F(X) = X*(1-EXP(X-10)) and initial velocities given by G(X) = -F(X).

Chapter 8 Programs

Program 8.1F Parabolic Equation Solve - Explicit Method

```
C EXAMPLE OF EXPLICIT METHOD OF SOLVING PARTIAL DIFFERENTIAL
C EQUATIONS. THIS SOLVES THE PROBLEM OF NONSTEADY HEAT
C CONDUCTION IN A PARALLEL-SIDED ENTITY.
C THE INITIAL TEMPERATURE DISTRIBUTION IS GIVEN BY THE FUNCTION
C F(X). THE BOUNDARY TEMPERATURES AT X = 0 AND X = 1 ARE
C GIVEN BY THE FUNCTIONS GO(DT) AND G1(DT) RESPECTIVELY.
C THE TEMPERATURES ARE PRINTED EVERY IFREQ TIME-STEPS, UNTIL THE
C CENTER TEMPERATURE EXCEEDS TMAX.

C......DEFINITIONS OF FUNCTIONS FOR COMPUTING
C INITIAL AND BOUNDARY CONDITIONS .....
      F(DIST) = 0.0
      GO(TIME) = 1.0
      G1(TIME) = 1.0
C
      DIMENSION TOLD(21), TNEW(21), X(21), ARRAY(2500)

C...... DATA INPUT
      READ (5,100) DDT, TMAX, M, IFREQ, IFRPLT
      INT = M/10
      FLOATM = M
      DX = 1.0/FLOATM
      RATIO = DDT/(DX*DX)
      WRITE (6,200) DX, DDT, TMAX, M, IFREQ, RATIO, IFRPLT
      CONST = 1.0 - 2.0*RATIO
C......SET AND PRINT INITIAL TEMPERATURES
      MP1 = M + 1
      DO 2 1 = 1, MP1
      FLOATI = I
      X(I) = (FLOATI - 1 0)/FLOATM
      TOLD(I) = F(FLOATI*DXO
      DT = 0.0
      WRITE (6,201)
      WRITE (6,202) DT, (TOLD(I)= 1, MP1, INT)
      ICOUNT = 0
C......COMPUTE TEMPERATURES FOR SUCCESSIVE TIME-STEPS
      DT = DT+DDT
      ICOUNT = ICOUNT+1
      DO 4 1 = 2, M
   4  TNEW(I) = RATIO*(TOLD(1-1) + TOLD(1+1)) + CONST*TOLD(I)
C......REPLACE NEW TEMPERATURES IN TOLD
      DO 5 1 = 2, M
   5  TOLD(I) = TNEW(I)
C......SET BOUNDARY CONDITIONS
      TOLD(1) = GO(DT)
      TOLD(MP1) = G1(DT)
C......PRINT T VALUES
      IF ((ICOUNT/IFREQ)*IFREQ .NE. ICOUNT) GO TO 3
      M2 = M/2
      WRITE (6,202) DT, (TOLD(I) = 1, MP1, INT)
      IF (TNEW(M2) .LE. TMAX) GO TO 3
C......FORMATS FOR INPUT AND OUTPUT STATEMENTS
 100 FORMAT (6X, F7.4, 10X, F5.2, 6X, I3, 10X, I3, 11X, I3)
 200 FORMAT (82HI UNSTEADY-STATE HEAT CONDUCTION IN A SLAB,
  1EXPLICIT METHOD, WITH THE PARAMETERS/ 12HO DX = F10.5/
```

```
     212H    DDT = , F10.5/ 12H    TMAX = , F10 5/ 12H   M=
     3I4/ 12H    IFREQ = , I4/ 12H    RATIO = , F10.5/
     4 12H IFRPLT = , I4)
C
 201 FORMAT (8H0   TIME, 18X, 39HVALUES OF TEMPERATURE AT ALL
    1GRIDPOINTS)
 202 FORMAT (1H , F7.3/(1H , 7X, 11F8.5))
 203 FORMAT (1H1)
 204 FORMAT (1H0, 38X, 25H DIMENSIONLESS DISTANCE X)

      END
```

Program 8.2F Parabolic Equation Solve - Implicit Method

```
C EXAMPLE OF IMPLICIT METHOD OF SOLVING A PARTIAL DIFFERENTIAL
C EQUATIONS. THIS PROBLEM IS THE SAME AS THE PREVIOUS PROBLEM.
C THE INITIAL TEMPERATURE DISTRIBUTION IS GIVEN BY THE FUNCTION
C F(X). THE BOUNDARY TEMPERATURES AT X = 0 AND X = 1 ARE
C GIVEN BY THE FUNCTIONS G0(DT) AND G1(DT), RESPECTIVELY.
C THE TEMPERATURES ARE PRINTED EVERY IFREQ TIME-STEPS, UNTIL THE
C CENTER TEMPERATURE EXCEEDS TMAX. THE TRIDIAGONAL SYSTEM OF
C EQUATIONS RESULTING AT EACH Time-STEP IS SOLVED BY THE
C SUBROUTINE TRIDAG.

C DEFINITIONS OF FUNCTIONS FOR COMPUTING
C INITIAL AND BOUNDARY CONDITIONS .....
      F(DIST)=0.0
      G0(TIME)=1.0
      G1(TIME)=1.0
      DIMENSION A(21), B(21), C(21), D(21), T(21)
C ..... CHECK INPUT PARAMETERS AND SET ARRAYS A, B, AND C
      READ (5,100) DDT, TMAX, M, IFREQ
      FLOATM=M
      DX=1.0/FLOATM
      RATIO = DDT/(DX*DX)
      WRITE (6,200) DX, DDT, TMAX, M, IFREQ, RATIO
      DO 2 1 = 2, M
      A(l) = - RATIO
      B(l) = 1.0+2.0*RATIO
      C(l) = - RATIO
C
C ......SET AND PRINT INITIAL TEMPERATURES
      MP1 = M + 1
      DO 3 1 = 1, MP1
      FLOATI=I
      T(l) = F(FLOATI*DX)
      DT=0.0
      Write (6,201)
      Write (6,202) DT, (T(l), I = 1, MP1)
      ICOUNT = 0
C
C PERFORM CALCULATIONS OVER SUCCESSIVE TIME-STEPS
    4 DT=DT+DDT
      ICOUNT=ICOUNT + 1
C SET BOUNDARY CONDITIONS
      T(1)=G0(DT)
      T(MP1)=G1(DT)
C COMPUTE RIGHT-HAND SIDE VECTOR D
      DO 5 I=2,M
    5        D(I)=T(I)
             D(2)= D(2)+RATIO*T(l)
             D(M)=D(M)+RATIO*T(MP1)
C ......COMPUTE NEW TEMPERATURES
      CALL TRIDAG (2, M, A, B, C, D, T)

C .....PRINT TEMPERATURES WHEN APPROPRIATE
      IF ((ICOUNT/IFREQ)*IFREQ .NE. ICOUNT) GO TO 4
      M2=M/2
      WRITE (6,202) DT, (T(I), I = 1, MP1)
      IF (T(M2).LE. TMAX GO TO 4
      GO TO 1
```

```
C
C .............FORMATS FOR INPUT AND OUTPUT STATEMENTS
  100 FORMAT (6X, F7.4, 10X, F5.2, 6X, I3, 10X, 13)
  200 FORMAT (82H1 UNSTEADY-STATE HEAT CONDUCTION IN A SLAB, IMPL
     1METHOD, WITH THE PARAMETERS/ 12H0 DX = , F10.5/
     212H DDT = , F10,5/ 12H TMAX = , F10.5/ 12H M =
     3I4/ 12H IFREQ = , I4/ 12H RATIO = , F10.5)
  201 FORMAT (8H0   TIME, 18X, 39HVALUES OF TEMPERATURE AT ALL
     1GRIDPOINTS)
  202 FORMAT (1H , F7.3/(1H , 7X, 11F8.5))

      END
C         SUBROUTINE FOR SOLVING A SYSTEM OF LINEAR SIMULTANEOUS
C         EQUATIONS HAVING A TRIDIAGONAL COEFFICIENT MATRIX.
C         THE EQUATIONS ARE NUMBERED FROM IF THROUGH L, AND THEIR
C         SUB-DIAGONAL, DIAGONAL, AND SUPER-DIAGONAL COEFFICIENTS
C         ARE STORED IN THE ARRAYS A, B, AND C. THE COMPUTED
C         SOLUTION VECTOR V(IF)...V(L) IS STORED IN THE ARRAY V.
C
      SUBROUTINE TRIDAG (IF, L, A, B, C, D, V)
      DIMENSION A(1), B(1), C(1), D(1), V(1), BETA(101), GAMMA(101)
C   ......COMPUTE INTERMEDIATE ARRAYS BETA AND GAMMA
      BETA(IF) = B(IF)
      GAMMA(IF)=D(IF)/BETA(IF)
      IFP1=IF + 1
      DO 1 I = IFP1, L
          BETA(I) - B(I) - A(I)*C(I-1)/BETA(I-1)
    1     GAMMA(I) = (D(I) - A(I)*GAMMA(I-1))/BETA(I)
C
C   ......COMPUTE FINAL SOLUTION VECTOR V
      V(L) = GAMMA(L)
      LAST = L - IF
      DO 2 K = 1, LAST
          I = L - K
   2     V(I) = GAMMA(I) - C(I)*V(I)/BETA(I)
      RETURN
      END
```

Program 8.3F Alternating Direction Implicit

```
C PROGRAM ADI ( INPUT, OUTPUT)
C THIS PROGRAM SOLVES THE UNSTEADY STATE HEAT FLOW EQUATION IN TWO
C SPACE DIMENSIONS USING THE A.D.I. METHOD
C
C PARAMETERS ARE :
C
C U - VECTOR OF TEMPERATURES AFTER ODD TRAVERSES
C V - VECTOR OF TEMPERATURES AFTER EVEN TRAVERSES
C UCF - MATRIX OF COEFFICIENTS FOR THE U'S
C VCF - MATRIX FOR THE V'S
C BCNDU - VECTOR OF BOUNDARY VALUES FOR THE U EQUATIONS
C BCUDV - SAME FOR THE V'S
C TOP - VECTOR TO HOLD BOUNDARY VALUES ACROSS TOP OF REGION
C BOT - SAME FOR THE VALUES ACROSS THE BOTTOM
C RT - HOLD THE RIGHT HAND SIDE VALUES
C LFT - HOLD VALUES FOR THE LEFT HAND EDGE
C M - NUMBER OF ROWS OF NODES IN THE GRID
C N - NUMBER OF COLUMNS OF NODES
C DIFF - THERMAL DIFFUSIVITY = K/C*DENSITY
C H - THE VALUE OF DELTA X - DELTA Y
C TIME - THE TIME VARIABLE
C TMAX - MAXIMUM VALUE OF TIME FOR WHICH COMPUTATIONS ARE
C DESIRED
C DT - TIME STEP SIZE, RELATED TO DELTA X THROUGH R
C
C ESTABLISH THE MATRICES AND VECTORS. PUT IN SOME VALUES WITH
C DATA STATEMENTS.
C
      REAL U(500),V(500),UCF(500,3),VCF(500,3)
      REAL BCNDU(500),BCNDV(500)
      REAL TOP(100),BOT(100),LFT(100),RT(100)
      DATA UCF,VCF / 3000 * -1.0 /
      DATA BCNDU,BCNDV / 1000 * 0.0 /
      DATA M,N,R / 7,15,0.5 /
      DATA H,DIFF / 0.125,0.152 /
      DATA TMAX / 20.0 /
      DATA RT,LFT,TOP,BOT / 100 * 100.0, 300 * 0.0 /
C
C FOR THE TEST CASE, SET TEMPERATURES EVERYWHERE EQUAL TO
C ZERO. ESTABLISH THESE BY A DATA STATEMENT.
C
      DATA V / 500 * 0.0 /
C
C SET UP THE COEFFICIENT MATRICES BY OVER-WRITING ON THE DIAGONAL
AND CERTAIN OFF DIAGONAL TERMS.
C
      MSIZE=M*N
      DO 10 I,MSIZE
           UCF(I,2)=1.0/R + 2.0
           VCF(I,2)=1.0/R + 2.0
      10 CONTINUE
      ML1=M - 1
      NL1=N - 1
      MSL1=MSIZE - 1
      DO 20 I=N,MSL1,N
           UCF(I,3)=0.0
           UCF(I+1,1)=0.0
```

```
      20 CONTINUE
      DO 30 I=M,MSL1,M
            VCF(I,3)=0.0
            VCF(I+1,1)=0.0
      30 CONTINUE
C
C GET VALUES INTO THE BCND VECTORS
C
      DO 40 I=1,N
            BCNDU(I)=TOP(I)
            J=MSIZE - N + I
            BCNDU(J)=BOT(I)
      40 CONTINUE
C
      DO 45 I=1,M
            J=(I-1)*N + 1
            BCNDU(J)=BCNDU(J) + LFT(I)
            J=I*N
            BCNDU(J)=BCNDU(J) + RT(I)
      45 CONTINUE
C
      DO 50 I=1,M
            BCNDV(I)= LFT(I)
            J=MSIZE - M + I
            BCNDV(J)=RT(I)
      50 CONTINUE
C
      DO 55 I=1,N
            J(I-1)*M+1
            BCNDV(J)=BCNDV(J) + TOP(I)
            J=MSIZE-M+1
            BCNDV(J)=BCNDV(J) + BOT(I)
      55 CONTINUE
C
C GET THE LU DECOMPOSITIONS OF UCF AND VCF
C
      DO 60 I=2,MSIZE
            UCF(I-1,3)=UCF(I-1,3)/UCF(I-1,2)
            UCF(I,2)=UCF(I,2) - UCF(I,1)*UCF(I-1,3)
            VCF(I-1,3)=VCF(I-1,3)/VCF(I-1,2)
            VCF(I,2)=VCF(I,2) - VCF(I,1)*VCF(I-1,3)
      60 CONTINUE
C DO ITERATIONS UNTIL TIME EQUALS TMAX.
      TIME=0.0
      DT=R/DIFF*H*H
   62 IF (TIM3.GT.TMAX) STOP
C COMPUTE THE R.H.S. FOR THE U EQUATIONS AND STORE IN THE U VECTOR.
C DO THE TOP AND BOTTOM ROWS AS A FIRST STEP.
C
      DO 65 1 = 1,N
            J = (1-1)*M + 1
            U(I) = ( 1.0/R - 2.0 )*V(J) + V(J+1) + BCNDU(I)
            K = MSIZE - N + I
            J = I * M
            U(K) = V(J-1) + ( 1.0/R - 2.0 )*V(J) + BCNDU(K)
      65 CONTINUE
C
C DO THE OTHER ROWS
C
      DO 75 I = 2,ML1
```

```
      DO 70 J = 1,N
            K=(I-1)*N + J
            L = I + (J-1)*M
            U(K) = V(L-1) + ( 1.0/R - 2.0 )*V(L) + V(L+1) + BCNDU(K)
      70 CONTINUE
      75 CONTINUE
C
C GET THE SOLUTION FOR THE ODD TRAVERSE
C
      U(1) = U(1) / UCF(1,2)
      DO 80 I = 2,MSIZE
            U(I) = ( U(I) - UCF(I,1)*U(I-1)) / UCF(I,2)
      80 CONTINUE
      DO 90 I = 1,MSL1
            JROW = MSIZE - I
            U(JROW) = U(JROW) - UCF(JROW,3)*U(JROW+1)
      90 CONTINUE
C
C COMPUTE THE R.H.S. FOR THE ODD TRAVERSE, STORE IN V. DO THE
C TOP AND BOTTOM ONES.
C
      DO 95 I=1,M
            J=(I-1)*N + 1
            V(I) - ( 1.0/R - 2.0 )*U(J) + U(J+1) + BCNDV(I)
            K=MSIZE - M + I
            J=I * N
            V(K) - U(J-1) + ( 1.0/R - 2.0 )*U(J) + BCNDV(K)
      95 CONTINUE
C NOW THE REST OF THE ROWS
      DO 105 I= 2,NL1
      DO 100 J=1,M
            K=(I-1) *M + J
            L=I + (J-1)*N
            V(K)=U(L-1) + (1.0/R-2.0)*U(L) + U(L+1) + BCNDV(K)
  100 CONTINUE
  105 CONTINUE
C DETERMINE THE SOLUTION FOR THE ODD TRAVERSE
      V(1)= V(1) / VCF(1,2)
      DO 110 I=2,MSIZE
            V(I)=(V(I) - VCF(I,1)*V(I-1)) / VCF(I,2)
  110   CONTINUE
      DO 120 I=1,MSL1
            JROW=MSIZE - I
            V(JROW)=V(JROW) - VCF(JROW,3)*V(JROW+1)
  120   CONTINUE
C PRINT OUT THE LAST RESULT.
      PRINT 202 TIME, ( V(I), I=1,MSIZE)
  202   FORMAT(//1X,' WHEN T = ',F6.3,' V VALUES

     1ARE'/(1X,15F8.4))
      GO TO 62
      END
```

Program 8.4F Coefficient Matrix Formation

```
SUBROUTINE COEFFMTX(A,NDIM,N,M)
C
C THIS SUBROUTINE FORMS THE COEFFICIENT MATRIX FOR THE LAPLACIAN ON
C A RECTANGULAR REGION THAT HAS N POINTS IN THE X DIRECTION AND M
C POINTS IN THE Y DIRECTION. IT ASSUMES THAT THE MATRIX ALREADY HAS
C ZEROS EVERYWHERE.
C PARAMETERS ARE :
C A - MATRIX THAT RETURNS COEFFICIENTS, MUST CONTAIN ZEROS TO
C START.
C NDIM - FIRST DIMENSION OF A IN THE MAIN PROGRAM
C N - NUMBER OF GRID POINTS IN THE X DIRECTION
C M- NUMBER OF GRID POINTS IN THE Y DIRECTION
C
      REAL A(NDIM,NDIM)
      INTEGER NDIM,N,M,NL1,NP1,MSIZE,MSLN,I
C
C     COMPUTE CONSTANTS
      NL1=N-1
      NP1=N+1
      MSIZE=N*M
      MSLN= MSIZE-N
C     PUT -4 ON THE DIAGONAL
      DO 10 I=1,MSIZE
            A(I,I) - -4.0
  10  CONTINUE
C     PUT 1 ABOVE AND BELOW THE DIAGONAL
      DO 15 I=2,MSIZE
            A(I-1,I)=1.0
            A(I,I-1)=1.0
  15  CONTINUE
C     REPLACE SOME ONES WITH ZEROS
      DO 20 I= N,MSLN,N
            A(I,I+1)=0.0
            A (I+1,I)=0.0
  20  CONTINUE
C     PUT IN THE SIDE BANDS AND GO BACK
      DO 30 I= NP1,MSIZE
            A(I,I-N)=1.0
            A(I-N,I)=1.0
  30  CONTINUE
      RETURN
      END
```

Program 8.5B Two-Dimensional Laplacian/Poisson Equation Solve

```
  5 REM PROGRAM POSOLV
     10 PRINT "TYPE 1 IF YOU WANT A PROGRAM EXPLANATION."
     12 PRINT
     13 PRINT"OTHERWISE, TYPE 2."
     14 INPUT L
     15 PRINT
     16 IF L=1 THEN 20
     18 IF L=2 THEN 40
     20 PRINT
     21 PRINT"THIS PROGRAM SOLVES THE TWO-DIMENSIONAL LAPLACE OR"
     22 PRINT"POISSON EQUATION FOR STEADY STATE CONDUCTION IN A"
     23 PRINT"SOLID IN A RECTANGULAR GEOMETRY SPECIFIED BY A 10X5"
     24 PRINT"GRID"
     25 PRINT"THE GOVERNING EQUATION IS EXPRESSED BY USE OF"
     26 PRINT"A FINITE DIFFERENCE APPROXIMATION."
     27 PRINT"
     28 PRINT"THE TOP AND BOTTOM SURFACES ARE ISOTHERMAL."
     29 PRINT
     30 PRINT"THE TWO VERTICAL SIDES ARE EITHER ADIABATIC OR"
     31 PRINT"ISOTHERMAL"
     32 PRINT"THE NUMBER OF ADIABATIC NODES, MEASURED DOWN FROM"
     33 PRINT"THE TOP ARE SPECIFIED BY THE INPUT VALUES OF I1 AND
I2"
     34 PRINT"FOR THE LEFT AND RIGHT SURFACES RESPECTIVELY."
     35 PRINT
     36  PRINT"A  LINEAR  TEMPERATURE  PROFILE  IN  THE  SOLID  IS
INITIALLY"
     37 PRINT"ASSIGNED"
     38 PRINT"THIS PROFILE IS MODIFIED UNTIL ALL BC'S"
     39 PRINT"AND THE DIFFERENTIAL EQUATION ARE SATISFIED. "
     40 DIM Y(10,5), T(10,5), X(10,5), Z(10,5)
     50 READ I1,I2,N1,D
     60 READ T1,T2
     95 REM ASSIGN GRID SIZE
     100 LET Y1=10
     120 LET X1=5
     190 REM INITIALIZE CONTROL CONSTANTS
     200 LET M=50
     202 LET P=0
     210 LET E1=0.001
     290 REM ASSIGN INITIAL TEMPERATURE DISTRIBUTION
     300 FOR I=1 TO Y1
     310 FOR J=1 TO X1
     320 LET A=I
     330 LET T(I,J)=T2+(T1-T2)*(A-1)/(Y1-1)
     340 NEXT J
     350 NEXT I
     355 PRINT"DO YOU WANT A PRINT OUT OF INITIALLY ASSIGNED"
     360 PRINT"LINEAR TEMPERATURE DISTRIBUTION"
     364 PRINT
     366 PRINT"IF YES TYPE 1. IF NO TYPE 2"
     370 INPUT B6
     372 IF B6=1 THEN 380
     374 IF B6=2 THEN 400
```

```
380 PRINT"THE ASSIGNED TEMPERATURE DISTRIBUTION IS"
382 FOR I=1 TO Y1
383 FOR J=1 TO X1
384 LET Y(I,J)=INT(100*T(I,J)+0.5)/100
385 NEXT J
387 NEXT I
388 MAT PRINT Y1
400 PRINT"BEGIN ITERATION"
404 REM INITIALIZE COUNTING VARIABLES
405 LET N=0
410 LET U=0
420 FOR J=1 TO X1
430 FOR I=2 TO Y1-1
440 IF J<>I THEN 500
450 IF I1>=I THEN 480
460 LET X(I,1 )=T1
470 IF I1<I THEN 570
480 LET X(I,J)=(T(I+1,J)+2*T(I,J+1)+T(I-1,J)+N1*(D^2))/4
490 IF I1>=I      THEN  570
500 IF      J<>X1 THEN  560
510 IF I2>=I      THEN  540
520 LET X(I,X1)=T1
530 IF I2<I THEN 570
540 LET X(I,J)=(T(I+1,J)+2*T(I,J-1)+T(I-1,J)+N1*(D^2 ))/4
550 IF I2>=I THEN 570
560LETX(I,J)=(T(I,J+1)+T(I,J-1)+T(I+1,J)+T(I-1,J)+N1*(D^2))4
570 LET Y(I,J)=X(I,J)-T(I,J)
580 IF ABS (Y(I,J))<E1 THEN 600
590 LET U=U+1
600 LET T(I,J)=X(I,J)
610 NEXT I
620 NEXT J
630 LET N=N+1
640 IF N>=M THEN 700
650 IF U=0 THEN 700
652 IF P>0.0 THEN 660
653 PRINT"AFTER EACH ITERATION THE U VALUE INDICATES THE
NUMBER"
654 PRINT"OF GRID POINTS WHERE A SOLUTION HAS NOT YET REACHED"
655 PRINT"THE REQUIRED ACCURACY."
656 LET P=P+1
657 PRINT
658 PRINT"IT. NO.         U VALUE"
659 PRINT
660 PRINT N,U
670 PRINT
690 IF U<>0.0 THEN 410
698 PRINT
700 PRINT N,U
710 PRINT
715 PRINT" THE FINAL SOLUTION IS"
720 FOR I=1 To Y1
722 FOR J=1 TO X1
724 LET Z(I,J)=INT(1000*T(I,J)+0.5)/1000
726 NEXT J
728 NEXT I
729 PRINT
730 MAT PRINT Z;
1032 PRINT"I1="I1,"I2="I2,"N1="N1
1033 PRINT
```

```
1034 PRINT"DELTA X="D
1035 PRINT
1040 DATA 0,0,0,.111
1045 DATA 0,1.0
2000 END
```

Program 8.6F Wave Equation Solver - One Dimensional

```
      WAVE PROGRAM (INPUT,OUTPUT)
C      THIS PROGRAM COMPUTES SOLUTIONS FOR THE ONE DIMENSIONAL WAVE
C       EQUATION. THE INITIAL DISPLACEMENTS OF THE VIBRATING STRING,
C       FROM X=0 TO X=XLEN ARE GIVEN BY F(X). THE INITIAL VELOCITIES
C       ARE GIVEN BY G(X). F(X) AND G(X) ARE DEFINED BY FUNCTION
C       SUBPROGRAMS. THE END POINTS ARE ASSUMED FIXED AT ZERO
C       DISPLACEMENT.
C
C       THE RELATION USED IS U(I,J-1) = U(I-1,J) + U(I+1,J) - U(I,J-1)
C       EXCEPT FOR THE FIRST TIME STEP, WHEN THE VALUE IS GIVEN BY
C       U(I,1) = 0.5*( U(I-1,J) + U(I+1,J) ) + .5/C*INTEGRAL OF INIT
C       VEL.
C
C     PARAMETERS ARE:
C
C       X       - DISTANCE ALONG THE STRING
C       DX      - INCREMENTS OF DISTANCE
C       XLEN    - TOTAL LENGTH OF STRING
C       N       - NUMBER OF SUBDIVISIONS
C       T       - TIME
C       TLAST   - FINAL VALUE OF TIME FOR WHICH SOLUTION IS DESIRED
C       F(X)   - INITIAL DISPLACEMENTS
C       G(X)   - INITIAL VELOCITIES
C       TDM    - VALUE OF TENSION / MASS * C SQUARED
C       U       - DISPLACEMENTS AT EVEN TIME INTERVALS
C       V       - DISPLACEMENTS AT ODD TIME INTERVALS
C
       REAL U(100), V(100),X,DX,XLEN,T,TLAST,F,G,TDM,SUBDX,XSUB
       INTEGER N,NP1,I,J
C
C      DEFINE DATA STATEMENTS FOR INITIAL VALUES
       DATA X,N,T/0.0,20,0.0/
       DATA TDM,XLEN,TLAST/3.0,10.0,1.000/
C      GET INITIAL DISPLACEMENTS
       NP1 = N + 1
       DX = XLCN / FLOAT (N )
       U(1) = 0.0
       U(NP1) = 0.0
       V(1) = 0.0
       V(NP1) = 0.0.
       DO 10 I = 2,N
             X = X + DX
             U (I ) = F(X)
       10 CONTINUE
C      WRITE OUT THE INITIAL DISPLACEMENTS
       PRINT 200, ( U(I), I = 1,NP1 )
C
C      GET DISPLACEMENTS AFTER FIRST TIME STEP. USE SIMPSON'S
C      RULE TO PERFORM INTEGRATIONS, USING 20 INTERVALS BETWEEN X(N-1
C      ) AND X(N+1). WE DO THIS FOR EACH INTERMEDIATE X VALUE.
       SUBDX = DX / 10.0
       XSUB = DX
       DO 30 I = 2,N
             SUM = 0. 0
```

```
            XSUB = XSUB - DX
            DO 20 J = 1,19, 2
      SUM=SUM+G(XSUB)+4.0*G(XSUB+SUBDX)+G(XSUB+2.0*SUBDX)
     1      XSUB=XSUB+2.0*SUBDX
   20       CONTINUE
      V(I) = 0.5*(U(I-1)+U(I+1))+0.5/SQRT(TDM)*SUBDX/3.0*SUM
   30       CONTINUE
C     WRITE OUT DISPLACEMENTS AFTER FIRST TIME STEP
      T=DX/TDM
      PRINT 201, T, ( V(I), I = 1,NP1 )
C     COMPUTE UNTIL TLAST IS REACHED
   35 IF (T.GE.TLAST ) STOP
      DO 40 I = 2,N
            U(I)=V(I-1)+V(I+1)-U(I)
   40 CONTINUE
      T=T+DX/TDM
      PRINT 201, T,(U(I),I=1,NP1)
      DO 50 I=2,N
            V(I) = U(I-1) + U(I+1) - V(I)
   50 CONTINUE
      T=T+DX/TDM
      PRINT 201, T,(V(I),I=1,NP1)
      GO TO 35
200   FORMAT(// ' SOLUTION TO VIBRATING STRING PROBLEM ',/,
     +       ' INITIAL DISPLACEMENTS ARE '//(1X,7F9.3))
201 FORMAT (/' AT T =  ' ,F8.3/(1X, 7F9.3))
      END
      REAL FUNCTION F(X)
      REAL X
      F=0.1*X*(X-6.0)*(X-10.0)
      RETURN
      END
C
      REAL FUNCTION G(X)
      REAL X
      G = X/3.0
      IF (X.GT.6.0) G=((10.0-X)/2.0
      RETURN
      END
```

9

Applications of Routines to the Finite Element Method

9.1 Introduction

In this text we have tried to present algorithms that are "useable". In addition examples have been presented that illustrate how the algorithms are used in practice. As the chapters have progressed, the material has gotten progressively more difficult but has been building on previous material.

This chapter attempts to utilize many of the previous techniques and in a very cursory manner give some background and insight into the finite element method through use of computer programs. It will be seen that finite element computations consist mainly of handling matrices and vectors. These operations include adding, multiplication, transposing, etc; a lot of what was covered in Chapter 4. The programs are written in FORTRAN and can be translated to the other languages as necessary.

The programs or subroutines given are not the most elegant or efficient, but they do illustrate how the method works.

9.2 What is the Finite Element Method?

The finite element method is a technique for solving partial differential equations by first discretising the equations describing the problem in their space dimensions. The discretisation is carried out over small regions of arbitrary shapes. This results in matrices that relate the input at specified points to the output at all points in the domain. To solve the equations over large regions, the matrix equations for the smaller regions are usually summed node by node. This will result in the development of the global matrix equations.

Many references are available that describe this technique. Perhaps the most notable is the text by Zienkiewicz and Taylor [95].

We will set up a simple application for the example in this chapter, the plate element. The techniques are applicable to all of the more complex elements.

The last two decades have seen an explosive growth in computer technology. This growth does not show signs of abating. It appears that the next few years will prove to be even greater, particularly with the introduction of novel forms of architecture (e.g., vector, array, and multiprocessor systems). The computer hardware developments are most noticeable at the two extremes of the spectrum. At one end the massive computer systems, usually referred to as supersystems, such as CRAY-2, and CDC CYBER 205, have radically different and novel architectures that result in high performance [computational speed of the order of 100 million floating point operations per second (100 MFLOPS) or more]. At the other end of the spectrum are various types of minicomputers, engineering workstations (microcomputers) microprocessors, and handheld computers.

The introduction of these new computing systems has made a strong impact on finite element technology.The supersystems have made possible new levels of sophistication in finite element modeling as well as in problem depth and scope which were not possible a short while ago. These platforms have allowed development of finite-element analysis software that allows the user to realize the full potential of the supersystems in finite-element computation using special parallel numerical algorithms and efficient programming strategies.

The small, low-cost computer systems (minis, micros, and engineering workstations) provide a high level of interaction and free the finite element analysts from the constraints often imposed by large central computation centers.

Computer-aided design and computer-aided manufacturing (CAD/CAM) are terms that have gained prominence in the late 1970s and early 1980s to describe the modern engineering process in which virtually all stages of the process are unified through a computerized information repository and data flow. Among the stages of engineering usually included in the CAD/CAM rubric are conceptual design of overall systems, technical analysis and design of various subsystems, resolution of subsystem-integration difficulties, production of manufacturing or construction drawings and documents, programming of production, and generation of manufacturing control information. The unified computer-assisted process encompasses the development and evaluation of alternative designs and the capability for redesigning the subsystems or the overall system. The intent is to maximize both engineering creativity and productivity by providing the engineer all the necessary information so that decisions and judgments can be made efficiently.

CAD/CAM systems are widely utilized in both research and industrial settings although not as extensively as might be expected within some segments of engineering and industry. It is clear that further advances will occur as several developments in hardware and software are increasingly combined to create an appropriate environment. Three chief developments are networking, which enables the sharing and rapid transfer of large volumes of data; engineering workstations, which provide individual designers access to data and computing tools; and integral interactive-graphics capabilities, which present the opportunity for efficient control and manipulation of the complex engineering process. The obvious common thread that ties together all CAD/CAM developments is the focus on data bases and data flow.

The finite-element method has been used in engineering analysis for many years. However, its use throughout the entire engineering design process has been limited by the cost of the computing resources and the workforce to synthesize and manage the data needed for multiple analyses required by design modifications. The drastically reduced expense of computing resources and the development of more integrated design software are increasing the cost-effectiveness of finite-element methods to the point that they can become the design process, not just a check on the final design or a tool for postfailure analysis.

The process of extracting the geometry of a design from a computer-aided design database, generating a finite-element model, analyzing that model, and retrieving the results for interpretation requires the transmittal of vast amount of information. Thus, data communication procedures between the various segments are paramount for the effective integration of the process. The two requirements placed on this data transfer are that it be as general, complete, and foolproof as possible and that it be as rapid as possible. Since there are no methods currently that meet these requirements.

9.3 Building the Basics for the Finite Element Method

The basic concept of the finite-element method is not new. It has been used throughout engineering for evaluating certain quantities (particularly area and volume) by adding or counting well-defined geometric figures (elements). Today's understanding of the finite-element however, is finding an approximate solution to a boundary- and initial-value problem by assuming that the domain is divided into well-defined subdomains (elements) and that the unknown function of the state variable is defined approximately within each element. With these individually defined functions matching each other at the element nodes or at its at the interfaces, the unknown function is approximated over the entire domain.

There are, of course, many other approximate methods for the solution of boundary-value problems, such as finite-difference methods, weighted-residual methods, Rayleigh-Ritz methods, Galerkin methods, and more. The primary difference between the finite-element method and most other methods is that in the finite-element method the approximation is confined to relatively small subdomains. It is, in a way, the localized version of the Rayleigh-Ritz method. Instead of finding an admissible function satisfying the boundary conditions for the entire domain, which, particularly for irregular domains, is often difficult, in the finite element methods the admissible functions are defined over element domains with simple geometry and pay no attention to complications at the boundaries. This is one of the reasons that the finite element method has gained superiority over the other approximate methods.

Since the entire domain is divided into numerous elements and the function is approximated in terms of its values at certain points (nodes), it is inevitable that the evaluation of such a function will require solution of simultaneous equations. Because of this, the finite element methods were not widely used until the middle of this century when powerful computers were conceived and developed.

9.3.1 Basic Steps in The Finite Element Method

Regardless of the physical nature of the problem, a standard finite element method involves the following steps. Depending upon the physical nature of the problem, the complexity of a solution may or may not require planning the solution in detail.

STEP 1. DEFINITION OF THE PROBLEM AND ITS DOMAIN. In finite element methods, there are three basic sources of approximation. The first one is the definition of the domain, both physically and geometrically; the second is the discretization of the physical domain; and the third is the solution algorithms. The approximations used in defining the physical characteristics of different regions of the domain are very much problem-oriented. The geometric definition of the domain, however, requires establishing global coordinate axes in reference to which the coordinates of certain points, which define the equations of the line and surfaces of elements, are to be described. This coordinate system need not be rectangular and cartesian as is often done; some curvalinear systems may actually be better suited to the specific problem.

The domain can be bounded or unbounded (some portions extend to infinity). For the bounded region of the domain, the idealization is done by using finite elements and for the unbounded portion by using infinite elements or boundary elements. Quite often, the entire domain is made up of subdomains. The interface conditions between subdomains must also be defined prior to discretization.

STEP 2. DISCRETION OF THE DOMAIN. Since the problem is usually defined over a continuous domain, the governing equations, with the exception of the essential boundary conditions are valid for the entirety of, as well as for any portion of, that domain. This allows idealization of the domain in the form of interconnected finite-sized domains (elements) of different size and shape. By doing this, certain approximations are introduced. Putting enough numbers of nodes between the elements (higher-order elements, etc.) also comes into the picture at this stage of the method. Here, one should be concerned with how well the idealized discrete domain represents the actual continuous domain. To a certain extent, it is true that the smaller elements (finer mesh) produce better results. But it is also true that the finer mesh results in a larger number of equations to be solved and decreases the accuracy, thereby contrasting with the purpose of using the method. The question then arises: what is the most efficient element type, size, and pattern? A partial answer to this question is given in the literature under the key word modeling. Adaptive processes or mesh refinements and automatic mesh generation are also techniques relevant to the discretization of the domain.

In finite-element idealization of the domain, we shall, in general, make reference to the following elements: (a) finite element and master element.

Finite elements are those when put together, result in the discrete version of the actual continuous domain. Finite elements are generally straight-sided (or -surfaced), particularly at the interior of the domain. They can be curved (as the higher-order elements) mainly at the boundaries or curved surfaces (as in shells). The curved elements, therefore, may contain geometric approximations in addition to physical approximations. Their geometric approximations are controlled by the number of nodes utilized at the exterior of the elements to define their shape. The physical approximations are controlled by the total number of nodes (exterior as well as interior) utilized in defining the trial functions (shape functions) for the state variable.

Master elements are those which are used in place of finite elements in order to facilitate computations in the element domain. The definition of shape functions and particularly of integration becomes simpler in master elements.

STEP 3. IDENTIFICATION OF STATE VARIABLE(S). Until this step, no reference has been made to the physical nature of the problem. Whether it is a heat-transfer problem, fluid- or solid mechanics problem, etc., comes into the picture at this stage. The mathematical description of steady-state physical phenomena, for instance, leads to an elliptic boundary-value problem in which the formula contains the state variable and the flux. These variables are related to each other by a constitutive equation representing a mathematical expression of a physical law.

Once the state variable and the flux have been identified, the formulation can take place containing either or both. The choice is usually dictated by the problem. Various finite-element methods, in general, are the result of such choices. Comprehensive discussions of alternative of finite element methods in mechanics may be found in [52].

STEP 4. FORMULATION OF THE PROBLEM. Very often a physical

problem is formulated either via a set of differential equations

$$Lu = f \tag{9.1}$$

with boundary conditions or by an integral equation (a functional)

$$\pi = \int_{\Omega} G(x,y,z,u)d\Omega + \int_{\Gamma} g(x,y,z,u)d\Gamma \tag{9.2}$$

subject to stationary requirement. While Equation (9.1) is referred to as the operational form of the physical problem, Equation (9.2) is referred to as the variational form of the same problem. Solution of either equation (inversion of L or minimization of π) yields the same results.

For the heat conduction problem in 2D space, for instance, Equations (9.1) and (9.2) take the following forms:

$$\begin{aligned} \nabla^2 u - c &= 0 \\ \pi &= \int_A (1/2u^2 + cu)dA \end{aligned} \tag{9.3}$$

In general, if a functional [Equation (9.2)] exists, the associated Euler-Lagrange differential equation (9.1) can be found. The reverse, however, is not necessarily true. While the differential equation may be approximated over a set of discrete points using finite differences, the associated functional can be minimized over a set of discrete domains by using the finite element method. A parallel treatment of the two classes of formulation can be found in Strang and Fix [80].

STEP 5. ESTABLISHING COORDINATE SYSTEMS. There are primarily two reasons for choosing special coordinate axes for the elements in addition to the global axes for the entire system. The first is the ease of constructing the trial functions for the elements and the second is ease of integration within the elements. However, since the elements will be assembled in the global frame, this step introduces additional computations in the form of coordinate transformations. Although the entire finite-element analysis can be carried out directly in global system, the benefit does not warrant the price paid for it. Since the coordinate transformations between any two coordinate systems is well defined, using the most suitable coordinate axes for each type of element and for the overall problem is highly recommended.

Depending upon the element shape, one usually chooses cartesian or curvalinear axes located within the element in reference to which the element matrix equation will be obtained. Other coordinate systems, known as natural coordinates such as area or volume coordinates are often employed in finite element analysis because the numerical integration is much simpler with respect to these coordinates.

The transformation of entities from an element to a corresponding master element involves mapping. Transformation from the element coordinate system to the global coordinates, however, involves only rotations, If both coordinate systems, local (x, y, z) and global (X, Y, Z), are chosen so as to be orthogonal, the rotation matrix is also orthogonal. This greatly facilitates the transformation between the two coordinate systems.

Once the coordinate axes are established, the element equations are ordinarily computed first in a master element. They are then finally transformed into the system for assembly. In some cases, however, the mapping of master element computations can be done directly into the global coordinate systems.

After the solution of simultaneous equations, the inverse transformation takes place in order to compute the physical entities required from the element domain.

STEP 6. CONSTRUCTING APPROXIMATE FUNCTIONS FOR THE ELEMENTS. Once the state variable(s) and the local coordinate system have been chosen, the function can be approximated in several ways. Only two entities need to be approximated. The first is physical (the state variable) and the second is geometrical (the shape of the element). If the element is actually made up of straight lines or planes, the coordinates of primary nodes (those at the extremes of the elements) will define the element shape accurately. In this case, the geometric approximation does not enter into the picture. Because of this, discretization of the entire domain is usually made by straight-line (linear) elements. For some problems, however, linear elements may introduce unacceptable errors, and discretization has to be accomplished by using isoparametric elements.

A similar argument is valid for the approximation of the state variable. It can be approximated in the form of a linear function or a higher-order function (i.e., quadratic, cubit, etc.). The user must decide whether to approximate physics (state variable) and geometry (shape) equally or to give preference to one or the other in various regions of the domain.

STEP 7. OBTAIN THE ELEMENT MATRICES AND EQUATIONS. At this point in the game it is assumed that the formulation and discretization of the domain with the desired element shapes and dimensions has been completed. The approximating function for the state variable is written in terms of "shape functions" (derived for each element geometry). This further approximation contains the approximation for the state variable and the coordinates for the element nodes that define the shape of the element. Substitution of this approximation into the variational approach to solving the system (Equation 9.2) yields, with considerable intensive integration, a system of equations to be solved for the state variables.

STEP 8. COORDINATE TRANSFORMATIONS. The determination of the system of equations for the nodal values involves the integration of the shape functions or their derivatives or both over the element. The integration is easier to evaluate when the interpolation equation is written in the terms of the element coordinate system (a coordinate system located on or within the boundaries of the element).

Coordinate transformations of physical entities such as vectors and matrices follow well-defined rules as those presented in Chapter Four.

STEP 9. ASSEMBLY OF THE ELEMENT EQUATIONS. The assembly of the element matrix equations is done according to the topological configuration of the elements after this equation is transformed into the global system. The assembly is done through the nodes at the interfaces which are common to the adjacent elements. At these nodes the continuities are established with respect to the state variable(s) and also with respect to the derivatives of the state variables.

The essential boundary conditions are introduced so that the final set of equations will be reduced or perhaps condensed.

STEP 10. SOLUTION OF THE FINAL SET OF EQUATIONS. The finite element methods yield the solution of a set of simultaneous equations. The solution procedure can be categorized into three parts: (1) direct, (2) iterative, and (3) stochastic.

The direct-solution techniques (originally proposed by Gauss a century ago) consist of a set of systematic steps and are used often in finite element solutions. The accuracy of the results is largely determined by the condition of equations (well or ill), the number of equations, and the computer (double precision, single precision, etc.). The symmetry and banded properties of equations can be well taken care of with these methods. The Gauss elimination and Cholesky's factorization (LU decomposition) are the most commonly used direct procedures. These methods are well suited to a small or moderate number of equations.

When systems are of a very large order, iterative procedures such as Gauss-Seidel or Jacobi iterations are more suited. Iterative methods are, in general, self-correcting, and the accuracy of the solution depends upon the number of iterations. Convergence is not always assured, but when convergence does occur, the sequence of equations plays an important role. The solution time is considerably less than that required by a direct procedure. On the other hand, the iterative methods are not suited for sets with multiple right-hand sides (multiple loading conditions in solid-mechanics problems, for example). When the set of equations to be solved is nonlinear, the Newton-Raphson iteration and its modified version appear to be the most commonly used methods. In recent years, the quasi-Newton method has also received considerable attention in nonlinear finite element analysis. A detailed presentation of the solution of simultaneous equations (linear and nonlinear) by using various procedures and with corresponding algorithms is presented in [52].

STEP 11. INTERPRETATION OF THE RESULTS. The previous step resulted in the approximate values of the state variable at discrete points (nodes) of the domain. Normally these values are interpreted and used for calculation of other physical entities, such as flux either throughout the domain or in certain regions of it.

We have elected to present a finite element solution to the two-dimensional heat transfer equations typical of that presented in Chapter 8.

Typical examples of this type of problem include steady seepage through soils and steady heat flow through a conductor. The program given is for two-dimensional planar conditions and uses four-node quadrilateral elements. Each node has only one unknown or degree of freedom associated with it. The unknown would represent, in the case of seepage problems, the fluid potential and, in the case of conduction problems, the temperature.

Systems that are governed by Laplace's equation require boundary conditions to be prescribed at all points around a closed domain. These boundary conditions commonly take the form of fixed values of the potential or values of the first derivative of the potential normal to the boundary. The problem amounts to finding the values of the fluid potential at points within the closed domain.

Being 'elliptic' in character, solution of Laplace's equation quite closely resembles the solution of the equilibrium equations in solid elasticity. Both methods ultimately require the solution of a set of linear simultaneous linear equations. The element `stiffness' matrices are formed numerically and assembled into a global stiffness matrix which is symmetrical and banded.

It is not the purpose of this text to go into detail behind the subroutines presented in the following sections. There are many texts that consider this subject.

9.3.2 General Flow Diagram for Solving Sample Problem

The following unstructured flow diagram is presented to illustrate the flow of the program.

Reserve space for all fixed and variable dimension arrays

Read data and initialize arrays

FOR ALL ELEMENTS

Find nodal coordinates

Determine the steering vector

Null the element matrices

Find the shape functions

Convert to global coordinates

Form the stiffness matrix

Add contributions from boundary conditions

Form global stiffness matrix

Solve equations

Print results

9.3.3 The Problem

The problem under consideration consists of a rectangular area with dimensions of 6x4. The top surface is held at a temperature of 0°F and the right side is fixed at a temperature of 50°F. The slab is insulated at the other two faces. Material properties for aluminum are used for this two-dimensional planar problem. The problem is illustrated in figure 9-1. The numbers shown at the corners are the node numbers and the numbers within the elements are the element numbers. Also shown are the top and right edge temperatures.

Problem specifics are listed below:

```
CONTROL INFORMATION

  number of node points             =          35
  number of element types           =           2
  number of load cases              =           1
  gravitational constant            =   3.8640E+02
```

NODAL DATA

NODE NO.	BOUNDARY CONDITION CODES DX	DY	DZ	RX	RY	RZ	NODAL POINT COORDINATES X	Y	Z	T
1	1	0	1	1	1	1	0.000E+00	0.000E+00	0.000E+00	0.000E+00
2	1	0	1	1	1	1	0.000E+00	1.000E+00	0.000E+00	0.000E+00
3	1	0	1	1	1	1	0.000E+00	2.000E+00	0.000E+00	0.000E+00
4	1	0	1	1	1	1	0.000E+00	3.000E+00	0.000E+00	0.000E+00
5	1	0	1	1	1	1	0.000E+00	4.000E+00	0.000E+00	0.000E+00
6	1	0	1	1	1	1	0.000E+00	5.000E+00	0.000E+00	0.000E+00
7	1	0	1	1	1	1	0.000E+00	6.000E+00	0.000E+00	0.000E+00
8	1	0	1	1	1	1	0.000E+00	0.000E+00	1.000E+00	0.000E+00
9	1	0	1	1	1	1	0.000E+00	1.000E+00	1.000E+00	0.000E+00
10	1	0	1	1	1	1	0.000E+00	2.000E+00	1.000E+00	0.000E+00
11	1	0	1	1	1	1	0.000E+00	3.000E+00	1.000E+00	0.000E+00
12	1	0	1	1	1	1	0.000E+00	4.000E+00	1.000E+00	0.000E+00
13	1	0	1	1	1	1	0.000E+00	5.000E+00	1.000E+00	0.000E+00
14	1	0	1	1	1	1	0.000E+00	6.000E+00	1.000E+00	0.000E+00
15	1	0	1	1	1	1	0.000E+00	0.000E+00	2.000E+00	0.000E+00
16	1	0	1	1	1	1	0.000E+00	1.000E+00	2.000E+00	0.000E+00
17	1	0	1	1	1	1	0.000E+00	2.000E+00	2.000E+00	0.000E+00
18	1	0	1	1	1	1	0.000E+00	3.000E+00	2.000E+00	0.000E+00
19	1	0	1	1	1	1	0.000E+00	4.000E+00	2.000E+00	0.000E+00
20	1	0	1	1	1	1	0.000E+00	5.000E+00	2.000E+00	0.000E+00
21	1	0	1	1	1	1	0.000E+00	6.000E+00	2.000E+00	0.000E+00
22	1	0	1	1	1	1	0.000E+00	0.000E+00	3.000E+00	0.000E+00
23	1	0	1	1	1	1	0.000E+00	1.000E+00	3.000E+00	0.000E+00
24	1	0	1	1	1	1	0.000E+00	2.000E+00	3.000E+00	0.000E+00
25	1	0	1	1	1	1	0.000E+00	3.000E+00	3.000E+00	0.000E+00
26	1	0	1	1	1	1	0.000E+00	4.000E+00	3.000E+00	0.000E+00
27	1	0	1	1	1	1	0.000E+00	5.000E+00	3.000E+00	0.000E+00
28	1	0	1	1	1	1	0.000E+00	6.000E+00	3.000E+00	0.000E+00
29	1	0	1	1	1	1	0.000E+00	0.000E+00	4.000E+00	0.000E+00
30	1	0	1	1	1	1	0.000E+00	1.000E+00	4.000E+00	0.000E+00
31	1	0	1	1	1	1	0.000E+00	2.000E+00	4.000E+00	0.000E+00
32	1	0	1	1	1	1	0.000E+00	3.000E+00	4.000E+00	0.000E+00
33	1	0	1	1	1	1	0.000E+00	4.000E+00	4.000E+00	0.000E+00
34	1	0	1	1	1	1	0.000E+00	5.000E+00	4.000E+00	0.000E+00
35	1	0	1	1	1	1	0.000E+00	6.000E+00	4.000E+00	0.000E+00

number of elements = 24
number of materials = 1
maximum temperatures per material = 1

MATERIAL PROPERTIES

weight density = 9.8000E-02

```
mass   density                     =            2.5360E-04
Stefan-Boltzmann constant          =            3.3025E-15
absolute temperature conversion    =            4.5970E+02
```

TEMPERATURE	K(N)	K(S)	K(T)	SP. HEAT
.0	2.315E-03	2.315E-03	2.315E-03	2.140E-01

ELEMENT CONNECTIVITY DATA

ELEM NO.	NODE I	NODE J	NODE K	NODE L
1	1	2	9	8
2	2	3	10	9
3	3	4	11	10
4	4	5	12	11
5	5	6	13	12
6	6	7	14	13
7	8	9	16	15
8	9	10	17	16
9	10	11	18	17
10	11	12	19	18
11	12	13	20	19
12	13	14	21	20
13	15	16	23	22
14	16	17	24	23
15	17	18	25	24
16	18	19	26	25
17	19	20	27	26
18	20	21	28	27
19	22	23	30	29
20	23	24	31	30
21	24	25	32	31
22	25	26	33	32
23	26	27	34	33
24	27	28	35	34

TEMPERATURE ELEMENTS

number of elements = 11

ELEMENT CONNECTIVITY DATA

ELEMENT NUMBER	NODE N	CODE KD	SPECIFIED TEMPERATURE	THERMAL STIFFNESS
1	7	1	5.0000E+01	1.0000E+01
2	14	1	5.0000E+01	1.0000E+01

3	21	1	5.0000E+01	1.0000E+01
4	28	1	5.0000E+01	1.0000E+01
5	29	1	0.0000E+00	1.0000E+01
6	30	1	0.0000E+00	1.0000E+01
7	31	1	0.0000E+00	1.0000E+01
8	32	1	0.0000E+00	1.0000E+01
9	33	1	0.0000E+00	1.0000E+01
10	34	1	0.0000E+00	1.0000E+01
11	35	1	0.0000E+00	1.0000E+01

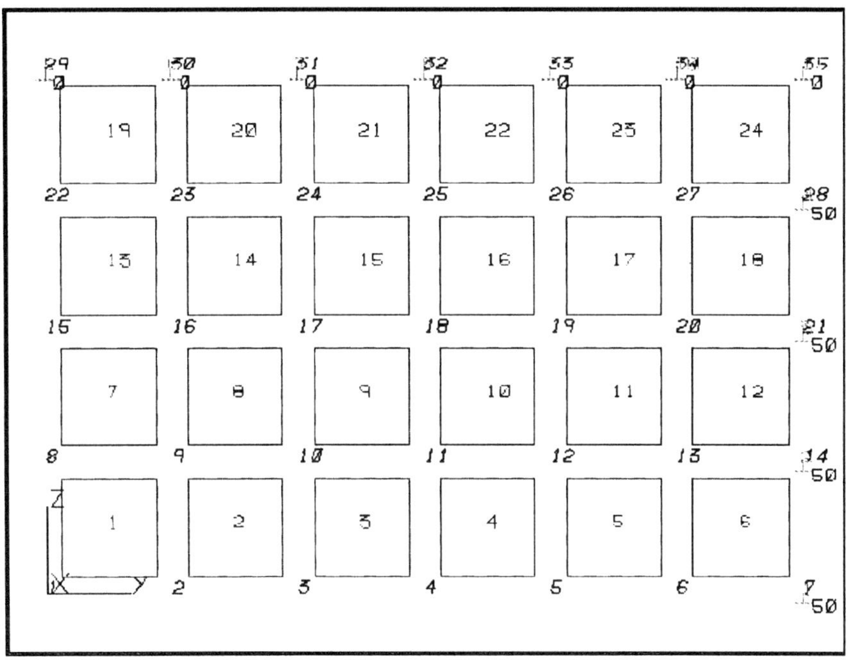

Figure 9-1: Node and Element Numbers For Example Problem

9.3.4 Variable Definitions for Sample FEA Program

Before giving the program solution to the problem it is worthwhile to list the variables used in the main program and subroutines with a short description of the variable.

NEX	number of elements in x direction
NEY	number of elements in y direction
N	number of degrees of freedom in mesh
IW	half-bandwidth of mesh
NN	number of nodes in mesh
NR	number of restrained nodes in mesh
NGP	number of integrating points per dimension
AA	dimension of elements in x direction
BB	dimension of elements in y direction
CONDX	conduction in x direction
CONDY	conduction in y direction
IFIX	number of prescribed freedoms
IDER	size of shape function (DER) matrix (local)
IDERIV	size of shape function (DERIV) matrix global)
IJAC	size of Jacobian matrix
IJACI	size of inverse Jacobian matrix
ICOND	size of conductivity matrix
IKDERV	size of product matrix COND*DERIV
IWGTSAMP	size of array for Gauss Quadrature weights
IDTKD	size of prod. array DERIVT*COND*DERIV
IKP	size of element stiffness matrix
ICOORD	size of element coordinate matrix
IDERVT	size of DERIV transpose
IKV	size of global stiffness matrix
ILOADS	size of global potentials
INF	size of nodal freedom array
INO	size of prescribed freedom array
IR	N*(IW+1)-working length of vectors KV and KVH
IT	dimensions of problem

NOD	number of nodes per element
NODOF	number of freedoms per node
DET	determinant of element Jacobian matrix
QUOT	scaled weighting coefficient
TEMP	nodal temperatures
WGTSAMP	quadrature abscissae and weights
JAC	Jacobian matrix
JACI	inverse of Jacobian Matrix
CON	conduction matrix
DTKD	product of DERIVT*CON*DERIV
KP	element stiffness matrix
COORD	element nodal coordinates
DER	derivatives of shape functions in local coordinates
DERIV	derivatives of shape functions in global coordinates
DERIVT	transpose of DERIV
KDERIV	product CON*DERIV
FUNCTION	element shape functions in local coordinates
G	element steering vector
KV	global `stiffness' matrix (IKV > IR)
KVH	copy of KV
LOADS	global potentials (ILOADS > N)
DISPS	calculated temperatures
NF	nodal freedom array (TNF > NN)
NO	prescribed freedoms (TNO > IFTX)
VAL	value to be fixed

9.3.5 The Finite Element Program

```
C       SOLUTION OF TWO-DIMENSIONAL HEAT CONDUCTION EQUATION OVER
C       A PLANE AREA USING 4-NODE QUADRILATERALS
C
        PARAMETER(IKV=1000,ILOADS=150,INF=50,INO=10)
C
      REAL JAC(2,2),JACI(2,2),CON(2,2),WGTSAMP(3,2),DTKD(4,4),KP(4,4),
     1COORD(4,2),DER(2,4),DERIV(2,4),DERIVT(4,2),KDERIV(2,4),FUNCTION(4),
     1VAL(INO),KVH(IKV),KV(IKV),LOADS(ILOADS),DISPS(ILOADS)
        INTEGER G(4),NO(INO),NF(INF,1)
        DATA IT,IJAC,IJACI,ICON,IDER,IDERIV,IKDERV/7*2/,IWGTSAMP/3/
        DATA IDTKD,IKP,ICOORD,IDERVT,NOD/5*4/,NODOF/1/
C
C      INPUT AND INITIALIZATION
C
        READ(5,*) NEX,NEY,N,IW,NN,NR,NGP,AA,BB,CONDX,CONDY
        CALL READNF(NF,INF,NN,NODOF,NR)
        IR=N*(IW+1)
        CALL NULVEC(KV,IR)
        CALL NULL(CON,ICON,IT,IT)
        CON(1,1)=CONDX
        CON(2,2)=CONDY
        CALL GAUSLEGN(WGTSAMP,IWGTSAMP,NGP)
C
C       ELEMENT INTEGRATION AND ASSEMBLY
C
        DO 10 IP=1,NEX
        DO 10 IQ=1,NEY
        CALL QUADGEO(IP,IQ,NEX,AA,BB,COORD,ICOORD,G,NF,INF)
        CALL NULL(KP,IKP,NOD,NOD)
        DO 20 I=1,NGP
        DO 20 J=1,NGP
        CALL SHAPE(DER,IDER,FUNCTION,WGTSAMP,IWGTSAMP,I,J)
        CALL MATMUL(DER,IDER,COORD,ICOORD,JAC,IJAC,IT,NOD,IT)
        CALL INV2X2(JAC,IJAC,JACI,IJACI,DET)
        CALL MATMUL(JAC1,IJAC1,DER,IDER,DERIV,IDERIV,IT,IT,NOD)
        CALL MATMUL(CON,ICON,DERIV,IDERIV,KDERIV,IKDERV,IT,IT,NOD)
        CALL MATRAN(DERIVT,IDERVT,DERIV,IDERIV,IT,NOD)
        CALLMATMUL(ERIVT,IDERVT,KDERIV,IKDERV,DTKD,IDTKD,NOD,IT,NOD)
        QUOT=DET*WGTSAMP(I,2)*WGTSAMP(J,2)
        CALL MSMULT(DTKD,IDTKD,QUOT,NOD,NOD)
  20    CALL MATADD(KP,IKP,DTKD,IDTKD,NOD,NOD)
  10    CALL GBLSTIF(KV,KP,IKP,G,N,NOD)
        CALL VECCOPY(KV,KVH,IR)
C
C       SPECIFY FIXED Potentials AND REDUCE EQUATIONS
C
        READ(5,*)IFIX,(NO(I),VAL(I),I.I,IFIX)
        CALL NULVEC(LOADS,N)
        DO 30 I=1,IFIX
        KV(NO(I))=KV(NO(I))+1.E20
  30    LOADS(NO(I))=KV(NO(I))*VAL(I)
        CALL BWREDUC(KV,N,IW)
C
C       SOLVE EQUATIONS AND RETRIEVE TEMPERATURES
C
        CALL BACSUB(KV,LOADS,N,IW)
```

```
      CALL PRINTV(LOADS,N)
      CALL MATVECMU(KVH,LOADS,DISPS,N,IW)
      TEMP=0.
      DO 40 I=1,IFIX
   40 TEMP=DISPS(NO(I))
      WRITE(6,'(E12.4)')TEMP
      STOP
      END

      SUBROUTINE BACSUB(BK,LOADS,N,IW)
C
C     THIS SUBROUTINE PERFORMS THE GAUSSIAN BACK-SUBSTITUTION
C
      REAL BK(*),LOADS(*)
      LOADS(I).LOADS(I)/BK(I)
      DO 1 I=2,N
      SUM=LOADS(I)
      I1=I-1
      NKB=I-IW
      IF(NKB)2,2,3
    2 NKB=I
    3 DO 4 K=NKB,II
      JN=(I-K)*N+K
      SUM=SUM-BK(JN)*LOADS(K)
    4 CONTINUE
      LOADS(I)=SUM/BK(I)
    1 CONTINUE
      DO 5 JJ=2,N
      I=N-JJ+1
      SUM=0.
      Il=I+1
      NKB=I+IW
      IF(NKB-N)7,7,6
    6 NKB=N
    7 DO 8 K=II,NKB
      JN=(K-I)*N+I
    8 SUM=SUM+BK(JN)*LOADS(K)
      LOADS(I)=LOADS(I)-SUM/BK(I)
    5 CONTINUE
      RETURN
      END

      SUBROUTINE BWREDUC(BK,N,IW)
C
C     THIS SUBROUTINE PERFORMS GAUSSIAN REDUCTION OF
C     THE STIFFNESS MATRIX STORED AS A VECTOR BK(N*(IW+1))
C
      REAL BK(*)
      DO 1 I=2,N
      IL1=I-1
      KBL=IL1+IW+I
      IF(KBL-N)3,3,2
    2 KBL.N
    3 DO 1 J=I,KBL
      IJ=(J-I)*N+I
      SUM=BK(IJ)
      NKB=J-IW
      IF(NKB)4,4,5
    4 NKB=I
    5 IF(NKB-IL1)6,6,8
```

```
    6     DO 7 M=NKB,IL1
          NI=(I-M)*N+M
          NJ=(J-M)*N+M
    7     SUM=SUM-BK(NI)*BK(NJ)/BK(M)
    8     BK(IJ).SUM
    1     CONTINUE
          RETURN
          END

          SUBROUTINE GBLSTIF(BK,KM,IKM,G,N,IDOF)
C         THIS SUBROUTINE FORMS THE GLOBAL STIFFNESS MATRIX
C         STORING THE UPPER TRIANGLE AS A VECTOR BK(N*(IW+1))
C
          REAL BK(*),KM(IKM,*)
          INTEGER G(*)
          DO 1 I=1,IDOF
          IF(G(I).EQ.0)GOTO 1
          DO 5 J=1,IDOF
          IF(G(J).EQ.0)GOTO 5
          ICD=G(J)-G(I)+1
          IF(ICD-1)5,4,4
        4 IVAL=N*(ICD-1)+G(I)
          BK(IVAL)=BK(IVAL)+KM(I,J)
        5 CONTINUE
        1 CONTINUE
          RETURN
          END

          SUBROUTINE SHAPE(DER,IDER,FUNCTION,WGTSAMP,IWGTSAMP,I,J)
C
C         THIS SUBROUTINE FORMS THE SHAPE FUNCTIONS AND
C         THEIR DERIVATIVES FOR 4-NODED QUAD ELEMENT
C
          REAL DER(IDER,*),FUNCTION(*),WGTSAMP(IWGTSAMP,*)
          ETA=WGTSAMP(I,I)
          XI=WGTSAMP(J,1)
          ETAM=.25*(1.-ETA)
          ETAP=.25*(1.+ETA)
          XIM=.25*(1.-XI)
          XIP=.25*(1.+XI)
          FUNCTION(1)=4.*XIM*ETAM
          FUNCTION(2)=4-*XIM*ETAP
          FUNCTION(3)=4.*XIP*ETAP
          FUNCTION(4)=4.*XIP*ETAM
          DER(1,1)=-ETAM
          DER(1,2)=-ETAP
          DER(1,3)=ETAP
          DER(1,4)=ETAM
          DER(2,1)=-XIM
          DER(2,2)=XIM
          DER(2,3)=XIP
          DER(2,4)=-XIP
          RETURN
          END

          SUBROUTINE MATVECMU(BK,DIS PS,LOADS,N,IW)
C         THIS SUBROUTINE MULTIPLIES A MATRIX BY A VECTOR
C         THE MATRIX IS SYMMETRICAL WITH ITS UPPER TRIANGLE
C         STORED AS A VECTOR
          REAL BK(*),DISPS(*),LOADS(*)
```

```
      DO 1 I=1,N
      X=0.
      DO 2 J=1,IW+1
      IF(I+J.LE.N+1)X=X+BK(N*(J 1)+I)*DISPS(I+J- 1)
    2 CONTINUE
      DO 3 J=2,IW+1
      IF(I-J +1.GE.1)X=X+BK((N-1)*(J-1)+I)* DISPS(I-J+1)
    3 CONTINUE
      LOADS(I)=X
    1 CONTINUE
      RETURN
      END

      SUBROUTINE MATMUL(A,IA,B,IB,C,IC,L,M,N)
C
C     THIS SUBROUTINE FORMS THE PRODUCT OF TWO MATRICES
C
      REAL A(IA,*),B(IB,*),C(IC,*)
      DO 1 I=1,L
      DO 1 J=1,N
      X=0.0
      DO 2 K.I,M
  2   X=X+A(I,K)*B(K,J)
      C(I,J)=X
  1   CONTINUE
      RETURN
      END

      SUBROUTINE MATRAN(A,IA,B,IB,M,N)
C
C     THIS SUBROUTINE FORMS THE TRANSPOSE OF A MATRIX
C
      REAL A(IA,*),B(IB,*)
      DO 1 I=1,M
      DO 1 J=1,N
    1 A(J,I)=B(I,J)
      RETURN
      END

      SUBROUTINE MVMULT(M,IM,V,K,I,Y)
C
C     THIS SUBROUTINE MULTIPLIES A MATRTX BY A VECTOR
C
      REAL M(IM,*),V(*),Y(*)
      DO 1 I=1,K
      X=0.
      DO 2 J=1,L
  2   X=X+M(I,J)*V(J)
      Y(I)=X
  1   CONTINUE
      RETURN
      END

      SUBROUTINE NULL(A,IA,M,N)
C
C     THIS SUBROUTINE NULLS A 2-D ARRAY
C
      REAL A(IA,*)
      DO 1 I=1,M
      DO 1 J=1,N
```

```
  1     A(I,J)=0.0
        RETURN
        END

        SUBROUTINE NULVEC(VEC,N)
C
C       THIS SUBROUTINE NULLS A COLUMN VECTOR
C
        REAL VEC(*)
        DO 1 I=1,N
  1     VEC(I).0.
        RETURN
        END

        SUBROUTINE MATADD(A,IA,B,IB,M,N)
C
C       THIS SUBROUTINE ADDS TWO EQUAL SIZED ARRAYS
C
        REAL A(IA,*),B(IB,*)
        DO 1 I=1,M
        DO 1 J=1,N
  1     A(I,J)=A(I,J)+B(I,J)
        RETURN
        END

        SUBROUTINE INV2X2(JAC,IJ AC,JACI,IJACI,DET)
C
C       THIS SUBROUTINE FORMS THE INVERSE OF A 2 BY 2 MATRIX
C
        REAL JAC(IJAC,*),JAC1(IJAC1,*)
        DET=JAC(1,1)*JAC(2,2)-JAC(1,2)*JAC(2,1)
        JAC1(1,1)=JAC(2,2)
        JAC1(1,2)=-JAC(1,2)
        JAC1(2,1)=-JAC(2,1)
        JACI(2,2)=JAC(1,1)
        DO 1 K=1,2
        DO 1 L=1,2
  1     JAC1(K,L)=JACI(K,L)/DET
        RETURN
        END


        SUBROUTINE VECCOPY(A,B,N)
C
C       THIS SUBROUTINE COPIES VECTOR A INTO VECTOR B
C
        REAL A(*),B(*)
        DO 1 I=1,N
  1     B(I)=A(I)
        RETURN
        END

        SUBROUTINE GAUSLEGN(WGTSAMP,IWGTSAMP,NGP)
C
C       THIS SUBROUTINE PROVIDES THE WEIGHTS AND SAMPLING POINTS
C       FOR GAUSS-LEGENDRE QUADRATURE
C
        REAL WGTSAMP(IWGTSAMP,*)
        GO TO(1,2,3,4,5,6,7),NGP
  1     WGTSAMP(1,1)=0
```

```
      WGTSAMP(1,2)=2.
      GOTO 100
2     WGTSAMP(1,1)=1./SQRT(3.)
      WGTSAMP(2,1).-WGTSAMP(I,I)
      WGTSAMP(1,2)=1.
      WGTSAMP(2,2)=1.
      GO TO 100
3     WGTSAMP(1,1)=.2*SQRT(15.)
      WGTSAMP(2,1)=.0
      WGTSAMP(3,1)=-WGTSAMP(1,1)
      WGTSAMP(1,2)=5./9.
      WGTSAMP(2,2)=8./9.
      WGTSAMP(3,2)=WGTSAMP(I,@)
      GO TO 100
4     WGTSAMP(1,1)=.861136311594053
      WGTSAMP(2,1)=.339981043584856
      WGTSAMP(3,1)=-WGTSAMP(2,1)
      WGTSAMP(4,1)=-WGTSAMP(I,I)
      WGTSAMP(1,2)=.347854845137454
      WGTSAMP(2,2)=.652145154862546
      WGTSAMP(3,2)=WGTSAMP(2,2)
      WGTSAMP(4,2)=WGTSAMP(1,2)
      GO TO 100
5     WGTSAMP(1,1)=.906179845938664
      WGTSAMP(2,1)=.538469310105683
      WGTSAMP(3,1)=.0
      WGTSAMP(4,1)=-WGTSAMP(2,1)
      WGTSAMP(5,1)=-WGTSAMP(I,I)
      WGTSAMP(1,2)=.236926885056189
      WGTSAMP(2,2)=.478628670499366
      WGTSAMP(3,2)=.568888888888889
      WGTSAMP(4,2)=WGTSAMP(2,2)
      WGTSAMP(5,2)=WGTSAMP(1,2)
      GO TO 100
6     WGTSAMP(1,1)=.932469514203152
      WGTSAMP(2,1)=.661209386466265
      WGTSAMP(3,1)=.238619186083197
      WGTSAMP(4,1)=-WGTSAMP(3,1)
      WGTSAMP(5,1)=-WGTSAMP(2,1)
      WGTSAMP(6,1)=-WGTSAMP(I,I)
      WGTSAMP(1,2)=.171324492379170
      WGTSAMP(2,2)=.360761573048139
      WGTSAMP(3,2)=.467913934572691
      WGTSAMP(4,2)=WGTSAMP(3,2)
      WGTSAMP(5,2)=WGTSAMP(2,2)
      WGTSAMP(6,2)=WGTSAMP(1,2)
      GO TO 100
7     WGTSAMP(1,1)=.949107912342759
      WGTSAMP(2,1)=.741531185599394
      WGTSAMP(3,1)=.405845151377397
      WGTSAMP(4,1)=.0
      WGTSAMP(5,1)=-WGTSAMP(3,1)
      WGTSAMP(6,1)=-WGTSAMP(2,1)
      WGTSAMP(7,1)=-WGTSAMP(I,I)
      WGTSAMP(1,2)=.129484966168870
      WGTSAMP(2,2)=.279705391489277
      WGTSAMP(3,2)=.381830050505119
      WGTSAMP(4,2)=.417959183473469
      WGTSAMP(5,2)=WGTSAMP(3,2)
      WGTSAMP(6,2)=WGTSAMP(2,2)
```

```
      WGTSAMP(7,2)=WGTSAMP(1,2)
  100 CONTINUE
      RETURN
      END

      SUBROUTINE QUADGEO(IP,IQ,NEX,AA,BB,COORD,ICOORD,G,N F,INF)
C
C     THIS SUBROUTINE FORMS THE COORDINATES AND STEERING VECTOR
C     FOR 4-NODE QUADS COUNTING IN X-DIRECTION
C     LAPLACE'S EQUATION 1-FREEDOM PER NODE
C
      REAL COORD(ICOORD,*)
      INTEGER NUM(4),G(*),NF(INF,*)
      NUM(1)=IQ*(NEX+1)+IP
      NUM(2)=(IQ-1)*(NEX+1)+IP
      NUM(3)=NUM(2)+1
      NUM(4)=NUM(1)+1
      DO 1 I=1,4
    1 G(I).NF(NUM(I),I)
      COORD(1,1)=(IP-1)*AA
      COORD(2,1)=(IP-1)*AA
      COORD(3,1)=IP*AA
      COORD(4,1)=IP*AA
      COORD(1,2)=-IQ*BB
      COORD(2,2)=-(IQ-1)*BB
      COORD(3,2)=-(IQ-1)*BB
      COORD(4,2)=-IQ*BB
      RETURN
      END
```

9.3.6 Solution Results

The temperature results are listed below and illustrated in Figure 9-2.

TEMPERATURES AT NODES

node number	temperature
1	1.1442D+01
2	1.2348D+01
3	1.5188D+01
4	2.0297D+01
5	2.8004D+01
6	3.8209D+01
7	5.0000D+01
8	1.0577D+01
9	1.1421D+01

10 1.4079D+01
11 1.8928D+01
12 2.6587D+01
13 3.7248D+01
14 5.0000D+01
15 8.1071D+00
16 8.7652D+00
17 1.0862D+01
18 1.4821D+01
19 2.1590D+01
20 3.3596D+01
21 5.0000D+01
22 4.3939D+00
23 4.7568D+00
24 5.9255D+00
25 8.1991D+00
26 1.2394D+01
27 2.0947D+01
28 4.9999D+01
29 5.3659D-05
30 1.1634D-04
31 1.4570D-04
32 2.0463D-04
33 3.2055D-04
34 6.4310D-04
35 3.5455D-04

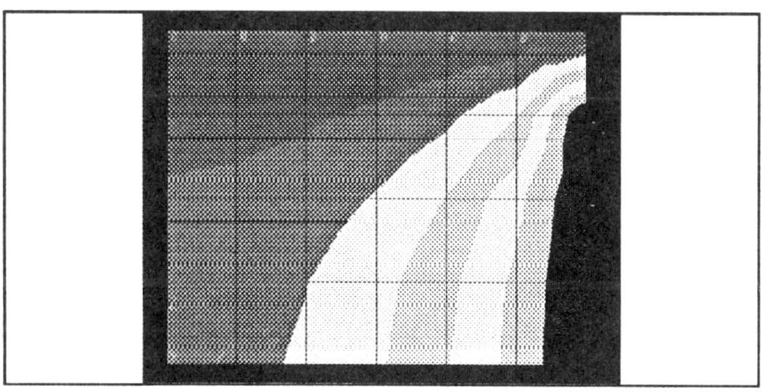

Figure 9-2: Analysis Results

9.4 Chapter Summary

The example presented in this chapter illustrates how to put the pieces together to solve seemingly complex problems. The above listed program can be used to solve two-dimensional potential flow problems by the following substitutions.:

Heat Conduction	Potential Flow
Temperature	Velocity Potential
Heat Flux	Normal Gradient
Insulated Edge	Zero Normal Flow

9.5 Problems

9-1) Prepare a flowchart of the computer code in the text and examine the possibility of a change of boundary conditions for the problem and how to implement the change(s).

9-2) Construct a problem analogous to the thermal problem in the chapter and calculate the flow field.

9-3) Cite other examples where you could solve flow problems using a thermal model and explain the analogy.

10

Fourier Transforms

10.1 Introduction

This chapter touches upon the subject of Fourier transforms and how data may be may be examined using this technique. The basic idea in a Fourier series is to express a periodic wave as a sum of complex exponentials all of which have the same period. To accomplish this, the complex exponentials satisfy the following orthogonality relation:

$$\frac{1}{P}\int_0^P e^{(i2\pi nx)/P}e^{(-i2\pi nx)/P}dx \begin{array}{l} = 0 \textit{ for } m \neq n \\ = 1 \textit{ for } m = n \end{array} \tag{10.1}$$

where P is the period.

A Fourier Transform is a common operation for analyzing various kinds of data to find the spectrum of the data. The idea is typically to pick out components of the data with particular frequencies or ranges of frequencies. This chapter is not intended to be anywhere near an exhaustive study of the subject but is included as an addition to the areas covered earlier in the text on interpolation and extrapolation. There are two uses for approximating functions. One is to replace complicated functions with a simpler function such that common operations such as differentiation and integration or evaluation may be performed more easily. The second lies in the fact that it is possible to recover a function from knowing just partial information about the data.

Many physical phenomena have periodic characteristics. Examples would be light and sound. A physical process can be described either in the time domain or in the frequency domain. The two representations are related by the Fourier transform equations developed below. These phenomena are described by functions, f(x), which are periodic and satisfy, for all x and some fixed number τ, the following:

$$f(x + \tau) = f(x) \tag{10.2}$$

Since the only periodic polynomials are the constant functions, one has to use other function classes for the effective approximation of periodic functions. Trigonometric polynomials offer themselves as an acceptable alternative.

Trigonometric polynomials of order n are usually written in the following format:

$$p(x) = \sum_{j=-n}^{n} c_j e^{ijx} \tag{10.3}$$

where the symbol i is the imaginary unit

$$i = \sqrt{-1} \tag{10.4}$$

Approximation by trigonometric polynomials is dominated by the Fourier series

$$f(x) \approx \sum_{j=-\infty}^{\infty} \hat{f}(j) e^{ijx} \tag{10.5}$$

with the Fourier coefficients calculated by

$$\hat{f}(j) = \frac{1}{2\pi} \int_0^{2\pi} f(x) e^{-ijx} dx \tag{10.6}$$

This series converges to f(x) for rather mild conditions. If f(x) is continuous with a piecewise-continuous first derivative, the series converges uniformly.

Using Parseval's relation that allows a conversion from time integration to frequency integration

$$\sum_j |\hat{f}(j)|^2 = \frac{1}{2\pi}\int_0^{2\pi} |f(x)|^2 dx \tag{10.7}$$

it can be shown that

$$f(x) = 2\sum_{j=0}^{\infty} |\hat{f}(j)| \cos\,(\theta_j + jx) \tag{10.8}$$

Therefore the periodic motion is now represented by a sum of simple harmonic oscillations. The amplitude is the first term, the frequency is $j/2\pi$, and the phase angle is θ_j.

The sequence of

$$|\,\hat{f}(0)\,|,\ |\,\hat{f}(1)\,|,\ \ldots \tag{10.9}$$

is called the power spectrum or just simply the spectrum.

It is usually not possible to calculate exactly the Fourier coefficients because the integral cannot be evaluated in closed form or the function f(x) is not known exactly. Numerical integration, which was discussed in Chapter 3, is used to facilitate the calculation process. Usually the composite trapezoidal rule is used. The N equally spaced points used for the integration are called the sampling points and the corresponding functional values are called the sample values. The number $2\pi/N$ is called the sampling interval and the inverse is called the sampling frequency.

For accuracy in applying this method, if you wish to observe a certain periodic phenomenon of frequency, then you must sample at a frequency twice as large as where you want to observe.

10.2 Fast Fourier Transforms

In discrete harmonic or Fourier analysis, the user calculates the Fourier coefficients by

$$c_j = \hat{f}_N(j) = \frac{1}{N}\sum_{k=0}^{N-1} f(x_k)e^{-ijx_k} \tag{10.10}$$

with

$$x_k = \frac{2\pi k}{N} \; for\ all\ k \tag{10.11}$$

to resolve the 2 π periodic motion described by f(x) into simple harmonics.

The evaluation of any particular function requires calculations on the order of N^2 operations (N is the number of sampling points). However, because of the strong interrelations between the numbers, the number of calculations is on the order of N $\log_2$ N. From this comes the name of fast Fourier transform or FFT. There are many FFTs and for the greatest efficiency in calculation it is necessary to write a different program for each different value of N that one wishes to use.

The algorithm of Danielson and Lanczos showed that a discrete Fourier transform of length N can be written as the sum of two discrete Fourier transforms of length N/2 each. One transform is composed of the even-numbered points of the original and the other is composed of the odd-numbered points of the original. In equation form,

$$\begin{aligned} F_k &= \sum_{j=0}^{N-1} f_j e^{2\pi ijk/N} \\ &= \sum_{j=0}^{N/2-1} f_{2j} e^{2\pi ikj/(N/2)} + W^k \sum_{j=0}^{N/2-1} f_{2j+1} e^{2\pi ikj/(N/2)} \end{aligned} \tag{10.12}$$

where

$$W^k = e^{2\pi i/N} \tag{10.13}$$

The above algorithm can be used recursively to generate the required values. To efficiently code such an algorithm, it is assumed here that the length of the input data is a power of 2. In cases where the length of the data is not a power of 2, the length may be augmented by adding zero samples. The basic advantages of using sample lengths based on a power of two are increased flexibility and speed in computation. The disadvantage is the increased complexity of the computational algorithm.

To provide computational efficiency in the algorithm, the data is sorted so that the input is bit-reversed order. A thorough discussion of this technique is found in Gold and Radner [43]. The result of the sorting of data is then in normal order. Therefore, the program that implements the algorithm has basically two sections. The first section sorts the data into bit-reversed order. The second section calculates transforms of length 2, 4, 8 ...,N. The one point transformations are the data points. The adjacent pairs are combined to obtain the two-point transforms. Next, the adjacent two-point transforms are combined to obtain the four-point transforms. This is continued until the first and second halves of the complete data set are combined into the complete transform. The loop controlling this is executed $\log_2$ N times. Computation of the sines and cosines of multiple angles occurs in recurrence relations in the inner loops. DATAIN is replaced by its discrete Fourier transform.

The algorithm is implemented in Program 10.1F.

The next subroutine, Program 10.2P written in PASCAL, reads in the raw data from a file and converts that data to complex numbers for the transformation. Capability is given for plotting and graphing the information on screen.

I have given two additional programs for a Fourier sine and a Fourier cosine series. These are special forms of the Fourier series for functions possessing either odd or even symmetry. If a function is odd on a given interval (-N,N) (i.e. f(-x) = -f(x)) then a Fourier sine series is used for calculational purposes. If a function is even on the interval (i.e. f(-x) = f(x)) then a Fourier cosine series is used on the interval.

The use of Fourier transforms and how to develop the transforms is an interesting subject. The programs presented in this chapter only give a glimpse into the power of this utility. The reader is prompted to explore this area further when considering how to analyze data.

10.3 Problems

10-1) Expand the function $f(x) = x^2 - 2x$ in a Fourier series for a period of 1 over the interval [0,1].

10-2) Expand the function $f(x) = 1$ on the interval [0,1] in a Fourier sine series.

10-3) Prove the orthogonality of the Fourier sine series:

$$\frac{2}{L}\int_0^L \sin\frac{(m\pi x)}{L}\sin\frac{(n\pi x)}{L}dx = 0 \ for\ m \neq n$$

$$= 1 \ for\ m = n$$

Chapter 10 Programs

Program 10.1F Fourier Transforms Using Danielson-Lanczos Algorithm

```
      SUBROUTINE FFT (DATAIN,N)
C     DATAIN -ARRAY OF NUMBERED PAIRS, REAL & COMPLEX
C            OF LENGTH 2*N.
C     N - NUMBER OF COMPLEX DATA POINTS
C
      REAL*8 OR,OI,ORT,OIT,OTEMP,ANG
      DIMENSION DATAIN(2*N)
      PI=3.141592519
      NN=2*N
      J=1
C     ******BIT REVERSAL OCCURS HERE******
      DO 10 I=1,NN,2
            IF (J.GT.I) THEN
                  TEMPREAL=DATAIN(J)
                  TEMPIMG=DATAIN(J+1)
                  DATAIN(J)=DATAIN(I)
                  DATAIN(J+1)=DATAIN(I+1)
                  DATAIN(I)=TEMPREAL
                  DATAIN(I+1)=TEMPIMG
            ENDIF
            M=NN/2
    1       IF ((M.GE.2) .AND. (J.GT.M)) THEN
                  J=J-M
                  M=M/2
            GOTO 1
            ENDIF
            J=J+M
   11  CONTINUE
C     ******DANIELSON-LANCZOS ROUTINE******
      MAXM=2
   40 IF (N.GT.MAXM) THEN
            2STEP=2*MMAX
            ANG=PI/MAXM
            OR=DCOS(ANG)
            OI=DSIN(ANG)
            ORT=1.DO
            ORI=0.DO
            DO 20 K=1,MAXM,2
                  DO 30 I=K,N,2STEP
                        J=I+MAXM
                        TEMPREAL=SNGL(OR)*DATAIN(J)
     1                        -SNGL(OI)*DATAIN(J+1)
                        TEMPIMG=SNGL(OR)*DATAIN(J+1)
     1                        +SNGL(OI)*DATAIN(J)
                        DATAIN(J)=DATAIN(I)-TEMPREAL
                        DATAIN(J+1)=DATAIN(I+1)-TEMPIMG
                        DATAIN(I)=DATAIN(I)+TEMPREAL
                        DATAIN(I+1)=DATAIN(I+1)+TEMPIMG
                  30 CONTINUE
            OTEMP=OR
            OR=OR*ORT-OI*OIT+OR
            OI=OI*ORT+OTEMP*OIT+OI
      20   CONTINUE
      MAXM=2STEP
      GO TO 40
```

```
      ENDIF
      RETURN
      END
```

Program 10.2P Fourier Transforms With Graph/Plotting Capability

```
Program FFTfile;
Uses Crt,Graph;
Type
   complex = object
               re,im : real;
               function mag: real;
               procedure equate(a,b : real);
               procedure EquateReal(a : real);
               procedure Neg;
             end;

   Function complex.mag: real;
   begin
      mag := sqrt( sqr(re) + sqr(im) );
   end;


   Procedure complex.equate(a,b : real);
   begin
      re := a;
      im := b;
   end;

   Procedure complex.equatereal(a: real);
   begin
      re := a;
      im := 0.0;
   end;

   Procedure complex.neg;
   begin
      re := -re;
      im := -im;
   end;

{**********************************}

Var
   distance,i,j     : integer;
   expon            : longint;
   i1,i2,r          : integer;
   Data             : array[0..1023] of integer;
   F                : array[0..1023] of complex;
   Result           : array[0..1023] of real;
   K                : array[0..512] of complex;
   Ktemp,R1temp,R2temp : complex;
   norm,tsample     : real;
   tscale           : real;
   graphdriver,graphmode : integer;
   xc,yc,xscale     : integer;
   textstr          : string[12];
   pixy1,pixy2      : integer;
   infile           : text;
   filename         : string[128];
   num,numsamples   : integer;
```

```
   datareal        : real;

{**********************************}

Procedure CADD     (arg1,arg2 : complex;
                    var result    : complex );

begin
   result.re := arg1.re + arg2.re;
   result.im := arg1.im + arg2.im;
end;

{**********************************}

Procedure Cmult    (arg1,arg2 : complex;
                    var result : complex );

begin
   result.re := arg1.re*arg2.re - arg1.im*arg2.im;
   result.im := arg1.re*arg2.im + arg2.re*arg1.im;
end;

{**********************************}

Function Bit_reverse(arg : integer) : integer;
Var
   temp,i : integer;

begin
   temp := 0;
   for i := 1 to r do begin
      temp := temp shl(1);
      if (arg and 1)>0 then temp := temp + 1;
      arg := arg shr(1);
   end;
   Bit_reverse := temp;
end;

{**********************************}
{**********************************}

Begin
   clrscr;
   writeln('FFT Program');
   numsamples := 0;
   write('Input File Name:  ');
   readln(filename);
   for   i   :=   1   to   length(filename)   do   filename[i]   :=
upcase(filename[i]);
   assign(infile,filename);
   reset(infile);
   writeln('Reading Data');
   readln(infile,tsample);
    readln(infile,numsamples);
   for i := 0 to numsamples-1 do begin
      readln(infile,data[i]);
   end;
   close(infile);
```

```
   num := numsamples DIV 1;
   writeln(numsamples,' samples spaced at ',tsample:8:6,' seconds.');
   r := trunc(ln(num)/ln(2.0));
   writeln('Initializing');
   for i := 0 to num div 2 do begin
      K[i].equate(cos(2*pi*i/num),sin(2*pi*i/num));
   end;

   writeln('Re-Ordering Data');
   for i := 0 to num-1 do begin
      F[i].equatereal(Data[bit_reverse(i)]);
   end;


   distance := 1;
   repeat
      writeln('Distance=',distance);
      for i := 0 to (num div (2*distance))-1 do begin
         for j := 0 to distance-1 do begin
            i1 := 2*distance*i+j;
            i2 := i1 + distance;
            expon := (j*(num div distance)) div 2;
{            writeln(i1:3,i2:3,expon:3);}

            {BUTTERFLY}

            CMULT(F[i2],K[expon],Ktemp);
            CADD(F[i1],Ktemp,R1temp);
            Ktemp.neg;
            CADD(F[i1],Ktemp,R2temp);
            F[i1] := R1temp;
            F[i2] := R2temp;

            {END BUTTERFLY}

         end;
      end;
      distance := distance*2;
   until distance > num div 2;



   norm := 0.0000;
   for i := 0 to num-1 do begin
      Result[i] := F[i].mag;
      if result[i] > norm then norm := result[i];
   end;
   for i := 0 to num-1 do begin
      result[i] := result[i]/norm;
      if result[i] < 1e-5 then result[i] := 1e-5;
      result[i] := 20*ln(result[i])/ln(10.0);
   end;


   {DISPLAY}

   DetectGraph(graphdriver,graphmode);
   InitGraph(graphdriver,graphmode,'c:\tp6\bgi');
   xc := 36;
   yc := 450;
```

```
   xscale := 512 div (num div 2);

   line(xc,yc-400,xc,yc);
   line(xc,yc,xc+512,yc);
   settextstyle(defaultfont,horizdir,1);
   settextjustify(centertext,toptext);
   tscale := 1/tsample/2/10;
   for i := 0 to 10 do begin
      line(xc+51*i,yc,xc+51*i,yc+4);
      str((i*tscale/1000):4:1,textstr);
      textstr := textstr + 'k';
      outtextxy(xc+51*i,yc+10,textstr);
   end;
   outtextxy(xc+256,yc+20,'FREQUENCY (Hz)');

   settextstyle(defaultfont,vertdir,1);
   settextjustify(bottomtext,centertext);
   for i := 0 to 10 do begin
      line(xc,yc-40*i,xc-4,yc-40*i);
      str(-10*(10-i),textstr);
      outtextxy(xc-5,yc-40*i,textstr);
   end;
   outtextxy(xc-15,yc-200,'RESPONSE db');

   settextstyle(defaultfont,horizdir,1);
   settextjustify(bottomtext,lefttext);
   outtextxy(320,10,filename);
   str(tsample:8:6,textstr);
   textstr := 'Ts = '+textstr;
   outtextxy(320,20,textstr);
   str(num,textstr);
   textstr := textstr + ' points';
   outtextxy(320,30,textstr);

   for i := 1 to num div 2 do begin
      pixy1 := trunc(yc-400-4*result[i-1]);
      pixy2 := trunc(yc-400-4*result[i]);
      line(xc+xscale*(i-1),pixy1,xc+xscale*i,pixy2);
   end;

   for i := 0 to (num div 2)-1 do begin
      putpixel(i,240-data[i] div 64,red);
   end;
End.
```

11

Ready-Made Math Software Packages

11.1 Introduction

In this chapter we will review four commercially available software packages for doing mathematics on personal computers. These programs represent the state-of-the-art in programming and user interface. The programs reviewed are Mathcad from Mathsoft, Inc., Derive from Soft Warehouse, Inc., Mathematica from Wolfram Research, Inc., and TK Solver Plus from Universal Technical Systems, Inc. Problems from the text are used to demonstrate the capabilities of each program as well as other problems.

Spreadsheet products such as Quattro Pro from Borland International, Excel from Microsoft Corporation, and Lotus 1-2-3 from Lotus Development Corporation have been a mainstay for engineering personnel wanting to do desktop calculations. Their ready-made tabular format is ideal for many repetitive calculations. The products reviewed herein provide a broader level of sophistication in handling mathematics. These programs, except for TK Solver Plus, allow the user to type in their equations, symbols, etc. as if he/she were using pencil and paper. The software's engine evaluates the expressions and/or crunches the numbers. Results are displayed as numbers, equations, or as graphs and animations. Of course the output depends upon the capabilities of the program.

As technical calculation software becomes more of an accepted tool, apart from the traditional programming languages of FORTRAN, C, PASCAL, and BASIC, the programs will become more feature oriented for specialized tasks.

11.2 DERIVE, A Mathematical Assistant

DERIVE is a menu-driven symbolic math system for personal computers ranging from an XT-class machine to the 80486 variety of computers available today. The program performs both numerical and symbolic equation solving and covers exact and approximate arithmetic, calculus, trigonometry, and vector and matrix algebra. DERIVE supports both two- and three-dimensional graphs. The user can plot more than one function on the same graph and can use multiple windows for easy comparison, zoom in and out on plots with a single function key, and can accurately locate coordinates graphically with a cross-hair feature. If your printer supports the extended ASCII character set, you can print the mathematical expressions. For the capture of plots, you will need a screen capture program such as Pizazz Plus from Application Techniques, Inc.

In using DERIVE, I found a very versatile program that will increase anyone's ability to solve math problems and explore "what-ifs" in looking for and at solutions. To illustrate DERIVE, problems from a variety of mathematics areas are selected and solved.

DERIVE has the combination of power, low cost, relatively modest hardware requirements (minimum of an XT-class machine), and a short learning curve. These factors make this program an excellent choice for students to engineering and scientific professionals.

DERIVE has all of the basic symbolic math: manipulations of polynomials, trigonometric and hyperbolic functions, integer factorialization, manipulations of symbolic sums, and symbolic integration and differentiation. Advanced functions such as Bessel, zeta, or hypergeometric functions are included in a utilities section. DERIVE does not use these definitions in symbolic calculations. Matrix and vector manipulation is extensive and allows the user to find eigenvalues and eigenvectors. The utility section provides methods for solving ordinary differential equations. Higher order differential equations can be handled by substitutions until the higher order equation is reduced to a equivalent series of first order equations.

On simple numerics, DERIVE is very quick. Very complex problems tend to slow the program somewhat. The package supports both two- and three-dimensional graphs. These plots convey the information well but lack the color and shading capability of the more expensive programs like Mathematica. Multiple plots can appear on the same graph and the user can specify multiple windows for comparison plotting.

There are no specific printer or plotter drivers supplied with DERIVE. Most printers that support the extended ASCII character set will allow the printing of the on screen information. Commercially available screen capture programs will provide the necessary avenue for inserting calculations into word processing documents and presentation programs.

Derives's interface does not provide text wrap. Long answers extend off the screen; however you can scroll horizontally to view the answer. Derive does not include direct support for special functions such as Bessel, Error, or Statistical. Functions like these are called from a utility routine. As an example, Bessel functions are defined as integrals. For numerical evaluation, this representation is most adequate. The symbolic processor within the Derive main program doesn't know about theses functions. Therefore, if, for example, a derivative of a Bessel function is needed, the user will get a sum of integrals instead of an answer given in terms of lower order Bessel functions.

Derive does not contain an extensive programming language per se. Two special functions, ITERATE and IF, allow programmability especially since IF supports recursion. Even with this ability, I found programming somewhat difficult. This could be due to the lack of an in-depth familiarity with all aspects of the program.

Graphics within the program are good when compared with the higher end programs such as Mathematica; in fact, given the outward simplicity of the program and the targeted operating platforms, the graphics are outstanding. Standard 2D line plots as-well-as 3D plots are available. These plots can be displayed in a window adjacent to a window for the mathematical expressions. While not exploiting the hardware of higher powered machines such as the 80386 and 80486 versions, the program will work on a garden variety of low powered XTs and ATs. The program has more direct competition from the program Mathcad v3.1. Still, for those persons looking for the most bang for the buck in a math software package, Derive is an outstanding program.

The first derive figure is a 3D plot of the function z=sin(x+sin(y).

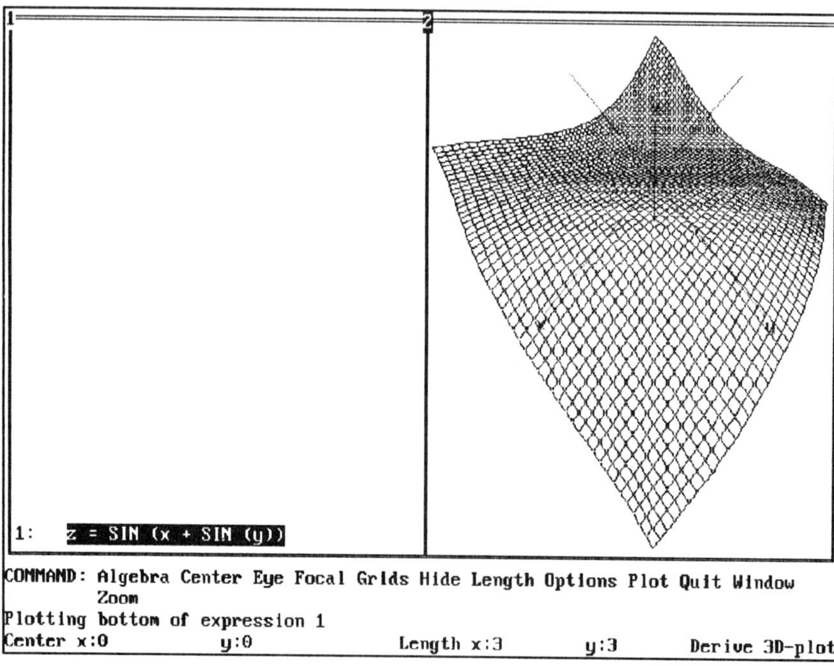

The second and third Derive figures illustrate some of the math capabilities of defining functions and integration and equation solving. Derive also does symbolic math as well.

```
                                               2           2
1:    DISTANCE (a, b, c, d) := √((c - a)  + (d - b) )

2:    DISTANCE (21, 11, -4, 10)

3:    √626

4:    25.0199

          1
      ─────────
            2
       1 + x

       ∞
      ⌠        1
      ⎮    ───────── dx
6:    ⎮          2
      ⌡     1 + x
       -∞

7:    ∞
COMMAND: Author Build Calculus Declare Expand Factor Help Jump soLve Manage
         Options Plot Quit Remove Simplify Transfer moVe Window approX
Compute time: 0.1 seconds
Simp(6)                                  Free:76%  Insert        Derive Algebra
```

```
1:    n

       k
2:     Σ   n
      n=1

       k (k + 1)
3:    ───────────
           2

4:    [x + y = 10, x - y = 5]

      ⎡      15        5 ⎤
5:    ⎢x = ─────, y = ───⎥
      ⎣       2        2 ⎦

6:    [x = 7.5, y = 2.5]

       2
7:    x  - 5 x + 6 = 0

8:    x = 2

9:    x = 3
COMMAND: Author Build Calculus Declare Expand Factor Help Jump soLve Manage
         Options Plot Quit Remove Simplify Transfer moVe Window approX
Compute time: 0.1 seconds
Solve(7)                                 Free:76%  Insert        Derive Algebra
```

11.3 Mathcad v3.1 (Windows)

Mathcad is the most widely used non-spreadsheet technical calculation software package. It is available on IBM-compatible PCs, Apple Macintoshes, and Unix workstations. Mathcad performs calculator type functions and has symbolic capabilities based on the Maple engine. Mathsoft has developed Mathcad as a technical publishing tool and has text as well as graphics import capabilities. The following section gives examples of how Mathcad is used. Some problems are taken from this text.

Mathcad runs a variety of PC (Windows and DOS), Macintosh, and UNIX platforms.

Mathcad has been the defacto standard for engineers, physicists, and others wanting to perform small to medium scale mathematical projects. Projects of this type include the computation of an integral, matrix manipulations, or solving a set of differential equations. The program, while lacking serious built-in programming capabilities has an interface designed for the person looking for a mathematical scratchpad. In fact, all advertisements point out the use of Mathcad as just that type of program. The user opens a window and enters equations, relations and graphics as if they were sitting down with pencil, paper, and a calculator of some description.

Mathcad has a limited symbolic calculation capabilities. The symbolic engine is licensed from Waterloo Maple Software, developers of Maple V, another high end mathematical package. This feature tends to round out Mathcad's capabilities although the symbolic processor is a very small subset of the Maple package. Mathcad's symbolic menu presents fourteen functions that were selected to cover most symbolic operations. These functions cover simplification and expansion, factorization, integration and differentiation,equation solving, series expansion and partial fractions, and the matrix operations of transposition, inverse, and determinant evaluation.

For everyday, quick calculations, Mathcad is an excellent package in the Windows format. The user can import and export data within the Windows environment and can make a nice presentation package.

To help in your work, Mathcad contains a Standard Electronic Handbook for instant access to many standard formulas, data, and useful formulas and commonly used equations. For the Windows environment, the user simply clicks on the formula or other information and pastes that information into the current document. There are available other electronic handbooks and application packages to aid in the rapid solution of problems.

The first demonstration comes from Chapter 6 and is the determination of eigenvectors and eigenvalues for a matrix.

Mathcad

Matrix Demonstration Using Equations 6.38 and 6.39.

$$M := \begin{pmatrix} 11 & 6 & -2 \\ -2 & 18 & 1 \\ -12 & 24 & 13 \end{pmatrix}$$

First, create the matrix by selecting Math from the menus and then select Matrices from the pull-down menus.

$$a := \text{eigenvals}(M)$$

$$a = \begin{pmatrix} 7 \\ 14 \\ 21 \end{pmatrix}$$

Fill in the matrix positions with the numbers as shown for the matrix M. The term "eigenvals( )" is a built-in function in Mathcad. Type a = and Mathcad calculates the eigenvalues.

Calculation of the corresponding eigenvectors is done by using the built-in function "eigenvec ( )". This is demonstrated below.

$$u := \text{eigenvec}(M, a_0) \qquad v := \text{eigenvec}(M, a_1) \qquad w := \text{eigenvec}(M, a_2)$$

$$u = \begin{pmatrix} 0.447 \\ 0 \\ 0.894 \end{pmatrix} \qquad v = \begin{pmatrix} -0.894 \\ -0.447 \\ 0 \end{pmatrix} \qquad w = \begin{pmatrix} 0 \\ 0.316 \\ 0.949 \end{pmatrix}$$

The following example shows how to solve higher a differential equation for transient heat conduction using the programming aids within Mathcad. This example is taken from the Numerical Handbook from Mathcad.

THE HEAT EQUATION

This application uses a finite-difference approach to solve the one-dimensional boundary-value problem known as the heat equation. We follow the algorithm and the example given in Section 8.2.1 of *Numerical Analysis* by Johnson and Riess (Addison-Wesley, 1982). The heat equation is

$$D_t(u(x,t)) := \alpha^2 \cdot D_{xx}(u(x,t)) \qquad Xi < x < Xf$$

$$0 < t < T$$

The function u(x, t) is to be zero at the endpoints of the space interval, and has values at time zero given by the function f(x).

$u(0,t) = u(L,t) = 0, \quad u(x,0) = f(x)$

Starting space coordinate...	$Xi := 0$
Ending space coordinate...	$Xf := 1$
Number of space subdivisions...	$SD := 20$
Time interval [0,T]...	$T := 1$
Number of iterations...	$L := 100$
Function u(x, 0) = f(x)...	$f(x) := \sin(\pi \cdot x) + 3 \cdot \sin(2 \cdot \pi \cdot x)$
α from equation...	$\alpha := 1$

The next sections set up the equations and propose a solution procedure.

The Solution

The matrix U contains the calculated values of u(x, t) as follows:

$$U := \text{augment}\left(z, \text{augment}\left(U^T, z\right)\right)^T$$

$$U_{i,j} := u\left(d_j, t_i\right)$$

where d and t are defined by:

$j := 1 .. SD + 1 \qquad d_j := Xi + (j - 1) \cdot Dx$

$k := 1 .. L + 1 \qquad t_k := 0 + (k - 1) \cdot Dt$

Note that the array origin is 1, so that the jth column of U gives the state after j - 1 time steps.

We can compare the calculated solution with the exact solution given by Johnson and Riess, which is

$$u(x,t) := \sin(\pi \cdot x) \cdot \exp(-\pi \cdot \pi \cdot t) + 3 \cdot \sin(2 \cdot \pi \cdot x) \cdot \exp(-4 \cdot \pi \cdot \pi \cdot t)$$

$Dx := \frac{Xf - Xi}{SD}$ $\quad Dt := T \cdot L^{-1}$ $\quad ORIGIN \equiv 1$

$s := \alpha^2 \cdot \frac{Dt}{2} \cdot Dx^{-2}$

$i := 1 .. SD - 1$

$B_{i,i} := 1 - 2 \cdot s$ $\quad W_i := f(Xi + i \cdot Dx)$

$A_{i,i} := 1 + 2 \cdot s$

$i := 1 .. SD - 2$

$B_{i+1,i} := s$ $\quad A_{i+1,i} := -s$

$B_{i,i+1} := s$ $\quad A_{i,i+1} := -s$

$C := A^{-1} \cdot B$

$n := 1 .. L - 1$

$U^{\langle 1 \rangle} := W$

$U^{\langle n+1 \rangle} := C \cdot U^{\langle n \rangle}$

$z_L := 0$ (setting the boundaries to 0)

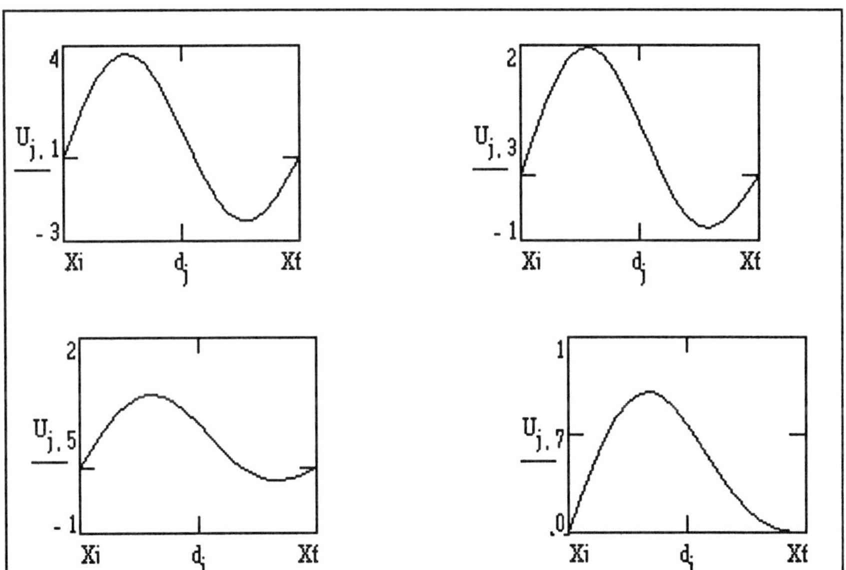

Additional output as well as error estimation is presented below.

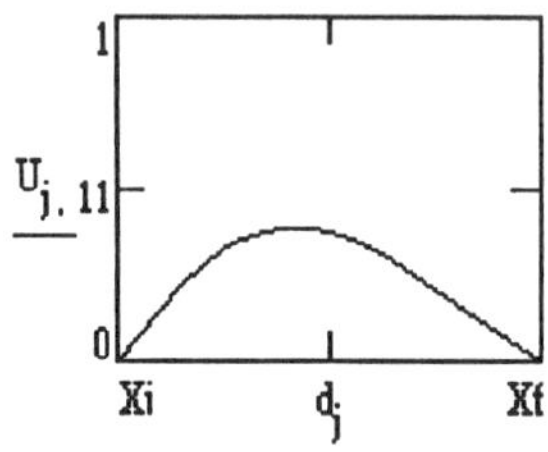

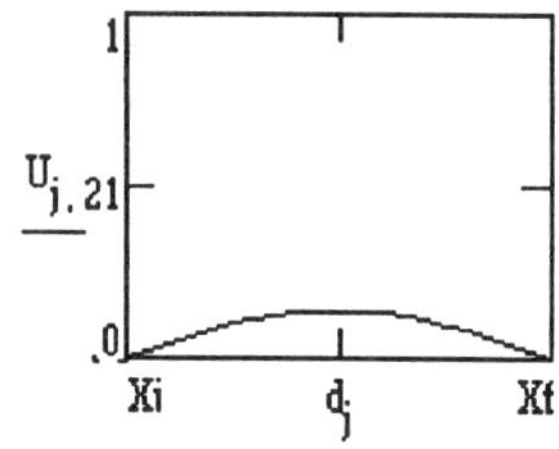

$x := Xi, Xi + .01 .. Xf$

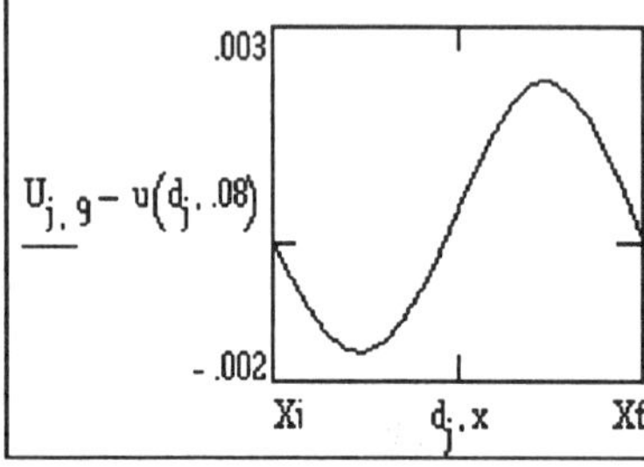

Error in the calculated solution at time .08.

The following example calculates the efficiencies for both regular and pin fins. Use is made of the hyperbolic functions built into Mathcad.

Determination of Circular Pin Fin and Straight Fin Efficiencies

hc - Convection coefficient [=] Btu/hr-ft^2 F
k - Fin thermal conductivity [=] Btu/hr-ft F
d - fin thickness,in
l - fin height, in

$dt := 55.2$

$c1 := .52 \quad c2 := .55 \quad c3 := .56$

$j := 1..4 \qquad k := 100$

$$d := \frac{.065}{12} \quad i := 1..100 \quad l_i := \frac{\frac{i}{100}}{12}$$

Estimation of Heat Transfer Coefficient

$$hhc_i := .52 \cdot c1 \cdot \left(\frac{dt}{l_i} \cdot 2\right)^{.25} \quad hvc_i := .52 \cdot c2 \cdot \left(\frac{dt}{l_i} \cdot 2\right)^{.25} \quad hvp_i := .52 \cdot c3 \cdot \left(\frac{dt \cdot 2}{l_i}\right)^{.25} \quad hhp_i := .52 \cdot c1 \cdot \left(\frac{dt}{l_i} \cdot 2\right)^{.25}$$

The following figure shows the results of the calculations.

$$mhc_i := \sqrt{4 \cdot \frac{hhc_i}{k \cdot d}} \quad mhvc_i := \sqrt{4 \cdot \frac{hvc_i}{k \cdot d}} \quad mhvp_i := \sqrt{2 \cdot \frac{hvp_i}{k \cdot d}} \quad mhhp_i := \sqrt{2 \cdot \frac{hhp_i}{k \cdot d}}$$

Efficiency Calculation Based on Fin Half Length (Fin Between Two Flat Plates)

$$\eta_i := \frac{\tanh\left(mhc_i \cdot \frac{l_i}{2}\right)}{mhc_i \cdot \frac{l_i}{2}} \quad \alpha_i := \frac{\tanh\left(mhvc_i \cdot \frac{l_i}{2}\right)}{(mhvc_i) \cdot \frac{l_i}{2}} \quad \beta_i := \frac{\tanh\left(mhvp_i \cdot \frac{l_i}{2}\right)}{mhvp_i \cdot \frac{l_i}{2}} \quad \gamma_i := \frac{\tanh\left(mhhp_i \cdot \frac{l_i}{2}\right)}{mhhp_i \cdot \frac{l_i}{2}}$$

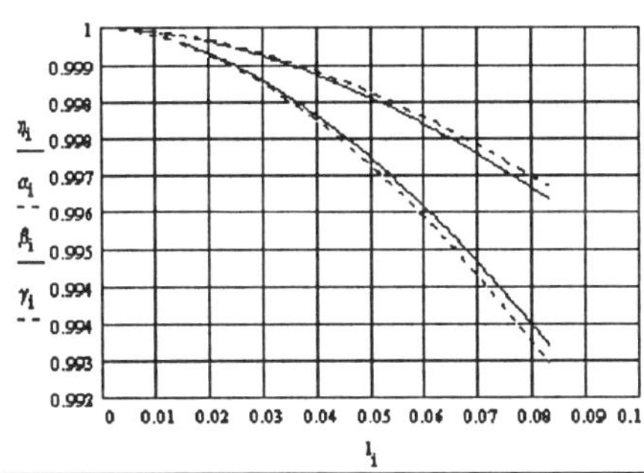

The next demonstration is the creation of a surface plot. As the reader can see, the problem requires definition of the range variables and a function description.

Creation of a Surface Plot

Set up range variables and define function: $i := 0, 1 .. 35$ $x_i := -3 + \left(\frac{6}{7}\right)\cdot\frac{i}{5}$ $y_i := -3 + \frac{6}{7}\cdot\frac{i}{5}$

$f(x, y) := \sin(x + \sin(y))$ $M_{(i,i)} := f(x_i, y_i)$

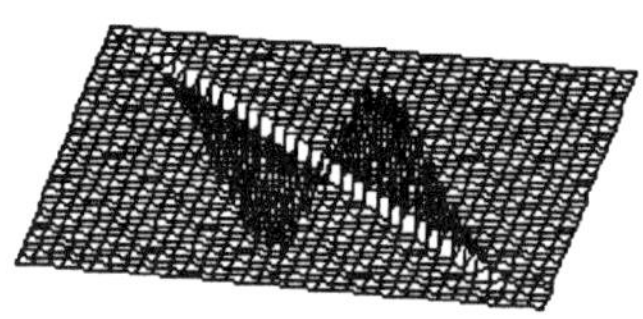

M

11.4 Mathematica 2.0 for Windows

Mathematica is the leading software for numeric, symbolic, and graphical computation used by professionals in almost every area of scientific computing. The Mathematica system is built in two parts: a kernel, that performs the computations, and a front end, that handles the user interaction. The functionality of the kernel is the same on all types of computers. Mathematica runs on Macintosh, MS-DOS based, DEC, DG, HP Apollo, IBM RISC, MIPS, NEC, NeXT, Silicon Graphics, Sony, Convex, and Stardent. The front end is built to take maximum advantage of the capabilities of a particular user interface system.

When running under Windows, Mathematica provides interactive documents known as notebooks in which text graphics, sound, and animation can be mixed with active Mathematica input. Mathematica notebooks running under Windows are interchangeable with notebooks created on Macintosh or NeXT computer systems. This ability allows Mathematica PC users a new and functional medium for doing calculations, rendering graphics, and presenting information.

Mathematica is positioned to be the complete tool for the exploration and evaluation of mathematical concepts and objects. To accomplish this task, Mathematica has robust solver and an extensive facility of built-in routines for problem solution and an impressive graphics capability. The programming language is one of the more complete that I have reviewed. The language contains low-level hooks so that, for example, one can access to the size and date of files. The Mathematica language uses case-sensitive keywords and brackets (rather than parentheses like other programs) for arguments. This is done to avoid ambiguity between functions and expressions. The program contains easy access to the many functions and extensive on-line help if the user mistypes or misspells a word or words. The number of mathematical routines is quite extensive. Mathematica contains about 850 different commands with a supplemental 143 files for other commands, a collection of statistical routines, and an extensive array of graphics functions.

Being the encompassing mathematical tool that Mathematica is requires substantial hardware. The minimum hardware requirements are at least a 80386 CPU with 4Mb RAM and at least 7Mb free on the hard disk. The more RAM available and a fast processor

The Windows version of Mathematica is a pleasure to use with the requisite pull-down menus and on-line help, and the text handling features the user expects from a Windows product. Unlike the DOS version, graphics are displayed along with the applicable text.

The text and graphics capabilities make Mathematica a terrific tool for the publication of technical documents. This feature, implemented through the concept of notebooks, allows for presentation of logic in an outline fashion.

In reviewing these included programs, I believe that Mathematica, though not perfect, is as close to where a symbolic processor should be when compared to other programs and future developments will further enhance its position as leader in the symbolic mathematics field.

The first area is to illustrate the graphics capability and ease of plotting functions within Mathematica. The following plot is offered on the following page.

```
Plot3D[Sin[x+Sin[y]],{x,-3,3},{y,-3,3}]
-SurfaceGraphics-
```

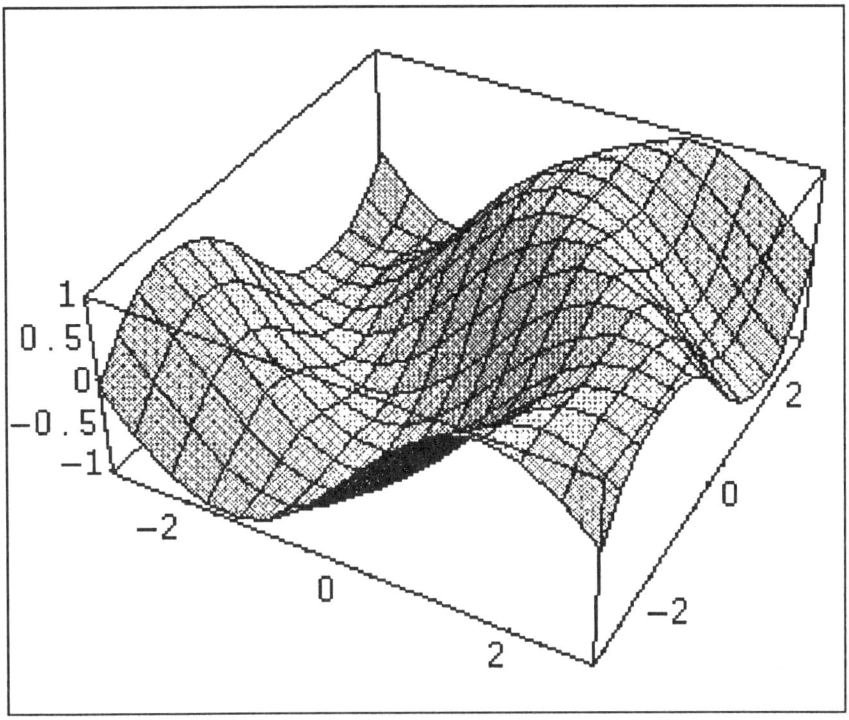

The next example illustrates the ease of integration of functions. Note that symbolic integration of indefinite integrals can be performed as well as definite integral integration.

The function to be integrated is

$$\int_{-\infty}^{\infty} \frac{1}{1+x^2} dx$$

The Mathematica solution is demonstrated below:

Demonstration of Integration Capabilities

```
Integrate[1/(1+x^2),{x,-Infinity,Infinity}]
Pi
```

The next demonstration is that of taking partial derivatives. Again, simple commands allow quick answers to be determined.

Derivatives - Partial

```
D[x^3+2x^y-3xy+4x,x]
```

$4 + 3\ x^2\ + 2\ x^{-1+y}\ -3\ \ y$

The next example is of indefinite integration and subsequent taking of the derivative to show how this is handled.

Indefinite Integrals

```
Integrate[1/(x^2+1),x]
ArcTan[x]
```

Differentiate Above Expression to Retrieve Original Expression

```
D[%,x]
  1
------
1 + x²
```

The next examples demonstrate the capability and ease in solving simultaneous equations. This extends to matrix manipulation as well.

```
Solve[{a x+b y==1,x - y==2},{x,y}]
          1      2 b          1      2 a
{{x -> ----- + -----, y -> ----- - -----}}
        a + b  a + b        a + b  a + b
Solve [{x^2 + y^2 ==1, x + y == a},{x,y}]
                     2                   2
        a - Sqrt[2 - a ]     2 a + 2 Sqrt[2 - a ]
{{x -> ----------------, y -> --------------------},
               2                      4

                     2                   2
        a + Sqrt[2 - a ]     2 a - 2 Sqrt[2 - a ]
 {x -> ----------------, y -> --------------------}}
               2                      4
```

Mathematica is an excellent program without peer.

11.5 TK Solver Release 2.0

TK Solver R2.0 is software designed to solve equations. The program handles algebraic equations, differential equations, complex variables, matrix and vector analysis, statistical analysis, and a host of other functions. "TK" stands for Tool Kit. The present release 2.0 can be characterized as a kit full of problem solving tools. Some of the tools are built-in such as the Direct and Iterative solvers, over 100 built-in functions, editors, calculators, and function and keystroke macro interpreters. An extensive library of supplied tools is included and will cover a great many of your needs and you can build models and modify existing models as you please.

Release 2.0 includes a menuing system for easy access to the built-in libraries for problem solving. In addition the program has a new feature called Presentation View that allows the creation and printing of presentation quality documents from any combination of TK sheets and subsheets.

The program comes with three manuals that cover basic operations to advanced operations of the program. What appears as a formable learning task at first is best approached first by working through the detailed examples and then strike out on your own while using the manuals as reference.

TK Solver uses its own language and format. Problems are posed as a series of rules in the "Rule Sheet". The rules need not be in any particular order to effect a solution. Addition of rules automatically sets up the "Variable sheet". Any variables may be solved for in the problem. If the problem cannot be solved by the Direct Solver, then the program utilizes its outstanding back-solving (iterative solution technique) capabilities.

TK Solver will act as a complete equation solving/model generating program in addition to improved graphics and output over earlier versions.

The following problems, supplied by Mr. Larry O'Keefe of Applied Engineering Concepts of Gainsville, Georgia shows an application of selecting fans based on system resistance and manufacturer supplied fan data and the estimation of the tipping speed of a trailer for a curve of a given radius.

The first problem illustrates the calculation resistance versus airflow for two electronic black boxes and the method of fan selection used. The equations for box system resistances were determined from measured data supplied from a vendor. The purpose of the analysis was to find a fan that would meet the required operating point for each of the boxes. The operating points required were 2 lb/min and 2.5 lb/min air which equated to approximately 51 cfm and 64 cfm. Fan data was supplied from a fan vendor. Various fan curves were inserted into the model until an appropriate fan was found that satisified the specification requirements. The illustrations below only give one iteration in the process. The next series of figures shows the Variable and Rule sheets for the creation of the model.

```
(c) Comment: FAN FLOW CURVE (ROTRON AXIMAX 501YH)                          7168/F9

==================== VARIABLE SHEET ================================================
St Input---- Name--- Output--- Unit----- Comment-----------------------------------
    .039     D                 #/FT3     DENSITY
L            CFM               FT3/MIN   VOLUMETRIC FLOW RATE
L            DPPPU             IN H2O    Box1 RESISTANCE
L            DPSPU             IN H2O    Box2 RESISTANCE
L            DPF               IN H2O    Vendor Supplied FAN CURVE
L            M                 LBS/MIN   MASS FLOW RATE
             CSPU    64.102564 FT3/MIN   2.5 LBS/MINUTE Required Flow Box2
             CPPU    51.282051 FT3/MIN   2 LBS/MINUTE Required Flow Box1
             DSTAR   1.9615385           DENSITY RATIO
==================== LIST FUNCTION: FAN ============================================
Comment:                 FAN FLOW CURVE (ROTRON AXIMAX 501YH)
Domain List:             CFM
Mapping:                 Table
Range List:              DPF
Element--- Domain---------------- Range------------------
1          0                      2.75
2          10                     2.35
3          20                     2.15
4          30                     1.95
5          40                     1.9
  F1 Help  F2 Cancel  F5 Edit  F9 Solve  / Commands  = Sheets  ; Window switch
```

The "lists" for the variables shown above with the "L" at the beginning of the line were created. Based on the fan information and the various rules for the calculations of the pressure drops through each device, the following graph was created.

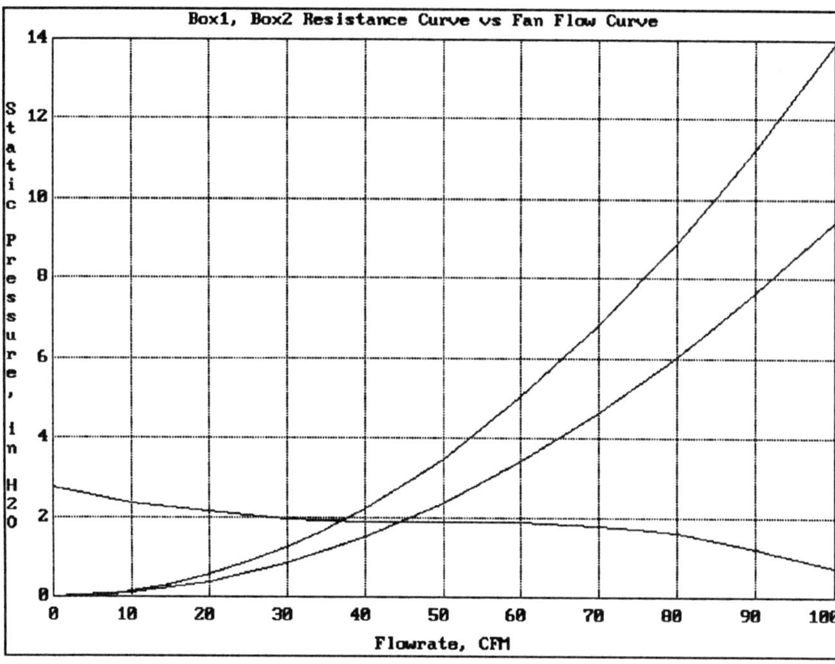

The next problem is the determination of the maximum speed of a trailer on a curve of a given radius. The solution is very straightforward and shows how even simple problems can be made interesting with the correct tools. In order to solve this problem, the variable and rule sheets have been created and are shown below.

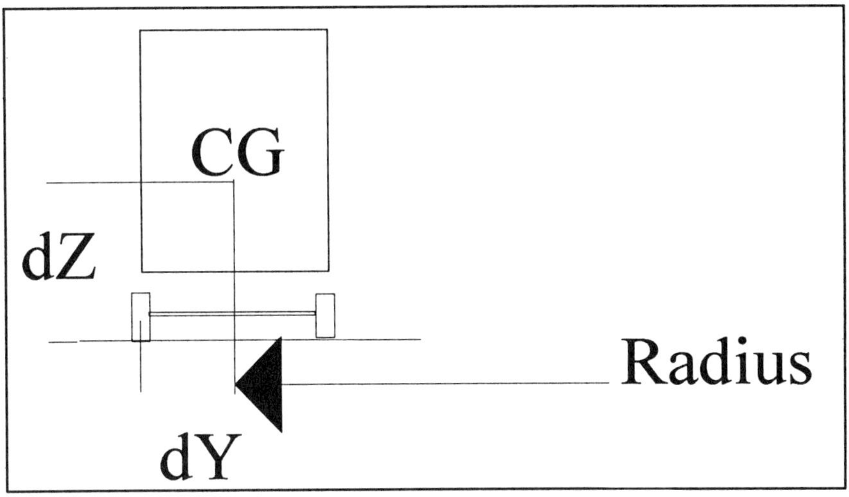

The variable and rule sheets are given below and is followed by the graphical output.

```
(3c) Comment: Date:                                              7172/F9

===================== VARIABLE SHEET ================================
St Input---- Name--- Output--- Unit----- Comment------------------------
                                         Trailer Tip Speed Calculation
                                         Outboard Spring 3 Inches Low
                                         Date:
L             Vmph    1.7649213 MPH      Tipping Speed
   20394.1    Wtrail            Lbs      Trailer/Module Weight
   20.21      Dy                In       Distance From C.G, To Outer Wheels
   97.12      Dz                In       Distance From Pavement To C.G.
L  1          Radius            Ft       Radius Of Road Curve
   32.2       G                 Ft/Sec2  Gravitational Constant
===================== RULE SHEET ====================================
S Rule------------------------------------------------------------------
* Wtrail*Dy=Ftip*Dz;Sum of Moments = 0 about outboard wheel/road contact.
* Vfps=(Ftip*Radius*G/Wtrail)^.5; F=ma
* Vmph=Vfps*3600/5280; Conversion to mph from fps.




 F1 Help  F2 Cancel  F5 Edit  F9 Solve  / Commands  = Sheets  ; Window switch
```

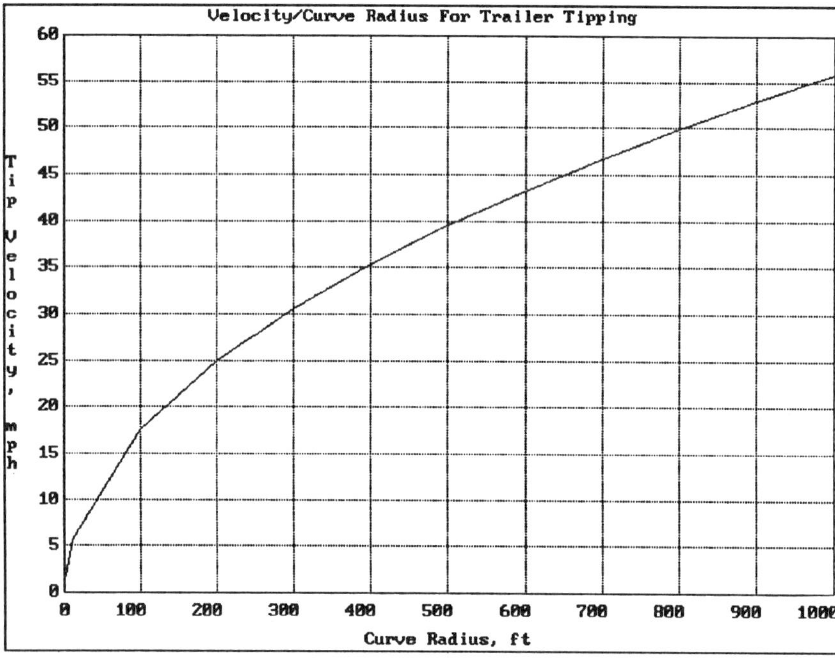
Velocity/Curve Radius For Trailer Tipping
Tip Velocity, mph
60
55
50
45
40
35
30
25
20
15
10
5
0
0
100
200
300
400
500
600
700
800
900
1000
Curve Radius, ft

The newest feature of TK Solver is the Presentation View format for reports. The example presented here is from the TK Solver Startup manual and covers the fundamental computations of loan analysis. The following figure shows the total amount of interest and payments on a $12000 loan for a 5 year period and for interest rates from 8 to 12 percent.

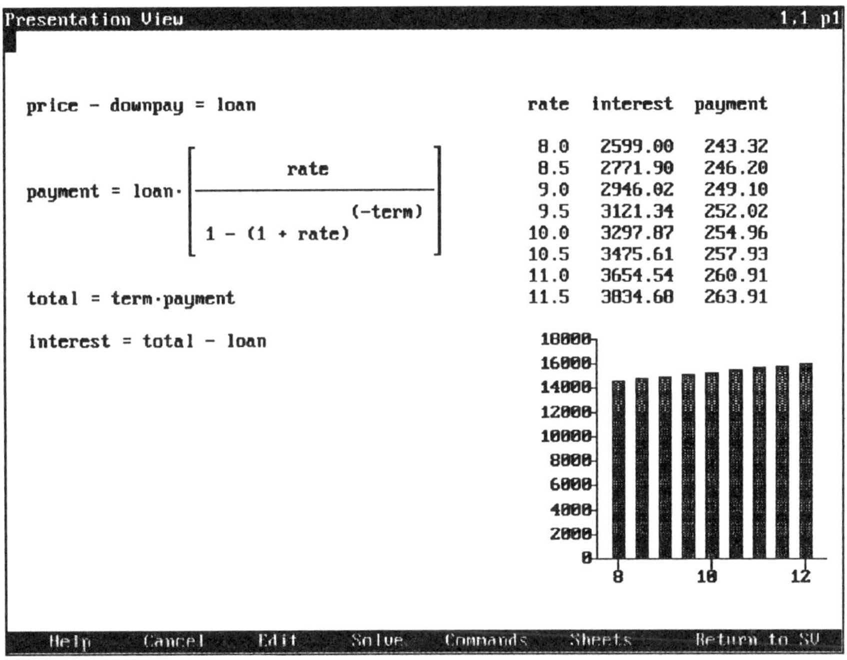

11.6 Math Software Vendors

The following addresses can be used to obtain more information about the programs reviewed in the previous sections.

Universal Technical Systems, Inc., 1220 Rock St., Rockford, IL 61101; (815) 963-2220. Program: TKSolver Release 2.0

Wolfram Research, Inc., 100 Trade Center Drive, Champaign, IL 61820; (217) 398-0747. Program: Mathematica Version 2.0

Soft Warehouse, Inc. 3660 Waialae Ave. Suite 304 Honolulu, HI 96816-3236; (808) 734-5801. Program: Derive (Version 2.51)

MathSoft, Inc. 201 Broadway St. Cambridge, MA 02139; (617) 577-1017. Program: Mathcad$^{3.1}$

Appendix

The computer programs in this book are available in FORTRAN, C, BASIC, or PASCAL and can be read with any machine able to read ASCII format. To purchase diskettes in either a 5¼" or 3½" format with one language version per diskette please use or copy the order form on the following page. The programs provide demonstration programs that illustrate the use of each subroutine and procedure in the book.

Unlicensed transfer of the programs from the above mentioned diskettes to any other format or to any other computer except that one computer for which the diskette is purchased is strictly prohibited. Licenses for authorized transfers to other computers are available from the address on the next page. Technical questions, corrections, and additional requests for information should be directed to the address on the following page also.

There are no cash refunds for diskettes. Only diskettes with manufacturing defects may be returned for replacement. All orders must be prepaid via personal check, cashier's check, or money order. Allow 10 working days for personal checks to clear. Ordinary postage for shipping is paid.

Order Form

Order From: Numerical Analysis Project, Order Department, P.O. Box 956683, Duluth, GA., 30136-9512. FAX#: (404) 975-0445.

Qty	Format 3½ or 5¼	Language	Price	Total
		FORTRAN	$21.95	
		BASIC	$21.95	
		PASCAL	$21.95	
		C	$21.95	

Total Enclosed---$

Name:____

Address:____

____

____

Selected and General Bibliography

1. Acton, Forman S. (1970), *Numerical Methods That Work*, Harper & Row, Publishers, Inc., New York.

2. Adby, P. R., and M. A. H. Dempster (1974), *Introduction to Optimization Methods*, Halsted Press, New York.

3. Aitkinson, Kendall E. (1985), *Elementary Numerical Analysis*, John Wiley & Sons, Inc., New York.

4. Aitkinson, L. V., and P. J. Harley (1983), *An Introduction to Numerical Methods with Pascal*, Addison-Wesley Publishing Company, Inc., Reading, Mass.

5. Aitkinson, Kendall E. (1978), *An Introduction to Numerical Analysis*, John Wiley & Sons, Inc., New York.

6. Anton, Howard (1984), *Elementary Linear Algebra*, 4th ed., John Wiley & Sons, Inc., New York.

7. Barakat, H.Z. and J.A. Clark, "On The Soution of The Diffusion Equations by Numerical Methods," J. Heat Transfer, 87-88, 421-427, 1966.

8. Bender, Edward A. (1978), *An Introduction to Mathematical Modeling*, John Wiley & Sons, Inc., New York.

9. Bender, Carl M., and Steven A. Orszag (1978), *Advanced Mathematical Methods For Scientists and Engineers*, McGraw-Hill Book Company, New York.

10. Binns, R.L. and Ku, D.N., "The Effect of Stenosis on the Wall Motion of Compliant Vessels", 1988.

11. Bitter, Gary G. (1983), *Microcomputer Applications for Calculus*, Prindle, Weber & Schmidt, Boston.

12. Blum, E. K. (1972), *Numerical Analysis and Computation: Theory and Practice*, Addison-Wesley Publishing Company, Inc., Reading, Mass.

13. Borse, G. J. (1985), *FORTRAN 77 and Numerical Methods for Engineers*, Prindle, Weber & Schmidt, Boston.

14. Brainerd, Walter S., and Lawrence H. Landweber (1974), *Theory of Computation*, John Wiley & Sons, Inc., New York.

15. Braun, Martin, Courtney S. Coleman, and Donald A. Drew, eds. (1983). *Differential Equation Models*, Springer-Verlag, New York.

16. Brent, Richard P. (1973), *Algorithms for Minimization without Derivatives*, Prentice Hall, Inc., Englewood Cliffs, N.J.

17. Buck, R. Creighton (1978), *Advanced Calculus, 3rd ed.*, McGraw-Hill Book Company, New York.

18. Bulirsch, R., and J. Stoer, (1980), *Introduction To Numerical Analysis*, Springer-Verlag, New York.

19. Bunday, Brian D. (1984), *Basic Linear Programming*, Edward Arnold (Publishers) Ltd., London.

20. Burden, Richard L., J. Douglas Faires, and Albert C. Reynolds, (1981), *Numerical Analysis, 2nd ed.*, Prindle, Weber & Schmidt, Boston.

21. Caranahan, B., et al (1969), *Applied Numerical Methods*, John Wiley and Sons, New York.

22. Cheney, Ward, and David Kincaid (1985), *Numerical Mathematics and Computing, 2nd ed.*, Brooks/Cole Publishing Co., Monterey, Calif.

23. Coddington, Earl A., and Norman Levinson (1955), *Theory of Ordinary Differerential Equations*, McGraw-Hill Book Company, New York.

24. Conte, S. D., and Carl de Boor (1980), *Elementary Numerical Analysis: An Algorithmic Approach*, McGraw-Hill Book Company, New York.

25. Cowley, S.J., "Elastic Jumps in Fluid-Filled Tubes", Journal of Fluid Mechanics, Vol. 116, 1982, pp. 459-473.

26. Crank, J., and P. Nicolson, "A Practical Method for Numerical Evaluation of Solutions of Partial Differential Equations of the Heat Conduction Type," Proc. Cambridge Phil.Soc., 43, 50-67, 1947.

27. Dahlquist, Germund, and Ake Bjorck (1974), *Numerical Methods*, Prentice-Hall, Inc. Englewood Cliffs, N.J.

28. Danielson, G.C. and C. Lanzcos,"Some Improvements In Practical Fourier Analysis and Their Application to X-Ray Scattering From Liquids," J. Franklin Inst., Vol. 233, pp. 365-380, 435-452.

29. Davis, Philip J., and Philip Rabinowitz (1984), *Methods of Numerical Integration 2nd ed.*, Academic Press, New York.

30. Davis, Philip J. (1963), *Interpolation and Approximation*, Blaisdell Publishing Company, Waltham, Mass.

31. Dodes, Irving Allen (1978), *Numerical Analysis for Computer Science*, North-Holland, New York.

32. Dorn, William S., and Daniel D. McCracken (1972), *Numerical Methods with FORTRAN IV Case Studies*, John Wiley & Sons, Inc., New York.

33. Douglas, J.,Jr., "On the Numerical Integration of $\partial^2u/\partial x^2 + \partial^2u/\partial y^2 = \partial u/\partial t$ by Implicit Methods," J. Soc. Ind. Appl. Math.,3, 42-65, 1955.

34. DuFort, E.C., and S. Frankel, "Stability Conditions in The Numerical Treatment of Parabolic Differential Equations," Mathematical Tables and Other Aids to Computation, Vol. 7,1953, pp. 135-152.

35. Elad, D., Kamm, R.D., and Shapiro, A.H., "Choking Phenomena in a Lung-Like Model", ASME Journal of Biomechanical Engineering, Vol. 109, 1987, pp. 1-9.

36. Fike, C. T. (1968), *Computer Evaluation of Mathematical Functions*, Prentice-Hall, Inc., Englewood Cliffs, N.J.

37. Flaherty, J.E., Keller, J.B., and Rubinow, S.I., "Post Buckling Behavior of Elastic Tubes and Rings with Opposite Sides in Contact", SIAM Journal of Applied Mathematics, 23(4), 1972, pp. 446-455.

38. Forsythe, George E., and Cleve B. Moler (1967), *Computer Solution of Linear Algebraic Systems*, Prentice-Hall, Inc., Englewood Cliffs, N.J.

39. Forsythe, George E., Michael A. Malcolm, and Cleve B. Moler (1977),

Computer Methods for Mathematical Computations, Prentice-Hall, Inc., Englewood Cliffs, NJ

40. Fox L., and D. F. Mayers (1968), *Computing Methods for Scientists and Engineers*, Clarendon Press, Oxford.

41. Gear, C. William (1971), *Numerical Initial Value Problems in Ordinary Differential Equations*, Prentice-Hall, Inc., Englewood Cliffs, N.J.

42. Gerald, Curtis F., and Patrick O. Wheatley (1984), *Applied Numerical Analysis, 3rd ed.*, Addison-Wesley Publishing Company, Inc., Reading, Mass.

43. Gold, B., and C.M. Radner, (1969), *Digital Processing of Signals*, McGraw-Hill Book Company, New York.

44. Greenspan, Donald (1971), *Introduction to Numerical Analysis and Applications*, Markham Publishing Co., Chicago.

45. Hamming, Richard W. (1973), *Numerical Methods for Scientists and Engineers, 2nd ed.*, McGraw-Hill Book Company, New York.

46. Henrici, Peter (1964), *Elements of Numerical Analysis*, John Wiley & Sons, Inc. New York.

47. Hildebrand, Francis B. (1974), *Introduction to Numerical Analysis, 2nd ed.*, McGraw-Hill Book Company, New York.

48. Hornbeck, Robert W. (1975), *Numerical Methods*, Quantum Publishers Inc., New York.

49. Householder, Alston S. (1953), *Principles of Numerical Analysis*, McGraw-Hill Book Company, New York.

50. Isaacson, Eugene, and Herbert Bishop Keller (1966), *Analysis of Numerical Methods*, John Wiley & Sons, Inc., New York.

51. Johnson, Lee W., and R. Dean Riess (1982), *Numerical Analysis, 2nd ed.*, Addison Wesley Publishing Company, Inc., Reading, Mass.

52. Kardestuncer, H. (Editor), Finite Element Handbook, McGraw-Hill, New

York,1987.

53. Lapidus L.,and J.H.Seinfeld (1971), *Numerical Solution of Ordinary Differential Equations*, Academic Press, Inc., New York.

54. Leon, Steven J. (1980), *Linear Algebra with Applications*, Macmillan Publishing Co., New York.

55. Maron, Melvin J. (1982), *Numerical Analysis: A Practical Approach*, Macmillan Publishing Co., New York.

56. McCalla, Thomas Richard (1967), *Introduction to Numerical Methods and FORTRAN Programming*, John Wiley & Sons, Inc., New York.

57. Morris, John L. (1983), *Computational Methods in Elementary Numerical Analysis*, John Wiley & Sons, Inc., New York.

58. Moursund, David G., and Charles S. Duris (1967), *Elementary Theory and Application of Numerical Analysis*, McGraw-Hill Book Company, New York.

59. Noble, Ben, and James W. Daniel (1977), *Applied Linear Algebra, 2nd ed.*, Prentice-Hall, Inc., Englewood Cliffs, N.J.

60. Oates, G.C., "Fluid Flow in Soft-Walled Tubes, I. Steady Flow", Medical and Biological Engineering 13, pp. 773-778.

61. Ortega, James M., and W. C. Rheinboldt (1970), *Iterative Solution of Nonlinear Equations in Several Variables*, Academic Press, Inc., New York.

62. Ortega, James M. (1972), *Numerical Analysis*, Academic Press, New York.

63. Peaceman, D.W. and H.H. Rachford, "The Numerical Solution of Parabolic and Elliptic Differential Equations," J. Soc. Appl. Math.,3, 28-41, 1955.

64. Pennington, Ralph H. (1970), *Introductory Computer Methods and Numerical Analysis, 2nd ed.*, Macmillan Publishing Co., New York.

65. Philips, G. M., and P. J. Taylor (1974), *Theory and Applications of Numerical Analysis*, Academic Press, Inc., New York.

66. Pizer, Stephen M. (1975), *Numerical Computing and Mathematical Analysis*,

Science Research Associates, Inc., Chicago, Ill.

67. Press, W.H.,et al (1989), *Numerical Recipes - The Art of Scientific Computing*, Cambridge, New York.

68. Ralston, Anthony, and Philip Rabinowitz (1978), *A First Course in Numerical Analysis 2nd ed.*, McGraw-Hill Book Company, New York.

69. Ralston, Anthony (1965), *A First Course in Numerical Analysis*, McGraw-Hill Book Company, New York.

70. Salvadori, Mario G., and Melvin L. Baron (1961), *Numerical Methods in Engineering 2nd ed.*, Prentice-Hall, Inc., Englewood Cliffs, N.J.

71. Scarborough, James B. (1966), *Numerical Mathematical Analysis*, The John Hopkins Press, Baltimore.

72. Scheid, Francis (1968), *Theory and Problems of Numerical Analysis*, McGraw-Hill Book Company, New York.

73. Scraton R. E. (1984), *Basic Numerical Methods: An Introduction to Numerical Mathematics on a Microcomputer*, Edward Arnold (Publishers) Ltd., London.

74. Shampine, L. F., and M. K. Gordon (1975), *Computer Solution of Ordinary Differential Equations*, W. H. Freeman and Company Publishers, New York.

75. Shapiro, A.H., "Steady Flow in Collapsible Tubes", ASME Journal of Biomechanical Engineering, Vol. 99, 1977, pp. 126-127.

76. Shoup, Terry E. (1983), *Numerical Methods for the Personal Computer*, Prentice-Hall, Inc., Englewood Cliffs, N.J.

77. Shoup, Terry E. (1979), *A Practical Guide to Computer Methods for Engineers*, Prentice-Hall, Inc., Englewood Cliffs, N.J.

78. Smith, W. Allen (1979), *Elementary Numerical Analysis*, Harper & Row, Publishers, Inc., New York.

79. Smith, G. D. (1978), *The Numerical Solution of Partial Differential Equations, 2nd ed.*, Oxford University Press, Oxford.

80. Strang, G. and G.F. Fix, *An Analysis of the Finite Element Method*, Prentice-Hall, Englewood Cliffs,N.J. 1973.

81. Snyder, Martin Avery (1966), *Chebyshev Methods in Numerical Approximation*, Prentice-Hall, Inc., Englewood Cliffs, N.J.

82. Stanton, Ralph G. (1961), *Numerical Methods for Science and Engineering*, Prentice-Hall, Inc., Englewood Cliffs, N.J.

83. Stark, Peter A. (1970), *Introduction to Numerical Methods*, The Macmillan Company, Toronto, Ontario.

84. Stroud, A. H., and Don Secrest (1966), *Gaussian Quadrature Formulas*, Prentice-Hall, Inc., Englewood Cliffs, N.J.

85. Thompson, William J. (1984), *Computing in Applied Science*, John Wiley & Sons, Inc., New York.

86. Todd, John (1977), *Basic Numerical Mathematics, Vol. 2 - Numerical Algebra*, Academic Press, Inc., New York.

87. Tompkins, Charles B., and Walter L. Wilson (1969), *Elementary Numerical Analysis*, Prentice-Hall, Inc., Englewood Cliffs, N.J.

88. Traub, J. F. (1964), *Iterative Methods for the Solution of Equations*, Prentice-Hall. Inc. Englewood Cliffs, N.J.

89. Vandergraft, James S. (1983), *Introduction to Numerical Computations*, Academic Press Inc., New York.

90. Vanlwaarden, John L. (1985), *Ordinary Differential Equations with Numerical Techniques*, Harcourt Brace Jovanovich, Publishers, San Diego, Calif.

91. Varga, Richard S. (1962), *Matrix Iterative Analysis*, Prentice-Hall, Inc., Englewood Cliffs, N.J.

92. Weizsacker, H.W., and Pinto, J.G., "Isotropy and Anisotropy of the Arterial Wall", Journal of Biomechanics, Vol. 21, No. 6, 1988, pp. 477- 487.

93. Wendrofl, Burton (1966), *Theoretical Numerical Analysis*, Academic Press, Inc., New York.

94. Young, David M. (1971), *Iterative Solution of Large Linear Systems*, Academic Press Inc., New York.

95. Zienkiewicz, O.C. and R.L. Taylor, *The Finite Element Method Vol 1, Fourth Edition*, McGraw-Hill, New York, 1989.

Index